M. GARVEY

Thermodynamic Behavior
of Electrolytes in Mixed Solvents

Thermodynamic Behavior of Electrolytes in Mixed Solvents

William F. Furter, EDITOR

Royal Military College of Canada

A symposium sponsored by

the Division of Industrial

and Engineering Chemistry at

the 170th Meeting of the

American Chemical Society,

Chicago, Ill.,

Aug. 27–28, 1975.

ADVANCES IN CHEMISTRY SERIES **155**

AMERICAN CHEMICAL SOCIETY

WASHINGTON, D. C.   1976

Library of Congress CIP Data

Thermodynamic behavior of electrolytes in mixed solvents.
(Advances in Chemistry Series; 155. ISSN 0065-2393)

   Includes bibliographic references and index.

   1. Electrolytes—Congresses. 2. Solvents—Congresses.
   I. Furter, William F., 1931-   . II. American Chemical Society. Division of Industrial and Engineering Chemistry. III. American Chemical Society. IV. Series: Advances in chemistry series; 155.

QD1.A355   no. 155   (QD565)   540'.8s
(541'.372)   76-54329
ISBN 0-8412-0302-4   ADCSAJ 155 1–404   (1976)

Copyright © 1976

American Chemical Society

PRINTED IN THE UNITED STATES OF AMERICA

Advances in Chemistry Series

Robert F. Gould, *Editor*

FOREWORD

ADVANCES IN CHEMISTRY SERIES was founded in 1949 by the American Chemical Society as an outlet for symposia and collections of data in special areas of topical interest that could not be accommodated in the Society's journals. It provides a medium for symposia that would otherwise be fragmented, their papers distributed among several journals or not published at all. Papers are refereed critically according to ACS editorial standards and receive the careful attention and processing characteristic of ACS publications. Papers published in ADVANCES IN CHEMISTRY SERIES are original contributions not published elsewhere in whole or major part and include reports of research as well as reviews since symposia may embrace both types of presentation.

CONTENTS

PREFACE

A major symposium on thermodynamic behavior of electrolytes in mixed solvents has been long overdue. This collection of papers attempts to draw together the wide range of effects that electrolytes exert in solvents consisting of two or more components. The papers reflect both the scope of the effects involved and the variety of work being done to elucidate them. Nine different countries are represented by the 25 contributions: the United States, Canada, Britain, Japan, Australia, Czechoslovakia, France, The Netherlands, and Spain. The book is divided into two sections; the first deals with the effects of electrolytes on the composition of the equilibrium vapor phase, and the second with effects on other thermodynamic and physicochemical properties of such solutions.

The topic covered in the 10 papers of the first section is commonly referred to as salt effect in vapor–liquid equilibrium and is potentially of great industrial importance. This salt effect leads to extractive distillation processes in which a dissolved salt replaces a liquid additive as the separating agent; the replacement often results in a greatly improved separating ability and reduced energy requirements. Two papers in this volume, those by Sloan and by Vaillancourt, illustrate the use of such processing to concentrate nitric acid from its aqueous azeotrope. Nevertheless, the effect has not been exploited by industry to nearly the extent that would seem to be merited by its scientific promise.

The papers in the second section deal primarily with the liquid phase itself rather than with its equilibrium vapor. They cover effects of electrolytes on mixed solvents with respect to solubilities, solvation and liquid structure, distribution coefficients, chemical potentials, activity coefficients, work functions, heat capacities, heats of solution, volumes of transfer, free energies of transfer, electrical potentials, conductances, ionization constants, electrostatic theory, osmotic coefficients, acidity functions, viscosities, and related properties and behavior.

In addition to the applications in extractive distillation referred to above, there are other industrial examples where electrolytes in mixed solvents occur. In many industrial situations nonvolatile electrolytes are either added to effect the separation of multicomponent process streams (e.g., the complexing agents added to enhance distribution coefficients in solvent extraction) or are present as a result of the process itself. Examples of the latter include the product streams from a wide range of chemical processes such as: neutralizations and certain esterifications and etherifications; systems such as those encountered in some fuel cells and batteries; and specialized processing such as certain distillations accompanied by chemical reaction. The properties and behavior of such mixtures must be well understood if separation operations involving them are to be designed effectively.

Finally I would like to thank Rose Boucher for stenographic support in the work of organizing and conducting the symposium and in preparing the book, and my wife Pamela for her patience and understanding.

Kingston, Ontario   WILLIAM F. FURTER
April 15, 1976

Effects on
Vapor–Liquid Equilibrium

Correlation of Vapor–Liquid Equilibrium Systems Containing Two Solvents and One Salt

ENRIQUE BEKERMAN[1] and DIMITRIOS TASSIOS

New Jersey Institute of Technology, Newark, N.J. 07102

Vapor–liquid equilibrium data for systems containing two solvents and one salt are correlated with the NRTL and LEMF equations. For the system methanol–water–LiCl the results are better than those of the Broul and Hala correlations. Good results were also obtained for the two systems containing CaCl₂ and potassium acetate, while for two systems containing HgCl₂—of questionable experimental accuracy—salting-out rather than salting-in was calculated. Prediction of the ternary behavior from the binary data appears possible, but additional experimental data are needed for the development of guidelines for the choice of the appropriate value of α for the solvent–solvent system.

The effect of salts on the vapor–liquid equilibrium of solvent mixtures has been of considerable interest in recent years. Introduction of a salt into a binary solvent mixture results in a change in the relative volatility of the solvents. This effect can be used to an advantage where the separation of the solvents is of interest. Furter and co-workers have demonstrated the potential importance of salts as separating agents in extractive distillation (*1, 2, 3*).

Several authors have attempted to correlate the vapor–liquid equilibrium (VLE) data for binary systems in the presence of salts at various concentrations. Johnson and Furter (*4*) successfully correlated a large number of systems consisting of an alcohol, water, and a salt at saturation, by the following equation:

$$\log \left(\frac{\alpha_s}{\alpha_0} \right) = k_3 x_3 \tag{1}$$

where k_3 is called the salt-effect parameter, α_s and α_0 are the relative volatilities with and without salt respectively, and x_3 is the mole fraction of the salt. The parameter k_3 was found to be surprisingly independent of the value of the liquid

---

[1] Current address: Tuck Industries, Inc., New Rochelle, N.Y. 10801

Table I. Data Sources

I. *Ternary Systems*	*References*
1. Methanol–water–lithium chloride @ 60°C	13
2. Methanol–water–calcium chloride @ 752 mmHg	4
3. Methanol–water–potassium acetate @ 764 mmHg	5
4. Methanol–water–mercuric chloride @ 758 mmHg	4
5. Ethanol–water–mercuric chloride @ 750 mmHg	4
II. *Binary Systems*	
1. Methanol–water @ 60°C	13
2. Methanol–water @ 760 mmHg	18, 21
3. Ethanol–water @ 760 mmHg	22
4. Water–lithium chloride @ 60°C	13
5. Water–calcium chloride @ 760 mmHg	23
6. Water–potassium acetate @ 760 mmHg	23
7. Water–mercuric chloride @ 760 mmHg	23
8. Methanol–lithium chloride @ 60°C	13
9. Methanol–calcium chloride @ 760 mmHg	23
10. Methanol–potassium acetate @ 760 mmHg	23
11. Methanol–mercuric chloride @ 760 mmHg	23
12. Ethanol–mercuric chloride @ 760 mmHg	23
III. *Antoine Constants*	22
IV. *Parameters for the Solubility Equation*	4
(Used for the mercuric chloride systems)	

phase mole fraction. Meranda and Furter (*5*) found that this independence of the parameter k_3 with x_2 does not apply to systems containing acetate salts at saturation. Yoshida (*6*) found that Equation 1 does not correlate water–acetic acid–salt systems.

Jaques and Furter (*7, 8*) successfully fitted the T-P-X data of various isobaric systems saturated with a salt by treating the systems as pseudobinaries and using the Wilson equation for correlation. Rousseau et al. (*9*) used a similar approach in correlating the data of Johnson and Furter by means of the van Laar (*10*), Wilson (*11*) and NRTL (*12*) equations.

Broul et al. (*13*) and Hala (*14*) developed a correlation scheme for systems containing two solvents and one salt, which they applied to several salt concentrations, not just to the saturation level as in the studies mentioned above. They utilized the binary VLE data for the three binaries (solvent 1–salt; solvent 2–salt; and solvent 1–solvent 2) along with the ternary data to correlate successfully the ternary results. They employed the Margules equation (*15*) with the addition of a term to account for the coulombic interactions.

Because of the limitations of the Margules equation—especially in predicting multicomponent VLE data—the Wilson, NRTL, and LEMF (*16*) equations are employed in this study. The experimental data on the systems presented in Table I were used in this work. These are the only systems for which both binary and ternary data could be found in the literature. As a matter of fact, uncertainties do exist about the accuracy of the two $HgCl_2$ systems. The maximum boiling point elevation is only 0.755°C for the EtOH–HgCl$_2$ and 1.27°C for the MeOH–HgCl$_2$ with most of the data points falling within 0.2°C. The data were obtained in 1896 and 1890 respectively. One additional point for each system consisting of the alcohol saturated with HgCl$_2$ was obtained from the Johnson and Furter data. The concentration of mercuric chloride in the ternaries had to be calculated from the Johnson and Furter solubility equation since the numerical data were not given. In addition, Jaques and Furter (7) indicate that the ethanol–water–mercuric chloride data failed to meet their thermodynamic consistency test. Therefore, the validity of the data for these two systems is somewhat questionable.

The Method

In the present study, systems composed of two solvents and a salt are treated as ternary systems. Data on the vapor pressure depression of the solvent by the salt for isothermal systems and on the boiling point elevation of the solvent in the presence of salt for isobaric systems are used to develop the parameters for the solvent–salt binaries. For such binaries only the activity coefficients for the solvent are considered. The parameters for all three binary sets are generated from the binary data by a regression subroutine.

Vapor phase ideality was assumed in all the computations. Because the pressure of the systems studied did not exceed atmospheric pressure, this assumption is acceptable. Hence, the activity coefficient for the solvent was defined by:

$$\gamma_i = \frac{y_i P}{x_i P_i{}^\circ} \tag{2}$$

where:

i = 1 and 2 for solvent 1 and solvent 2
x_i = liquid phase analytical composition (assuming no salt dissociation when a salt is present)
y_i = vapor phase mole fraction (equal to unity for solvent salt binaries)
P = total pressure
$P_i{}^\circ$ = saturation pressure of pure i at system temperature.

The following three equations were used for the correlation of the activity coefficient to solution composition. The Wilson (11) equation:

$$\ln\gamma_k = -\ln\left[\sum_{j=1}^{m} \alpha_j \Lambda_{kj}\right] + 1 - \sum_{i=1}^{m} \frac{x_i \Lambda_{ik}}{\sum_{j=1}^{m} x_j \Lambda_{ij}} \tag{3}$$

where:

$$\Lambda_{ij} = \frac{v_j}{v_i} \exp\left[-\frac{\lambda_{ij} - \lambda_{ii}}{RT}\right]$$

As seen from Equation 3 only binary parameters are needed for the determination of the ternary activity coefficients, and two parameters per binary system are needed.

The NRTL equation (12):

$$\ln\gamma_i = \frac{\sum\limits_{j=1}^{n} \tau_{ji}G_{ji}x_j}{\sum\limits_{k=1}^{n} G_{kj}x_k} + \sum\limits_{j=1}^{n} \frac{x_jG_{ij}}{\sum\limits_{k=1}^{n} G_{kj}x_k}\left(\tau_{ij} - \frac{\sum\limits_{l=1}^{n} x_l\tau_{lj}G_{lj}}{\sum\limits_{k=1}^{n} G_{kj}x_k}\right) \qquad (4)$$

where

$$\tau_{ji} = (g_{ji} - g_{ii})/RT$$

$$G_{ji} = \exp(-\alpha_{ji}\tau_{ji})$$

$$\alpha_{ij} = \alpha_{ji} \text{ and } g_{ij} = g_{ji}$$

Again, only binary parameters are needed for the determination of the ternary. For a given binary system the equation contains two adjustable parameters ($g_{12} - g_{11}$) and ($g_{12} - g_{22}$). The equation also contains the parameter α_{12}, which according to Renon and Prausnitz must assume a positive value. Renon and Prausnitz indicate that, on the basis of the type of system under consideration, a value of α_{12} can be determined a priori equal to 0.20, 0.30, or 0.47. On the other hand, Marina and Tassios (16) showed that α can assume negative values and recommended $\alpha = -1.0$. This is known as the LEMF equation. In general α can be considered an adjustable parameter, and some authors establish its value by regressing the experimental data when correlating binary VLE data or predicting multicomponent ones.

The Wilson equation was considered first. Great problems were encountered with this equation. Negative values for one or both of the parameters Λ_{12} and Λ_{21} were often obtained by regressing the solvent–salt data. As it can be seen from Equation 3, a negative value for Λ_{ij} is unacceptable.

Following the failure of the Wilson equation the NRTL and LEMF equation were considered. Preliminary studies using analytical mole fractions, i.e. assuming no salt dissociation, gave good results for the isothermal system (water–methanol–LiCl); for isobaric data, however, no results could be obtained since the bubble point temperature subroutine failed to converge. It was decided, therefore, to assume complete salt dissociation, as suggested by Broul (13) and Hala (14) and to define the mole fractions and activity coefficients as follows:

$$\left.\begin{array}{l} x_i' = \dfrac{x_i}{x_i + \nu x_3} \\[2ex] x_3' = 1 - x_i' \\[2ex] \gamma_i' = \dfrac{P}{x_i'P_1^{\circ}} \end{array}\right\} \quad \text{Binary System, } i = 1 \text{ and } 2 \qquad (5)$$

$$\left.\begin{aligned}
x_1' &= \frac{x_1}{x_1 + x_2 + \nu x_3} \\[2mm]
x_2' &= \frac{x_2}{x_1 + x_2 + \nu x_3} \\[2mm]
x_3' &= 1 - x_1' - x_2' \\[2mm]
\gamma_i' &= \frac{y_i P}{x_i' P_i^\circ}
\end{aligned}\right\} \quad \text{Ternary System} \qquad (6)$$

where:

ν = number of ions resulting from the complete dissociation of one molecule of salt, and

x_i = analytical mole fraction.

This definition of x' and γ' is more realistic at low and moderate salt concentrations and is in agreement with that of Sada and Morisue (*17*). Broul and Hala also assumed complete salt dissociation. The assumption of full dissociation of the salt may not be entirely valid at high salt concentrations, especially where the concentration of the nonaqueous solvent is also high. However, even in those instances where the assumption of full dissociation of the salt may be invalid, it appears to describe the system better than ignoring salt ionization completely. The terms x' and γ' are referred to hereafter as ionic mole fraction and ionic activity coefficient, respectively. These should not be confused with the mean ionic terms used by Hala which are also based on complete salt dissociation, but are defined differently. No convergence problems were encountered when the ionic quantities were employed.

The Values of α_{13}, α_{23}, and α_{12}.

First the value of α for the solvent–salt binaries was considered. No precedent for using the NRTL equation for such systems is known to the writers except for the work of Rousseau et al., where α was obtained by regression of the data, but this was applied to solvent–solvent–salt systems treated as pseudobinaries. Determination of α by regression of the solvent–salt data lead to the observation that, even though elimination of one data point resulted often in a drastic change of the value of α, the quality of the obtained fit was always good and, for practical purposes, independent of the value of α. It was decided therefore to use the value $\alpha_{13} = \alpha_{23} = 0.30$. The values of the ionic activity coefficients for certain binaries did not deviate from unity by more than 10%. These binaries consisted of those where the salt is not very soluble in the solvent and the effect on the boiling point of the solvent is very small. Treating such binaries as ideal for two of the systems resulted in comparable ternary correlations, as shown in Table II. Assuming the MeOH–CaCl$_2$ system to behave ideally yielded better ternary results than those obtained by treating it as real. For the potassium acetate system the assumption of ideality results in reduction of the maximum error in y_1. This assumption used

Table II. Effect of Assuming Ideal Behavior for the Binary System Where γ's Are Within 10% From Unity[a]

	System							
	MeOH (1)–Water (2)–CaCl₂(3)				MeOH (1)–Water (2) –Potass. Acet.(3)			
α_{13}		0.30		0.30		0.30		0.30
$g_{13}\text{-}g_{11}$	Ideal	1681	Ideal	1681	Ideal	1786	Ideal	1786
$g_{13}\text{-}g_{33}$		16647		16647		−1970		−1970
α_{23}	0.30	0.30	0.30	0.30	0.30	0.30	0.30	0.30
$g_{23}\text{-}g_{22}$	−1602	−1602	−1602	−1602	1558	1558	1558	1558
$g_{32}\text{-}g_{33}$	−1566	−1566	−1566	−1566	−1735	−1735	−1735	−1735
α_{12}	−0.10	−0.10	0.30	0.30	−0.10	−0.10	0.30	0.30
$g_{12}\text{-}g_{11}$	1431	1431	−171	−171	1431	1431	−171	−171
$g_{12}\text{-}g_{22}$	1194	1194	763	763	1194	1194	763	763
$\lvert\Delta y_1\rvert_{mean}$	0.029	0.039	0.037	0.065	0.040	0.026	0.074	0.056
$\lvert\Delta y_1\rvert_{max}$	0.048	0.091	0.069	0.140	0.069	0.085	0.137	0.264

[a] Comparison is based on Perry's data for the 1–2 binary.

in all four sets of isobaric data, will also simplify the procedure where binary data are not readily available. In such a case, a measurement of the boiling point at saturation would indicate whether or not the binary system can be described as behaving ideally within the solubility range. Rousseau and co-workers took a similar approach when they neglected the effect of the salt on the vapor pressure of the alcohol in alcohol–water–salt systems.

Turning now to the solvent–solvent binary, the effect of the value of α_{12} on the quality of the obtained fit is well established (12, 16). Since this binary had the largest number of experimental activity coefficients—for the solvent–salt binaries only the γ of the solvent is used—it was decided to let α_{12} vary between +1.0 and −3.0 with the best fit of the ternary data as criterion for its optimum value. The possibilities of varying the other two α's (α_{13} and α_{23}) to obtain the best ternary fit was rejected; although it would probably lead to better correlation of the ternary results, it could not lead to any predictive scheme. The number of available systems is simply too limited for the establishment of optimum α values for all three binaries.

Results and Discussion

Figures 1–5 present plots of the average absolute error in y_1 vs. α_{12} for the five ternary systems considered in this study. For three of the four isobaric systems, two sources of VLE data for the MeOH–water binary have been used to demonstrate the impact on the correlation of the ternary data. While the difference is not drastic, the better data of Ramalho et al. (18) also provide better correlation, as evidenced by Figures 1–3. The ensuing discussion is based on the latter data. Table III presents the optimum α_{12} in the positive and negative regions. Considering the diversity of the systems studied—e.g., three involved

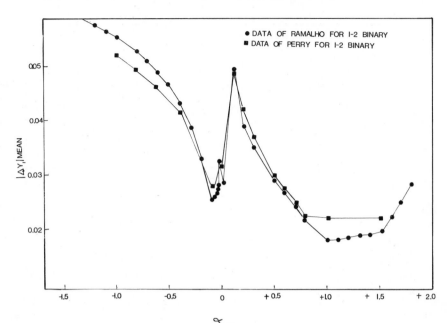

Figure 1.   Plot of $|\Delta y_1|_{mean}$ vs. α_{12}: methanol–water–CaCl$_2$ @ 752 mmHg; $\alpha_{13} = 0.0$ (ideal), $\alpha_{23} = 0.30$

salting-out and two salting-in—this is not unexpected. On the other hand, examination of more data could lead to a set of optimum values for α_{12} as a function of the system involved. This, of course, was done for binary systems by Renon and Prausnitz (*12*).

As far as correlation of ternary data is concerned, the method has provided results that range from poor to excellent: for the MeOH–water–HgCl$_2$ system no salt effect is predicted; for the EtOH–water–HgCl$_2$ system, salting-out instead of salting-in is predicted. The reservations concerning the quality of the binary data for these two systems have been mentioned earlier. For the MeOH–

Table III.   Optimum Values of α_{12} in the Positive and Negative Regions Based on Ternary Vapor Mole Fraction Calculation

System	Negative Region		Positive Region	
	α_{12}	$|\Delta y_1|_{mean}$	α_{12}	$|\Delta y_1|_{mean}$
MeOH–water–CaCl$_2$	−0.1	0.024	1.0	0.019
MeOH–water–Pot. acet.	−0.2	0.038[a]	0.2	0.040[a]
MeOH–water–HgCl$_2$	−1.0	0.037	0.4	0.042
MeOH–water–LiCl	−1.1	0.010	—	—
EtOH–water–HgCl$_2$	−0.3	0.056	—	—

[a] For $\alpha_{12} = 0.0$, $|\Delta y_1|_{mean} = 0.036$.

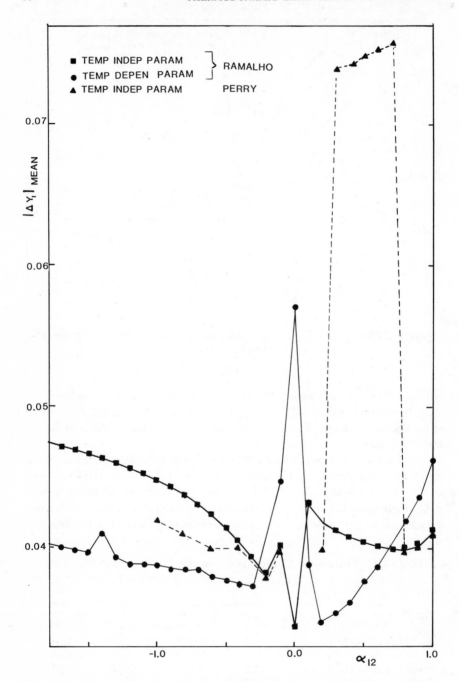

Figure 2.   Plot of $|\Delta y_1|_{mean}$ vs. α_{12}: methanol–water–potassium acetate @ 764 mmHg; $\alpha_{13} = 0.0$ (ideal), $\alpha_{23} = 0.30$

water–potassium acetate system salting-out is correctly predicted, as shown in Figure 6. Good results were also obtained for the MeOH–water–calcium chloride system, as shown in Figure 7. Finally, excellent results were obtained for the MeOH–water–lithium chloride system, as shown in Table IV.

For the isobaric data the values of the average absolute deviation in temperature is about 2°C except for the potassium acetate system where the value is very high at about 17.5°C (*19*). For the latter system, use of parameters that were linear functions of temperature to account for the large temperature variations encountered in the three binary and the ternary systems (up to 61°C), gave

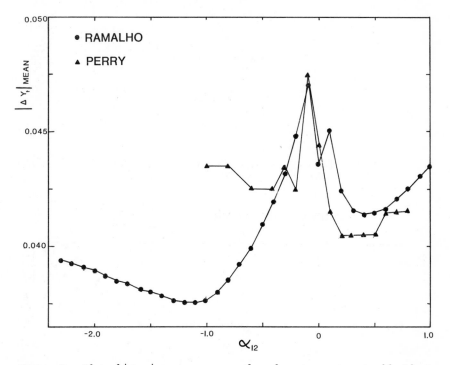

Figure 3.   Plot of $|\Delta y_1|_{mean}$ vs. α_{12}: methanol–water–mercuric chloride @ 758 mmHg; $\alpha_{13} = 0.30$, $\alpha_{23} = 0.0$ (ideal)

a slightly improved y_1 fit but essentially no improvement in the temperature fit (*19*). The studies of Rousseau et al. and Jaques and Furter (*7, 8, 9, 20*) did not include this system. Meranda and Furter (*5*) report serious difficulties in obtaining the data for this system because of the high solubility of potassium acetate. They also report that the correlation of Johnson and Furter, Equation 1 in this paper, which applies to a large variety of salts, failed for the MeOH–water–potassium acetate system. Considering the difficulties encountered in obtaining and correlating the data for this system, the performance of our correlation should be considered good.

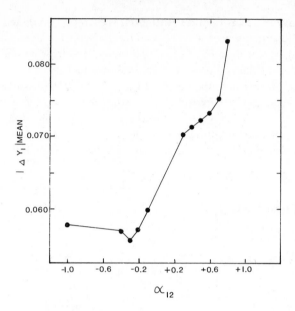

Figure 4.   Plot of $|\Delta y_1|_{mean}$ vs. α_{12}: ethanol–water–mercuric chloride @ 750 mmHg; $\alpha_{13} = 0.30$, $\alpha_{23} = 0.0$ (ideal)

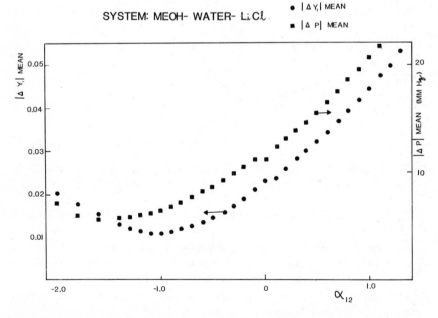

Figure 5.   Plot of $|\Delta y_1|_{mean}$ vs. α_{12}: plot of $|\Delta P|_{mean}$ vs. α_{12}; methanol–water–lithium chloride @ 60°C; $\alpha_{13} = 0.30$, $\alpha_{23} = 0.30$

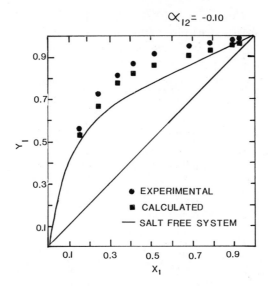

Figure 6.   Plot of y_1 vs. x_1 (salt-free basis): methanol–water–potassium acetate @ 764 mmHg; $\alpha_{12} = -0.20$, $\alpha_{23} = 0.30$, $\alpha_3 = 0.0$ (ideal)

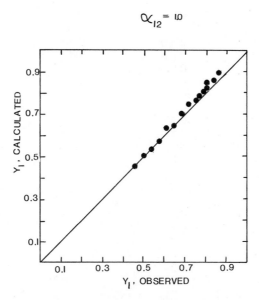

Figure 7.   Plot of y_{cal} vs. y_{obs}: methanol–water–CaCl$_2$ @ 752 mmHg; $\alpha_{12} = -0.10$, $\alpha_{23} = 0.30$, $\alpha_{13} = 0.0$ (ideal)

Table IV. Comparison of Results Obtained for the System
Methanol–Water–Lithium Chloride to the Results Obtained by
Broul and Hala

| | | Correlation of | |
| | $\alpha_{12} = -1.00$ | Broul | Hala |
y_1 obs.	y_1 calc.	y_1 calc.	y_1 calc.
0.093	0.079	0.083	0.083
0.114	0.104	0.107	0.107
0.042	0.070	0.067	—
0.103	0.110	0.104	0.105
0.166	0.163	0.161	0.162
0.155	0.164	0.157	0.158
0.245	0.232	0.263	0.229
0.156	0.168	0.170	—
0.243	0.239	0.229	0.230
0.335	0.339	0.332	0.334
0.414	0.412	0.407	0.408
0.340	0.362	0.351	—
0.340	0.351	0.353	—
0.499	0.496	0.492	0.493
0.432	0.448	0.431	0.441
0.725	0.667	0.748	0.742
0.499	0.512	0.506	—
0.588	0.587	0.589	0.597
0.490	0.494	0.495	—
0.571	0.579	0.575	0.575
0.801	0.778	0.847	—
0.673	0.676	0.687	0.686
0.616	0.622	0.620	0.620
0.646	0.653	0.653	0.653
0.729	0.738	0.754	—
0.709	0.710	0.715	0.714
0.849	0.846	0.904	—
0.776	0.791	0.812	0.809
0.761	0.773	0.785	0.783
0.752	0.764	0.766	—
0.877	0.885	0.933	—
0.808	0.822	0.843	—
0.810	0.825	0.843	—
0.826	0.839	0.857	0.855
0.850	0.861	0.882	0.878
0.900	0.910	0.943	—
0.876	0.887	0.915	—
0.846	0.856	0.861	0.860
0.895	0.904	0.928	—
0.942	0.948	0.967	0.965
$|\Delta y_1|_{mean}$	0.011	0.019	0.012

Table IV presents the results for the MeOH–water–LiCl system, along with those of Broul et al. and Hala.   The Hala study omitted some points of high salt concentration where the Broul calculations gave the largest errors.   The results are definitely better than those of Broul and slightly better than those of Hala.

Conclusions

The approach presented in this study correlates the available experimental results with accuracy ranging from poor to excellent.   On the basis of the results for the MeOH–water–LiCl system, the method provides better accuracy than the more complicated methods of Broul et al. and Hala.   The method is applicable to all salt concentrations up to saturation.   Use of the method for prediction purposes will require additional experimental data to establish the optimum value of α_{12} as a function of the system type.

Nomenclature

g_{ii}, g_{ij} = parameters of NRTL equation

G_{ij} = function of NRTL parameters per Equation 4

k_3 = salt effect parameter in Equation 1

P = total pressure

P_i° = saturated vapor pressure for pure solvent i at the temperature of the system

R = ideal gas constant

T = temperature, $^\circ K$

v_i = molar volume of component i in Wilson equation

x_i = analytical mole fraction of component i in the liquid phase; number of moles of i/total number of moles

x_i' = ionic mole fraction of component i, the liquid phase; defined by Equations 5 and 6

y_i = mole fraction of component i in the vapor phase

α_0 = relative volatility in the absence of salt in Equation 1

α_s = relative volatility in the presence of salt in Equation 1

α_{ij} = parameter in the NRTL equation

γ_i = activity coefficient of solvent i defined by Equation 2

γ_i' = ionic activity coefficient of solvent i defined by Equations 5 and 6

Δ = difference between observed and calculated values

ν = number of ions resulting from the dissociation of one mole of salt in Equations 5 and 6

Literature Cited

1. Cook, R. A., Furter, W. F., *Can. J. Chem. Eng.* (1968) **46**, 119.
2. Furter, W. F., "Extractive and Azeotropic Distillation," *Adv. in Chem. Ser.* (1972) **115**, 35.
3. Furter, W. F., Cook, R. A., *Int'l J. Heat Mass Transfer* (1967) **10**, 1, 23.
4. Johnson, A. I., Furter, W. F., *Can. J. Chem. Eng.* (1960) **38**, 78.
5. Meranda, D., Furter, W. F., *AIChE. J.* (1971) **17**, 1, 38.
6. Yoshida, F., Yasunishi, A., Hamada, Y., *Kagaku Kogaku* (English) (1964) **2**, 2, 162.
7. Jaques, D., Furter, W. F., *AIChE. J.* (1972) **18**, 2, 343.
8. Jaques, D., Furter, W. F., "Extractive and Azeotropic Distillation," Adv. in Chem. Ser. (1972) **115**, 159.
9. Rousseau, R. W., Ashcraft, D. L., Shoenborn, E. M., *AIChE. J.* (1972) **18**, 4, 825.
10. Van Laar, J. J., *Z. Physik Chem.* (1910) **72**, 723.
11. Wilson, G. M., *J. Amer. Chem. Soc.* (1964) **86**, 127.
12. Renon, H., Prausnitz, J. M., *AIChE. J.* (1968) **14**, 135.
13. Broul, M., Hlavaty, K., Linek, J., *Collect. Czech. Chem. Commun.* (English) (1969) **34**, 3428.
14. Hala, E., AIChE.—*Symp. Ser.* (1969) **32**, 5.
15. Larson, D., Tassios, D., *Ind. Eng. Chem. Process Des. Dev.* (1972) **11**, 35.
16. Marina, J. M., Tassios, D., *Ind. Eng. Chem. Process Des. Dev.* (1973) **12**, 67.
17. Sada, E., Morisue, T., J. of Chem. Eng. of Japan (English) (1973) **6**, 5, 385.
18. Ramalho, R. S., Tiller, F. M., James, W. J., Bunch, D. W., *Ind. Eng. Chem.* (1961) **53**, 67.
19. Bekerman, E. M., M.S. Thesis, N.J. Inst. of Tech., Newark, N.J., 1976.
20. Jaques, D., Furter, W. F., The Can. J. of Chem. Eng. (1972) **50**, 502.
21. Perry, J. H., "Chemical Engineer's Handbook," 4th ed., McGraw-Hill, N.Y., 1963.
22. Hala, E., Wichterle, I., Polak, J., Boublik, T., "Vapor–Liquid Equilibrium Data at Normal Pressures," Pergamon Press, N.Y., 1968.
23. Timmermans, J., "Physico–Chemical Constants of Binary Systems," Vol. 3, Interscience Publ. Inc., N.Y., 1960.

RECEIVED July 28, 1975.

A New Approach to the Calculation of Liquid–Vapor Equilibrium Data for Partially Miscible Systems Containing Salts at Saturation

DEREK JAQUES

Department of Applied Chemistry, Royal Melbourne Institute of Technology, Melbourne, Victoria, 3000, Australia

A method for interpolation of calculated vapor compositions obtained from Π-T-x data is described. Barker's method and the Wilson equation, which requires a fit of raw T-x data, are used. This fit is achieved by dividing the T-x data into three groups by means of the miscibility gap. After the mean of the middle group has been determined, the other two groups are subjected to a modified cubic spline procedure. Input is the estimated errors in temperature and a smoothing parameter. The procedure is tested on two ethanol- and five 1-propanol-water systems saturated with salt and found to be satisfactory for six systems. A comparison of the use of raw and smoothed data revealed no significant difference in calculated vapor composition.

There are two basic approaches to the calculation of vapor compositions from boiling point–liquid composition data or vapor pressure–liquid composition data: (a) the coexistence equation (1) which requires the smoothing of experimental T-x or Π-x data first, or (b) a correlating equation which relates the excess free energy with liquid composition. Various equations have been proposed, but Barker (2), who pioneered this method, employed Scatchard's equation (3). Raw or smoothed data are used, but the smoothing process may introduce unwarranted errors.

The present author (4) has previously preferred to use raw isobaric data coupled with Barker's method (2) and the Wilson equation (5), but interpolation of the calculated discrete vapor composition values requires smoothing of boiling point–liquid composition data at some stage.

The problems of smoothing are most acute with partially miscible systems, and it is with these that the present paper deals.

Procedure

The following method allowed comparison of the use of raw and smoothed data.

Step 1.   The function $\Sigma(\Pi - \Pi_c)^2$ is minimized where the total pressure is given by:

$$\Pi_c = xp_1'\gamma_1\phi_1 + (1 - x)p_2'\gamma_2\phi_2 \tag{1}$$

The vapor pressures of the pure liquid components are replaced by the vapor pressures of the liquids saturated with salts.   The Wilson equation (5) in its three-constant form is employed as the correlating equation.   This yields values of A_{21}, A_{12}, and C and the corresponding vapor compositions.

Step 2.   The common tangent gives the miscibility gap when molar free energy of mixing under isothermal conditions is plotted against liquid composition (6).   For the isobaric data of the present investigation, molar free energy of mixing divided by the absolute temperature is plotted against liquid composition because the T-x fit has not been applied at this stage.   (There was no significant difference in calculated miscibility gap between neglecting the temperature variation and using the smoothed temperatures from Step 3.)   An algorithm employed terminates when the slopes on either side of the miscibility gap differ by less than 0.0002.   This step is omitted if experimental data are available.

Step 3.   The T-x data are divided into three groups by means of the values from Step 2.   The arithmetic mean of the middle group is found.   Each of the other groups, together with either the datum for water saturated with salt and for the first equilibrium phase, or the datum for alcohol saturated with salt and for the second equilibrium phase, are subjected to cubic spline smoothing with the use of the Larkin method (7).   The normal spline smoothing process puts a cubic between every adjacent pair of data points and also requires that the first and second derivatives at each knot (data point) be equal.   In Larkin's modification each measured temperature is assumed to consist of signal and noise which are subject to a Gaussian (bell-shaped) distribution.   The temperatures are weighted by dividing by their estimated standard deviation.   This ensures that all adjusted temperatures come from the same distribution, which is a normal one of zero mean and unit variance.   A smoothing parameter, which is input, is used as a measure of how much each of the curves for signal and noise differ from a normal distribution of mean zero and unit variance.

The choice of error in temperature measurement and smoothing parameter was made as follows.   The error in temperature was assumed to be ±0.2°C for all liquid mixtures, and ±0.05°C for single liquids.   The rationale for the different treatment lies in lumping all the error in the ordinate.   Hence, in the case of single liquids one source of error is removed. In practice, the different treatment is generally necessary because the ends of the spline are not so rigidly defined. The optimum smoothing parameter was determined from a table of first and second derivatives given by runs of the program with various smoothing parameters by selecting the lowest parameter value which did not show discontinuity in the table. This was confirmed by examination of the T-x graph itself.

Finally, any spikes at the intersections of the three sections were removed by extrapolation of the center section.

Step 4. The smoothed temperatures were used to interpolate vapor compositions.

Step 5. Steps 1–4 were repeated in the second run with replacement of the experimental temperatures by the smoothed values.

Application to Experimental Data

The above procedure is now applied to two ethanol–water (8, 9) and five 1-propanol–water systems (9) which have been saturated with an inorganic salt and which show partial miscibility. The vapor pressures and molar volumes (10), and second virial coefficients of water (11), ethanol (12), and 1-propanol (13) were obtained by interpolation of literature data. The vapor pressures of water saturated with salts over a temperature range are available for all salts (14) except lead nitrate. Such data are unavailable for both alcohols saturated with salt. Hence a correction to the saturation vapor pressure is made by multiplying by the ratio of the vapor pressure of alcohol saturated with salts to the vapor pressure

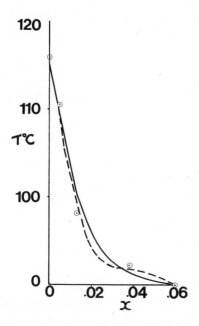

Figure 1. Comparison of experimental and calculated boiling points for 1-propanol–water–ammonium chloride system. Experimental: ⊙. *Calculated:* $Q_1 = 5 \times 10^{-6}$, ——; $Q_1 = 10^{-6}$, - - - -.

Table IA.   Effect of Smoothing Parameters for the Excess Water
Region of the 1-Propanol–Water–Ammonium Chloride System

Experimental		$Q_1 = 0.5 \times 10^{-5}$			$Q_1 = 10^{-6}$		
x	$T°C$	$T°C$	dT/dx	d^2T/dx^2	$T°C$	dT/dx	d^2T/dx^2
0.0	115.90	115.83	−1273	0	115.91	−1333	0
0.0050	110.40	109.56	−1216	22688	109.24	−1336	−1402
0.0130	98.20	100.95	−888	59260	99.45	−993	+87354
0.0380	92.20	91.48	−101	3718	91.85	−20	−9521
0.0591	89.98	89.90	−62	0	90.02	−120	0

Table IB.   Effect of Smoothing Parameters for the Excess Alcohol
Region of the 1-Propanol–Water–Ammonium Chloride System

Experimental		$Q_2 = 10^{-3}$			$Q_2 = 0.5 \times 10^{-3}$		
x	$T°C$	$T°C$	dT/dx	d^2T/dx^2	$T°C$	dT/dx	d^2T/dx^2
0.635	89.98	89.89	2.2	0	89.92	1.4	0
0.690	90.00	90.04	3.8	60	90.04	3.5	77
0.753	90.4	90.42	8.3	83	90.42	8.8	90
0.803	91.0	90.95	13.4	119	90.98	13.6	104
0.892	92.9	92.67	25.9	163	92.61	23.3	114
0.949	93.6	94.53	41.3	376	94.36	42.0	542
1.0	97.0	96.96	50.9	0	96.97	55.8	0

of pure alcohol at the boiling point of the salt solution.   This ratio is assumed
independent of temperature.   For lead nitrate in water a similar correction is
applied.

Because of uncertainties about the presence of an azeotrope close to the
water-free datum, the standard temperature error of 0.2°C was applied to this
datum for both ethanol systems.

Tables IA and IB show how the smoothing parameters are selected for the
1-propanol–water–ammonium chloride system (9).   For the water-rich region
the smaller of the two parameters gives an unwarranted inflection point which

Table II.   Comparison of the Use

		Run 1, raw data			
Alcohol	Salt	σ_π	σ_y	x'	x''
Ethanol	KNO_3	29.1	0.0201	0.110	0.419
Ethanol	$(NH_4)_2SO_4$	52.1	0.0518	0.042	0.428
1-Propanol	NH_4Cl	42.3	0.0253	0.059	0.633
1-Propanol	NaCl	12.0	0.0389	0.057	0.573
1-Propanol	$NaNO_3$	41.0	0.1381	0.296	0.700
1-Propanol	KCl	23.2	0.0325	0.060	0.561
1-Propanol	$Pb(NO_3)_2$	24.8	0.0164	0.092	0.455

is clearly visible on Figure 1. With the alcohol-rich data the position is not so clearcut, but the second derivatives for the smaller Q_2 are changing much less smoothly. This choice is somewhat subjective.

Comparison of columns 3 and 7 of Table II shows, as expected, that for all systems except the first, the sample deviations of total pressure are reduced after temperature smoothing. At this point in time no clearcut explanation can be given for the unexpected result for the potassium nitrate system (8). However, the paper reveals that an Othmer still was used to give x and y values only, and these data were subjected to smoothing. The boiling points of synthetic liquid measures were determined separately in a three-necked flask under total reflux. With the other systems of Table II all data were measured in a modified Othmer still.

A comparison of columns 4 and 8 reveals no clear pattern, which is perhaps of greater significance. The use of raw data yields smaller values of the vapor composition sample deviations in four out of six cases, but the effects are small and could be masked by errors in the vapor compositions themselves. It seems likely that the greatest source of error lies in determination of vapor composition. Thus there is very little difference in using raw or smoothed data. A typical example of the fit is shown in Figure 2. The optimum smoothing parameters used in run 1 were found to be the same as required for run 2, and are listed in columns 11 and 12 of Table II.

Comparison of the calculated miscibility gap data for the two runs reveals no significant difference except for sodium nitrate. This system also exhibits much higher values of σ_y in both runs. We have considered two explanations for the poor correlation.

The elevation of boiling point in this case is much greater than for any of the other systems. The assumptions that the heat term in the Gibbs–Duhem equation can be neglected and that the effect of the salt can be expressed in terms of its effect on the vapor pressure of each solvent independently become less viable as the boiling point elevation increases.

of Raw and Smoothed T-x Data

| Run 2, smoothed data | | | | Smoothing parameters | |
| | | | | $Q_1 \times 10^5$ | $Q_2 \times 10^3$ |
σ_π	σ_y	x'	x''		
35.1	0.0218	0.119	0.413	1	50
26.3	0.0608	0.051	0.432	1	50
19.9	0.0236	0.057	0.631	0.5	1
9.3	0.0389	0.057	0.568	0.1	5
39.6	0.1542	0.369	0.718	50	1
16.6	0.0316	0.059	0.560	0.05	1
16.2	0.0177	0.102	0.452	5	5

A second explanation involves the relative accuracy of the first three data points. Examination of the Δy values (Figure 3) reveals that the contribution of these three values distorts the final result. By ignoring them we can reduce the sample deviation from 0.1381 to 0.0387. The $\Delta \Pi$ values (Figure 4) show no particular bias; hence we might conclude that a likely source of error is in the experimental vapor compositions. In practice, the attainment of true equilibrium

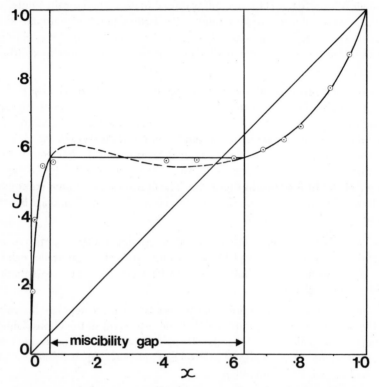

Figure 2. Comparison of experimental and calculated vapor compositions for 1-propanol–water–ammonium chloride system. Experimental: ⊙. Calculated in run 1: ———.

at these low alcohol concentrations and high salt concentrations is difficult. At this stage, it is not possible to decide in favor of either explanation.

A complete listing of m, c, ϵ, δ, A_{21}, A_{12}, and C values for raw data are available elsewhere (15).

Discussion

When given ordinate data represent approximate values, the problem can be tackled in three steps:

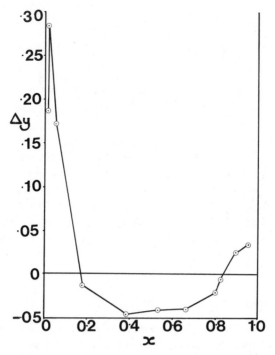

Figure 3.   *Difference between experimental and calculated vapor compositions vs. liquid composition for 1-propanol–water–sodium nitrate system*

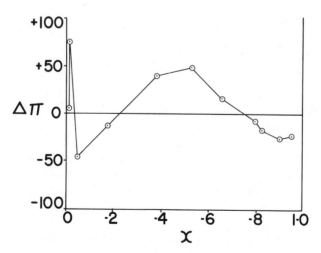

Figure 4.   *Difference between experimental and calculated total pressure vs. liquid composition for 1-propanol–water–sodium nitrate system*

(a) Smooth the given data.
(b) Fit a function to the smoothed data.
(c) Use the function to give best estimates of the ordinate values.
The best known method of attack, which uses least squares and either a polynomial of suitable degree or a cubic spline, combines the first two steps. This approach has been used successfully by Klaus and Van Ness (*16*). Larkin (*7*) notes that the problem belongs to the mathematical theory of probability, so we wish to extract the most probable signal from noisy data (step a) and then use a cubic spline to fit the extracted signals (step b). This approach provides a meaningful estimate of the errors in the ordinate values and a reasonable degree of smoothing. The ad hoc nature of the least squares approach is replaced by a method which is based upon sound probability theory. In Klaus and Van Ness's method weighting is employed only to ensure that all data points contribute equally. By the imposition of a purely arbitrary weighting, their approach might be expected to give a result similar to the one obtained in this paper, but this weighting is more difficult to justify mathematically.

Acknowledgments

The author wishes to thank Q. van Abbe of the Computer Centre at R.M.I.T. for many helpful discussions about spline functions, Bruce Bacon of the Australian Bureau of Statistics for assistance with the program for finding the miscibility gap, and the Computer Centre at R.M.I.T. for provision of facilities. The author would also like to express his gratitude to the Royal Melbourne Institute of Technology for the provision of a travel grant which made possible the presentation of this paper at the 170th ACS meeting in Chicago.

Nomenclature

Subscripts:

$1 =$ alcohol
$2 =$ water
$c =$ calculated

A_{21}, A_{12}, C	= empirical constants in Wilson equation
m, c	= empirical constants in the equation $\log p_2' = m \log p_2^0 - c$
p_i'	= vapor pressure of component i saturated with salt
Q_1, Q_2	= smoothing parameter in water-rich and alcohol-rich region respectively
T	= temperature ($^\circ$C)
x	= mole fraction of alcohol in the liquid phase, calculated on salt-free basis
x', x''	= composition of the two equilibrium liquid phases
γ_i	= activity coefficient of component i

δ, ϵ	= ratio of vapor pressure of water (alcohol) saturated with salt to the vapor pressure of pure solvent at the salt solution boiling point
Δy	$= y - y_c$
$\Delta \Pi$	$= \Pi - \Pi_c$
Π	= total pressure
σ_y	= sample deviation of the vapor composition
σ_{Π}	= sample deviation of the total pressure in Equation 1
ϕ_i	= correction term for non-ideality of component i in an ideal gaseous solution

Literature Cited

1. Van Ness, H. C., "Classical Thermodynamics of Non-Electrolyte Solutions," Pergamon Press, Oxford, 1964, p 137.
2. Barker, J. A., *Aust. J. Chem.* (1953) **6**, 207.
3. Scatchard, G., Raymond, C. L., *J. Am. Chem. Soc.* (1938) **60**, 1278.
4. Jaques, D., Furter, W. F., ADVAN. CHEM. SER. (1972) **115**, 159.
5. Wilson, G. M., *J. Amer. Chem. Soc.* (1964) **86**, 127.
6. Rowlinson, J. S., "Liquids and Liquid Mixtures," 2nd ed., Butterworth, London, 1969, p 144.
7. Larkin, F. M., "Optimal Estimation of Bounded Linear Functionals from Noisy Data," *Inf. Process. Proc. L.F.I.P. Congr., 1971*, Vol. 2. Ljubljana, North-Holland, pp 1335–1345.
8. Rieder, R. M., Thompson, A. R., *Ind. Eng. Chem.* (1950) **42**, 379.
9. Johnson, A. I., Furter, W. F., *Can. J. Chem. Eng.* (1960) **38**, 78.
10. Prausnitz, J. M., Eckert, C. A., Orye, R. V., O'Connell, J. P., "Computer Calculations for Multicomponet Vapor–Liquid Equilibrium," Prentice-Hall, Englewood Cliffs, N.J., 1967.
11. Rowlinson, J. S., *Trans. Faraday Soc.* (1949) **45**, 974.
12. Kretschner, C. B., Wiebe, R., *J. Am. Chem. Soc.* (1954) **76**, 2579.
13. Cox, J. D., *Trans. Faraday Soc.* (1961) **57**, 1674.
14. "International Critical Tables," Vol. III, McGraw-Hill, New York, 1929, pp 362–374.
15. Jaques, D., *Ind. Eng. Chem.* Process Des. Dev. (1976) **15**, 236.
16. Klaus, R. L., Van Ness, H. C., AIChE J. (1967) **13**, 1132.

RECEIVED July 14, 1975.

3

Correlation and Prediction of Salt Effect in Vapor–Liquid Equilibrium

WILLIAM F. FURTER

Department of Chemical Engineering, Royal Military College of Canada, Kingston, Ontario, Canada K7L 2W3

A review is presented of techniques for the correlation and prediction of vapor–liquid equilibrium data in systems consisting of two volatile components and a salt dissolved in the liquid phase, and for the testing of such data for thermodynamic consistency. The complex interactions comprising salt effect in systems which in effect consist of a concentrated electrolyte in a mixed solvent composed of two liquid components, one or both of which may be polar, are discussed. The difficulties inherent in their characterization and quantitative treatment are described. Attempts to correlate, predict, and test data for thermodynamic consistency in such systems are reviewed under the following headings: correlation at fixed liquid composition, extension to entire liquid composition range, prediction from pure-component properties, use of correlations based on the Gibbs–Duhem equation, and the recent "special binary" approach.

The use of a dissolved salt in place of a liquid component as the separating agent in extractive distillation has strong advantages in certain systems with respect to both increased separation efficiency and reduced energy requirements. A principal reason why such a technique has not undergone more intensive development or seen more than specialized industrial use is that the solution thermodynamics of salt effect in vapor–liquid equilibrium are complex, and are still not well understood. However, even small amounts of certain salts present in the liquid phase of certain systems can exert profound effects on equilibrium vapor composition, hence on relative volatility, and on azeotropic behavior. Also extractive and azeotropic distillation is not the only important application for the effects of salts on vapor–liquid equilibrium; while used as examples, other potential applications of equal importance exist as well.

For simplicity, the discussion in this review will be limited to the simplest system of this type possible: one consisting of two volatile components and a single salt, the latter considered to be completely nonvolatile. Hence, while the liquid phase will consist of all three components, the equilibrium vapor will contain only the two volatile species. Most previous investigators who have measured vapor–liquid equilibrium data in such systems have used saturated salt concentrations throughout to determine the maximum salt effect possible at each value of liquid composition. In such a case, of course, salt concentration changes as liquid composition is varied (except in the unlikely case of a salt being equally soluble in the two liquid components). However, a few investigators, recognizing that salt concentration would tend to remain essentially constant from tray to tray in an actual distillation column (as constant, in fact, as the concept of constant molal overflow is valid), have preferred to use salt concentrations held constant at values below saturation. The majority of the data reported in the literature are for boiling systems under isobaric conditions. The topics of salt effect in vapor–liquid equilibrium and extractive distillation by salt effect have been reviewed recently. Their literature is treated by Furter and Cook (*1*); their theory and practice by Ciparis, Dobroserdov, and Kogan (*2*), and by Furter (*3*); and the state of the art by Furter (*4*). Ciparis (*5*) has published a compilation of data for 188 systems.

The selective effect that a salt can have on the volatilities of the two liquid components, and hence on the composition of the equilibrium vapor, comes about primarily through effects exerted by the salt ions and/or molecules on the structure of the liquid phase. The most likely effect to be expected is that the salt would induce formation of association complexes, or clusters, of molecules of the volatile components about its ions. This effect would lower both of their volatilities but by differing amounts depending on the degree of selectivity of the particular salt in the preference of its ions for clustering with the molecules of one volatile component over those of the other. A preference for associating with the less volatile component would result in an increase in relative volatility and hence in ease of separation, and a preference for the more volatile component would have the opposite effect.

This is not the only structural mechanism possible. For instance, the salt may alter an existing liquid structure by promoting, destroying, or otherwise affecting interactions between the two volatile components, in some cases increasing rather than decreasing the volatility of a component. Still other effects are possible. All, of course, are functions of the relative amounts of all components present. Also, the types of short-range forces involved in liquid structure and in its promotion or other alteration by a salt may differ from system to system and from salt to salt.

The great increase in complexity in solution thermodynamics which occurs when a salt is dissolved to substantial concentration in a mixture of two liquid components becomes fully apparent in the realization that the liquid phase so created is a concentrated solution of an electrolyte whose degree of dissociation is a function of the relative proportions of the other two components present, and hence of liquid composition. Unless the salt is either fully associated or fully dissociated at all liquid compositions, it exists as a mixture of molecules and two species of ions, the relative proportions of which change with liquid composition. Undoubtedly, all three species play parts in determining the overall effect of the salt on the activities of the individual liquid components. However, their respective contributions, while possibly interrelated, probably all differ from each other. In one region of liquid composition, one effect may predominate, while in another region of the same system, it is quite possible that a different effect may prevail. Meranda and Furter (6) and others have observed the existence of such crossovers in salt effect in certain systems.

Such complexities tend to explain why progress has been relatively slow, at least until recently, in the formulation of effective relations and techniques for the representation of salt effect in vapor–liquid equilibrium.

Correlation at Fixed Liquid Composition

The original equation for salt effect in vapor–liquid equilibrium, proposed by Furter (7) and employed subsequently by Johnson and Furter (8), described the effect of salt concentration on equilibrium vapor composition under the condition of a fixed ratio of the two volatile components in the liquid phase. The equation, derived from the difference in effects of the salt on the chemical potentials of the two volatile components, with simplifying approximations reduces to the form

$$\ln (\alpha_s/\alpha) = kN_3 \tag{1}$$

In other words, the equation defines an improvement factor, which consists of the ratio of relative volatility with salt present (calculated using liquid composition on a salt-free basis for direct comparison purposes) to relative volatility at the same liquid composition but without salt present. It relates the logarithm of this improvement factor in a direct proportionality with N_3, the mole fraction of salt present in the liquid on a ternary basis. Jaques and Furter (17) tested the equation with data taken at several constant liquid compositions in four alcohol–water–inorganic salt systems, and observed good agreement.

The constancy of k with changing salt concentration is predicted only when the ratio of volatile components in the liquid is held constant. Considering that the salt effect is believed to be a complex function of interactions and self-interactions between all system components (which in turn are functions not only of composition but also of degree of salt dissociation, which itself is composition-dependent), there would be little reason to expect that k should have a single value for an entire system. That is, k would not be expected to remain constant over the entire liquid composition range of a given system. Nevertheless, in order to represent vapor–liquid equilibrium data for systems containing salts, an equation which holds over the entire composition range is required. Because of the absence of any such effective relation at the time, and also because of the undoubted attractiveness of using a single-constant equation for representing such complex phenomena, various investigators have employed Equation 1 empirically to correlate the data for entire systems.

Extension to Entire Liquid Composition Range

Furter (7) and Johnson and Furter (8, 9) tested Equation 1 with vapor–liquid equilibrium data over the entire concentration range for 24 alcohol–water– inorganic salt systems, in all of them discovering unexpectedly that a single best value of k for each system permitted correlation of the data to within average absolute deviations of one percent or less. Kogan et al. (44) observed good agreement with data for 14 salt-containing systems. When Johnson and Furter separated k into two individual salting parameters, one for salt–alcohol and the other for salt–water, it was observed that while the individual parameters varied considerably in value with alcohol–water proportionality, they tended to do so in a manner such that their difference, k, remained remarkably constant. In these particular systems the various interaction and self-interaction mechanisms comprising salt effect, all of which are composition-dependent, tended somehow to balance each other as liquid composition was varied. More recent investigations have shown that this apparent balancing effect (that is, the insensitivity of k to liquid composition) is not universal. Although Ramalho and Edgett (10) observed it in certain propionic acid–water–salt systems; and Ohe, Yokoyama, and Nakamura (11) observed it in the methanol–ethyl acetate–calcium chloride system; Yoshida, Yasunishi, and Hamada (12) observed it with certain methanol–water–salt systems but not with the acetic acid–water–salt systems which they tested. Meranda and Furter (6, 13, 14, 15), who experimented with a wide range of organic and inorganic salts and salt pairs in alcohol–water systems, observed little variation of k with liquid composition in some systems and large variations in others. The nature and use of Equation 1 have been discussed not only by Furter (7) and by Johnson and Furter (8) but more recently by Furter and Cook (1), by Meranda and Furter (13, 14), and by Jaques and Furter (16, 17). Ramalho and Edgett (10) described a reference diagram method for using Equation 1 with data obtained at constant salt concentration. Various investigators, particularly in the USSR, have experimented with similar relations with varying degrees of success, often under constraints such as infinite dilution and others. Their efforts have been reviewed elsewhere (1).

The effect of a given salt on vapor composition in a given system is, of course, a function of the relative proportions of the two volatile components in the liquid as well as of salt concentration, and an equation for correlation of salt effect at other than fixed liquid composition should contain liquid composition as a factor. Hashitani and Hirata (18) reported some success with a purely empirical equation which related the improvement factor of Equation 1 both to salt concentration and to liquid composition. Guyer, Guyer, and Johnsen (19) proposed an empirical relationship between vapor composition change and the concentration of one volatile component in the liquid while maintaining a fixed ratio of the concentrations of the other volatile component and salt. Sada, Kito, Yamaji, and Kimura (20), by considering vapor pressure lowering caused by the salt, derived an equation similar otherwise to Equation 1 but with the right hand side modified to an expression of salt concentration allowing for the numbers of cations and anions per salt molecule produced by dissociation, and containing no empirical parameter. They reported test results for the equation, which is restricted to systems in which the salt is soluble in only one of the two volatile components, with data for three benzene–ethanol–salt systems tested. Bedrossian and Cheh (21) proposed two empirical correlations and tested them with data for one system. Lu (22) proposed a correlation involving modified expressions of liquid composition which were intended to relate the relative contributions of the two volatile components to the overall salt solution. Liquid composition was expressed in pseudo mole fractions based on relative vapor pressure lowerings by the salt in question. Alvarez et al. (23, 24) have proposed and tested various empirical equations, in certain cases adopting the pseudo mole fraction approach of Lu.

The classical work relating liquid phase activity coefficients to the concentrations of electrolyte and nonelectrolyte species present in aqueous solution, and to the respective electrostatic interactions involved, was published by Long and McDevit (25). Hala (26) developed a generalized expression for characterization of vapor–liquid equilibrium in multicomponent systems containing both electrolyte and nonelectrolyte components. In his method a ternary system, for instance, is treated as three binaries. Correlation with excess free energy of mixing is achieved through extrapolation of dilute solution behavior to standard states. Jaques and Furter (16, 27) derived an equation relating the salt effect to both liquid composition and salt concentration by subtracting from the Long–McDevit equation in its ten-constant form the Redlich–Kister equation written in a four-constant form. The resulting six-constant equation was tested with data for a total of 12 systems, and, not surprisingly, yielded a better fit than did the one-constant Equation 1.

Prediction from Pure-Component Properties

Pure-component properties from which prediction of salt effect in vapor–liquid equilibrium might be sought, include vapor pressure lowering, salt solubility, degree of dissociation and ionic properties (charges and radii) of the salt, polarity, structural geometry, and perhaps others.

It has been generally held, at least until recently, that a salt dissolved in the liquid phase would enrich the equilibrium vapor in the component in which it was less soluble and impoverish it in the component in which it was more soluble. It was also assumed that the magnitude of the effect on vapor composition depended not only on how much salt was present but also on the degree of difference between the solubilities of the salt in the two liquid components taken separately. Various investigators, including Tursi and Thompson (28) and Fogg (29), have tried to relate the salt effect to this solubility difference alone, but their success has not been marked since such other factors as degree of dissociation and the ionic properties of the salt are neglected. Furter (7) and Johnson and Furter (8) demonstrated that the prediction that a salt would alter vapor composition in favor of the component in which it was less soluble was mathematically rigorous in the limited case of systems possessing festoon-like solubility curves. They also proposed an empirical equation expressing the relationship between salt saturation concentration and liquid composition in terms of the two pure-component salt solubilities and one other parameter. The expectations for prediction of the salt effect from solubility factors alone have been laid to rest by more recent discoveries of systems which behave anomalously in respect to the earlier-held generalities relating magnitude of salt effect to pure-component solubilities. Meranda and Furter (6, 13, 14) and Newstead and Furter (30), for instance, reported systems in which a reversal in the salt effect takes place at some point in the liquid composition range even though the salt is clearly more soluble in one component than in the other, others in which a salt was observed to enrich the vapor throughout in the component in which it is more, rather than less, soluble, and still others in which a large change in vapor composition is caused by a salt having little difference between its pure-component solubilities in the two liquid components.

The possibilities for predicting the salt effect from ionic properties alone are likewise improbable. General orders of effectiveness for anions and cations have been observed to exist by various investigators, including Long and McDevit (25), Prausnitz and Targovnik (31), Johnson and Furter (7, 8), Ciparis and Smorigaite (32), and others. In general, for similarly charged ions, the order of decreasing effectiveness follows the order of increasing ionic radius. Although the anion order is reasonably independent of the cation and vice versa in some systems, the uniformity decreases considerably when the liquid components are polar. In general, orders of ion effectiveness are only approximate and tend to exhibit some variance from system to system. For relating salt effect to ion radius and charge, the degree of dissociation, the number of ions per molecule, and the salt concentration must be considered so that ion parameters are isolated from the latter factors. The literature pertaining to ion order is reviewed in more detail elsewhere (1).

Over the years, various other theories and models have been proposed for predicting salt effect in vapor–liquid equilibrium, including ones based on hydration, internal pressure, electrostatic interaction, and van der Waals forces. These have been reviewed in detail by Long and McDevit (25), Prausnitz and Targovnik (31), Furter (7), Johnson and Furter (8), and Furter and Cook (1). Although the electrostatic theory as modified for mixed solvents has had limited success, no single theory has yet been able to account for or to predict salt effect on equilibrium vapor composition from pure-component properties alone.

Use of Correlations Based on the Gibbs–Duhem Equation

Empirical relations which work well for a variety of systems can be useful for correlating and, in limited cases, even predicting data but cannot be considered as criteria for judging their thermodynamic consistency. The criterion for thermodynamic consistency of vapor–liquid equilibrium data is that they must be consistent with the Gibbs–Duhem equation. Therefore only those relations which themselves are consistent with the Gibbs–Duhem equation can be deemed reliable for judging the thermodynamic consistency of such data or for correlating them in a thermodynamically consistent manner. The principal correlations and consistency tests for vapor–liquid equilibrium in use today, including those of van Laar, Margules, Redlich–Kister, Scatchard, Renon, Wilson, and others, can all be considered approximations to the integration of the Gibbs–Duhem equation.

For systems of the type under consideration, that is, consisting of two volatile components and a salt, there has been controversy over whether binary or ternary forms of correlating equations should be used, and over whether the presence of the salt should be included in the liquid mole fraction data used to calculate liquid activity coefficient values for the two volatile components. One point, however, is absolutely clear. It would be thermodynamically incorrect not to acknowledge the presence of the salt in calculating liquid-phase activity coefficients.

If the two volatile components are designated A and B respectively, and component A is used as an example, the mole fraction of component A in the liquid expressed on a salt-free basis is

$$x_A = \frac{\text{moles A}}{\text{moles A} + \text{moles B}} \tag{2}$$

However, if activity coefficient data were calculated for component A using the pure component vapor pressure and liquid composition data on a salt free basis, the activity coefficient values would not normalize (i.e., approach unity as x_A approaches unity) unless the salt were insoluble in component A. A better liquid composition expression would be

$$x_A' = \frac{\text{moles A}}{\text{moles A} + \text{moles B} + \text{moles salt}} \tag{3}$$

However the question of whether the salt should be considered as a molecular or ionic constituent is raised. The laws of solution theory suggest the latter. Hence, unless the salt is either fully associated or fully dissociated over the entire liquid composition range, the varying degree of salt dissociation over this range is important. In other words, since both species of ion and salt molecules contribute to the total effect caused by a partially dissociated salt, the total number of salt particles (ions and molecules) present should be considered. This would suggest that an even more correct expression of liquid composition for use in calculating liquid phase activity coefficients would be

$$x_A'' = \frac{\text{moles A}}{\text{moles A} + \text{moles B} + n \text{ moles salt}} \tag{4}$$

where n is a factor accounting for the degree of dissociation of the salt and is the statistical average number of particles (ions and molecules) of the salt in solution per molecule of salt dissolved.   For example, for the salt NaCl, $n = 1$ fully associated, $n = 2$ fully dissociated, and n lies between these limits for partial dissociation.   (Note that n is not a constant but is rather a function of the relative proportions of the volatile components A and B present in the liquid phase.) However even Equation 4 is not an ideal expression; an even more sophisticated approach would be to have the value of n also take account of ion radii and charges.   The real problem, of course, is in knowing the degree of salt dissociation as a function of liquid composition in a boiling system, and is a major reason why so little progress has been made over the years on thermodynamic correlation of salt effect in vapor–liquid equilibrium.

In summary, to be thermodynamically rigorous, the salt presence must be recognized in calculating activity coefficients for use in correlating equations; its degree of dissociation as a function of liquid composition, among other factors, must be considered also.   Conversely, it may be possible to apply data that are believed to be consistent to a consistency test in order to calculate degree of dissociation as a function of liquid composition.

Various investigators have encountered the difficulties of attempting to correlate data for systems containing salts by one or another of the common correlating equations or consistency tests for vapor–liquid equilibrium data, often in their binary versions, using activity coefficients computed from salt-free liquid composition data.   Among them are Lindberg and Tassios (33), Rius and Alvarez (34), Costa and Moragues (35), and other investigators such as Kogan, Rozen, and their associates whose work has been reviewed elsewhere (1).   Recently, Sada and Morisue (36) attempted to derive a relationship consistent with the Gibbs–Duhem equation by expressing salt effect on vapor composition as a function of liquid composition.   The liquid composition expression which they used acknowledged both the presence of the salt and the number of ions per salt molecule, but not ionic charges or radii, or variation in degree of salt dissociation with liquid composition.

The Special Binary Approach

Jaques and Furter (37, 38, 39, 40) devised a technique for treating systems consisting of two volatile components and a salt as "special binaries" rather than as ternary systems.   In this pseudo binary technique the presence of the salt is recognized in adjustments made to the pure-component vapor pressures from which the liquid-phase activity coefficients of the two volatile components are calculated, rather than by inclusion of the salt presence in liquid composition data.   In other words, alteration is made in the standard states on which the activity coefficients are based.   In the special binary approach as applied to salt-saturated systems, for instance, each of the two components of the binary is considered to be one of the volatile components individually saturated with the salt. The pure-component vapor pressures used to calculate liquid-phase activity coefficient values for the volatile components are the vapor pressures of the volatile components each saturated with salt at the temperature in question, rather than of the volatile components alone. Instead of defining the reference fugacity as the saturated vapor pressure of the pure component, it is defined as that of the pure component as depressed by the presence of the salt. In other words, the activity coefficients of the volatile components are based on standard states consisting of each volatile component saturated individually with the salt. Although only data for salt-saturated systems have been tested so far, the technique is not limited to saturation; for salt concentrations below saturation the standard-state vapor pressures would be those of the volatile components each depressed individually by the salt, at the salt concentration and temperature in question. The temperature in question is, of course, the boiling point of the salt-containing liquid phase at the composition for which an activity coefficient is calculated. The advantage of the approach is that any of the common correlating equations or thermodynamic consistency tests based on approximations to the Gibbs–Duhem equation can then be used in their standard binary forms, and with liquid compositions calculated on a salt-free basis. Hence the method obviates the necessity for knowing the relationship between degree of salt dissociation and liquid composition that would otherwise be required to calculate activity coefficients, information which for most systems is not generally known. The special binary approach, while semi-empirical, has been employed successfully by Jaques and Furter both in testing for thermodynamic consistency and in the correlation of data.

Jaques and Furter (37, 39) tested the thermodynamic consistency of literature data for 23 alcohol–water–salt systems and Jaques (41) studied 17 additional systems using the Herington method as adapted to their special binary technique.

Jaques and Furter (38, 40) also applied the special binary approach to the correlation and prediction of salt effect in vapor–liquid equilibrium using the Wilson equation as the correlating relation. For a total of 22 systems tested, the average deviations between observed and calculated vapor compositions ranged from <0.5% to as high as 3%, with most falling around the 1% mark. However, in all cases unsmoothed data from the literature were used; hence, these deviations include experimental scatter. More recently, Ashcraft (42) and Rousseau, Ashcraft, and Schoenborn (43) have reported success with a similar technique, employing as correlating relations the van Laar, Wilson, and Renon equations.

Acknowledgment

The research programs on extractive distillation by salt effect and on salt effect in vapor–liquid equilibrium at the Royal Military College of Canada are supported by the Defence Research Board of Canada, Grant No. 9530-142.

Literature Cited

1. Furter, W. F., Cook, R. A., *Int. J. Heat Mass Transfer* (1967) **10**, 23.
2. Ciparis, J. N., Dobroserdov, L. L., Kogan, V. B., "Salt Rectification," Khimiya, Leningrad, USSR, 1969.
3. Furter, W. F., *Chem. Eng.* (*I. Chem. E., London*) (1968) **219**, CE173.
4. Furter, W. F., "Extractive and Azeotropic Distillation," ADV. CHEM. SER. (1972) **115**, 35.
5. Ciparis, J. N., "Data of Salt Effect in Vapour–Liquid Equilibrium," Lithuanian Agricultural Academy, Kaunas, Lithuania, USSR, 1966.
6. Meranda, D., Furter, W. F., *A.I.Ch.E. J.* (1974) **20**, 103.
7. Furter, W. F., Ph.D. Thesis, University of Toronto, Toronto, Ontario, Canada, 1958.
8. Johnson, A. I., Furter, W. F., *Can. J. Chem. Eng.* (1960) **38**, 78.
9. Johnson, A. I., Furter, W. F., *Can. J. Chem. Eng.* (1965) **43**, 356.
10. Ramalho, R. S., Edgett, N. S., *J. Chem. Eng. Data* (1964) **9**, 324.
11. Ohe, S., Yokoyama, K., Nakamura, S., *J. Chem. Eng. Data* (1971) **16**, 70.
12. Yoshida, F., Yasunishi, A., Hamada, Y., *Kagaku Kogaku* (1964) **28**, 133.
13. Meranda, D., Furter, W. F., *Can. J. Chem. Eng.* (1966) **44**, 298.
14. Meranda, D., Furter, W. F., *A.I.Ch.E. J.* (1971) **17**, 38.
15. Meranda, D., Furter, W. F., *A.I.Ch.E. J.* (1972) **18**, 111.
16. Jaques, D., Furter, W. F., *Can. J. Chem. Eng.* (1972) **50**, 502.
17. Jaques, D., Furter, W. F., *Ind. Eng. Chem. Fundam.* (1974) **13**, 238.
18. Hashitani, M., Hirata, M., *J. Chem. Eng. Japan* (1969) **2**(2), 149.
19. Guyer, A., Guyer, A., Jr., Johnsen, B. K., *Helv. Chim. Acta* (1942) **38**, 946.
20. Sada, E., Kito, S., Yamaji, H., Kimura, M., *J. Appl. Chem. Biotechnol.* (1974) **24**, 229.
21. Bedrossian, A. A., Cheh, H. Y., *A.I.C.h.E. Symp. Ser.* (1974) **70** (140), 102.
22. Lu, B. C.-Y., *Ind. Eng. Chem.* (1960) **52**, 871.
23. Alvarez Gonzales, J. R., Bueno Cordero, J., Galan Serrano, M. A., *An. Quim.* (1974) **70**, 262.
24. Alvarez Gonzales, J. R., Galan Serrano, M. A., *An. Quim.* (1974) **70**, 271.
25. Long, F. A., McDevit, W. F., *Chem. Rev.* (1952) **51**, 119.
26. Hala, E., *Int. Chem. Eng. Symp.* Ser. (*I. Chem. Engin., London*) (1969) **32** (3), 8.
27. Furter, W. F., Jaques, D., 1970 Annual Report, Grant No. 9530-40, Appendix B, Defence Research Board of Canada, Ottawa, Ontario, 1970.
28. Tursi, R. R., Thompson, A. R., *Chem. Eng. Prog.* (1951) **47**, 304.
29. Fogg, E. T., University Microfilms, Publication 5589, Ann Arbor, Michigan, 1953.
30. Newstead, W. T., Furter, W. F., *A.I.Ch.E. J.* (1971) **17**, 1246.
31. Prausnitz, J. M., Targovnik, J. H., *Chem. Eng. Data* Ser. (1958) **3**, 234.
32. Ciparis, J. N., Smorigaite, N., *Zh. Obshch. Khim.* (1964) **34** (12), 3867.
33. Lindberg, G. W., Tassios, D., *J. Chem. Eng. Data* (1971) **16**, 52.
34. Rius Miro, A., Alvarez Gonzales, J. R., Uriarte Hueda, A., *An. Quim.* (1960) **56B** (6), 629.
35. Costa Novella, E., Moragues Tarraso, J., *An. Quim.* (1952) **48B** (*6*), 441.
36. Sada, E., Morisue, T., *J. Chem. Eng. Japan* (1973) **6**,(5), 385.
37. Furter, W. F., Jaques, D., 1970 Annual Report, Grant No. 9530-40, Appendix A, Defence Research Board of Canada, Ottawa, Ontario, 1970.
38. Furter, W. F., Jaques, D., 1971 Annual Report, Grant No. 9530-40, Appendix D, Defence Research Board of Canada, Ottawa, Ontario, 1971.
39. Jaques, D., Furter, W. F., *A.I.Ch.E. J.* (1972) **18**, 343.
40. Jaques, D., Furter, W. F., "Extractive and Azeotropic Distillation," ADV. CHEM. SER. (1972) **115**, 159.
41. Jaques, D., *A.I.Ch.E. J.* (1974) **20**, 189.
42. Ashcraft, D. L., M.S. Thesis, North Carolina State University at Raleigh, Raleigh, N.C., 1972.
43. Rousseau, R. W., Ashcraft, D. L., Schoenborn, E. M., *A.I.Ch.E. J.* (1972) **18**, 825.
44. Kogan, V. B., Bulushev, S. F., Safronov, V. M., Moskovets, O. F., *Zh. Prikl. Khim.* (1959) **32**, 2409.

RECEIVED June 9, 1975.

4

The Correlation of Vapor–Liquid Equilibrium Data for Salt-Containing Systems

J. E. BOONE, R. W. ROUSSEAU, and E. M. SCHOENBORN

Department of Chemical Engineering, North Carolina State University, Raleigh, N. C. 27607

A procedure is presented for correlating the effect of non-volatile salts on the vapor–liquid equilibrium properties of binary solvents. The procedure is based on estimating the influence of salt concentration on the infinite dilution activity coefficients of both components in a pseudo-binary solution. The procedure is tested on experimental data for five different salts in methanol–water solutions. With this technique and Wilson parameters determined from the infinite dilution activity coefficients, precise estimates of bubble point temperatures and vapor phase compositions may be obtained over a range of salt and solvent compositions.

Separation processes which involve non-volatile salts arise in two situations. First, as an alternative to extractive or azeotropic distillation, salts may be added to a system to alter the vapor–liquid equilibrium behavior. Second, there are cases where a salt is generated in the process before final product purification. For example, product streams from processes involving esterification, etherification, or neutralization contain salts and are often fed to separation units such as distillation or stripping columns.

An accurate representation of the phase equilibrium behavior is required to design or simulate any separation process. Equilibrium data for salt-free systems are usually correlated by one of a number of possible equations, such as those of Wilson, Van Laar, Margules, Redlich–Kister, etc. These correlations can then be used in the appropriate process model. It has become common to utilize parameters from such correlations to obtain insight into the fundamentals underlying the behavior of solutions and to predict the behavior of other solutions. This has been particularly true of the Wilson equation, which is shown below for a binary system.

$$\ln \gamma_1 = -\ln (x_1 + \Lambda_{12} x_2) + x_2 \left[\frac{\Lambda_{12}}{x_1 + \Lambda_{12} x_2} - \frac{\Lambda_{21}}{\Lambda_{21} x_1 + x_2} \right] \tag{1a}$$

$$\ln \gamma_2 = -\ln (x_2 + \Lambda_{21} x_1) - x_1 \left[\frac{\Lambda_{12}}{x_1 + \Lambda_{12} x_2} - \frac{\Lambda_{21}}{\Lambda_{21} x_1 + x_2} \right] \tag{1b}$$

$$\Lambda_{ij} = \frac{v_j{}^L}{v_i{}^L} \exp \left[- \frac{\lambda_{ij} - \lambda_{ii}}{RT} \right] \tag{1c}$$

$$\lambda_{ij} = \lambda_{ji} \tag{1d}$$

An advantage of the Wilson equation is that it involves only two parameters per binary and may be extended, without further information, to estimate multi-component phase equilibrium behavior.

Two approaches have been used in correlating the phase equilibrium behavior of complex mixtures involving a non-volatile salt dissolved in a binary solvent mixture. Johnson and Furter (1) developed what appears to be the most popular approach by correlating the ratio of relative volatilities of the solvents as a function of salt concentration. Meranda and Furter (2) review this approach and present experimental determinations of the necessary parameters as a function of mole fraction of one of the solvents.

An alternative approach is to estimate activity coefficients of the solvents from experimental data and correlate these coefficients using, for example, the Wilson equation. Rousseau et al. (3) and Jaques and Furter (4) have used the Wilson equation, as well as other integrated forms of the Gibbs–Duhem equation, to show the utility of this approach. These authors found it necessary, however, to modify the definitions of the solvent reference states so that the results could be normalized.

The data against which these two approaches have been tested have for the most part been on salt-saturated systems. There have been no tests of the second approach in which the solution was not salt-saturated. Many applications, however, would be concerned with systems having salt concentrations below saturation; in any case, to limit a technique to salt-saturated systems means that an available degree of freedom has been unnecessarily removed.

Therefore, the objectives of this study were to investigate the influence of salt concentration on the vapor–liquid equilibrium behavior of aqueous solutions of methyl alcohol and to develop a fundamentally sound approach to correlating the influence of salt on the behavior.

Pseudobinary Systems

To apply the binary form of the Wilson equation, Jaques and Furter (4, 5) and Rousseau et al. (3) treated the salt–solvent systems as pseudobinaries by expressing the solvent compositions on a salt-free basis. In addition, reference fugacities were defined to adjust the vapor pressure of the liquids by an amount equal to the vapor pressure lowering caused by the presence of the salt. While this approach resulted in successful correlations, it was clearly empirical.

For this study, a different pseudobinary approach was adopted: solvent 1, which salted out, was designated as component 1* while the mixture of solvent 2 and the salt in a constant mole ratio was designated component 2*. Defining the system in this manner means that it can be treated as a binary and the equilibrium relationships governing the behavior of the system can then be written as

$$\gamma_{i*} \, x_{i*} \, f_{i*}{}^0 = \phi_{i*} \, y_{i*} \, P \tag{2}$$

where

$$x_{1*} = n_1/(n_1 + n_2 + n_3) \tag{3a}$$

$$x_{2*} = (n_2 + n_3)/(n_1 + n_2 + n_3) \tag{3b}$$

Since the salt is nonvolatile

$$y_{i*} = y_i \tag{4}$$

Choosing the reference state of each component in the pseudobinary solution to be the pure component (1* and 2*), the reference state fugacities are given by

$$f_{1*}{}^0 = P^s{}_1 \tag{5a}$$

$$f_{2*}{}^0 = P^s{}_{2*} = P^s{}_2 - \Delta P^s{}_2 \tag{5b}$$

Infinite Dilution Activity Coefficients

Shreiber and Eckert (6) developed the concept of estimating Wilson parameters for binary systems by examination of infinite dilution activity coefficients. Their technique was adapted to the pseudobinary salt–solvent system and used to determine the effect of salt on activity coefficients and the corresponding Wilson parameters. This was done as outlined in the following paragraphs.

Long and McDevitt (7) express the salt effect, S_i, as the ratio of the fugacity of a component in the salt-containing system to the fugacity of that component in the salt-free system at the same ratio of solvent 1 to solvent 2. Using a superscript (′) to denote the salt-free system

$$S_i = \frac{f_i}{f_{i'}} \tag{6}$$

where

$$\frac{x_i}{x_j} = \frac{x_{i'}}{x_{j'}} \tag{7}$$

Substituting the usual expressions for fugacity into Equation 6

$$S_i = \frac{x_i \, \gamma_i \, P^s_i}{x_{i'} \, \gamma_{i'} \, P^s_{i'}} \tag{8}$$

Recall that solvent 1 is identical to component 1* and therefore $f_1 = f_1{}^*$. Substituting this expression into Equation 8 and rearranging

$$\frac{\gamma_{1*}}{\gamma_{1'}} = \frac{x_{1'}}{x_{1*}} \frac{P^s_{1'}}{P^s_{1*}} S_1 \tag{9}$$

Taking the limit of both sides of Equation 9 as x_{1*} goes to zero

$$\lim_{x_{1*} \to 0} \frac{\gamma_{1*}}{\gamma_{1'}} \equiv \frac{\gamma_{1*}{}^\infty}{\gamma_{1'}{}^\infty} = \lim_{(x_{1*} \to 0)} \left[\frac{x_{1'}}{x_{1*}} \frac{P^{s*}_{1'}}{P^s_{1*}} S_1 \right] \tag{10}$$

From Equations 3b and 4

$$\lim_{x_{1*} \to 0} \frac{x_{1'}}{x_{1*}} = \frac{1}{x_2} \tag{11}$$

King (8) points out that the salt effect has been successfully correlated by Setschenow's equation in which

$$S = \exp \left(k_s \, C_e \right) \tag{12}$$

Furthermore,

$$\lim_{x_{1*} \to 0} \frac{P^s_{1'}}{P^s_{1*}} = \frac{P^s_1 \text{ (at boiling point of 2)}}{P^s_{1*} \text{ (at boiling point of 2*)}} = \frac{P^s_1 \, (2)}{P^s_1 \, (2*)} \tag{13}$$

Substituting Equations 11, 12, and 13 into Equation 10

$$\frac{\gamma_{1*}{}^\infty}{\gamma_{1'}{}^\infty} = \frac{1}{x_2} \frac{P^s_1 \, (2)}{P^s_1 \, (2*)} \exp \left(k_s \, C_e \right) \tag{14}$$

Now considering solvent 2

$$\frac{\gamma_{2*}}{\gamma_{2'}} = \frac{y_{2*} \, x_{2'}}{y_{2'} \, x_{2*}} \frac{P^s_{2'}}{P^s_{2*}} \tag{15}$$

In the limit as x_{2*} approaches zero, $y_{2*}/y_{2'}$ is approximately 1. It is also shown easily that $x_{2'}/x_{2*}$ is approximately 1. Therefore, taking the limit of both sides of Equation 15 as x_{2*} goes to zero,

$$\lim_{x_{2*} \to 0} \frac{\gamma_{2*}}{\gamma_{2'}} = \frac{\gamma_{2*}{}^\infty}{\gamma_{2'}{}^\infty} \frac{P_{2'} \text{ (at boiling point of 1)}}{P^s_{2*} \text{ (at boiling point of 1)}} = \frac{P^s_2}{P^s_2 - \Delta P^s_2} \tag{16}$$

On examination of Equations 14 and 16, it is clear that estimates of vapor pressure lowering and k_s are all that are required to determine the influence of a given salt on the vapor–liquid equilibrium behavior of a binary solvent mixture.

　　To test the validity of the concept outlined in the previous paragraphs, experimental data were obtained for the effect of each of five salts on the equilibrium properties of methanol–water mixtures; each data set was treated as a pseudobinary and values for k_s determined from observed infinite dilution activity coefficients.

Experimental Equipment and Procedure

Included among the salts chosen for study were those that cause salting-out (NaBr, NaF, KCl, Li Cl) and salting-in (HgCl$_2$) of methanol in aqueous solutions. To test the technique described above, the vapor–liquid equilibria of systems of constant ratios of salt to solvent 2 were measured. For example, in cases where methanol is salted out, the experiments were done at constant salt-to-water ratios, and when methanol is salted in (salting-out of water), constant salt-to-methanol ratios were used. This was done by preparing a solution of a fixed salt molality and using it as component 2* in the equilibrium still. Thus, references to molality refer to the ratio moles of salt to 1000 g of solvent 2.

Vapor–liquid equilibrium experiments were performed with an improved Othmer recirculation still as modified by Johnson and Furter (2). Temperatures were measured with Fisher thermometers calibrated against boiling points of known solutions. Equilibrium compositions were determined with a vapor fractometer using a type W column and a thermal conductivity detector. The liquid samples were distilled to remove the salt before analysis with the gas chromatograph; the amount of salt present was calculated from the molality and the amount of solvent 2 present. Temperature measurements were accurate to ±0.2°C while compositions were found to be accurate to 1% over most of the composition range. The system pressure was maintained at 1 atm. ± 1 mm Hg.

Table I. Equilibrium Data for Methanol–H$_2$O–HgCl$_2$

Salt Molality	x_2*	y_2*	$T(°C)$
2.0	0.033	0.225	94.1
2.0	0.081	0.395	88.5
2.0	0.185	0.530	83.7
2.0	0.274	0.655	80.0
2.0	0.413	0.701	77.1
2.0	0.538	0.799	73.6
2.0	0.753	0.879	71.5
2.0	0.919	0.955	66.9
3.0	0.044	0.192	95.3
3.0	0.107	0.401	89.6
3.0	0.212	0.543	84.8
3.0	0.298	0.621	81.0
3.0	0.373	0.654	78.0
3.0	0.569	0.800	73.9
3.0	0.748	0.871	71.0
3.0	0.921	0.956	67.6

Table II. Equilibrium Data for Methanol–H₂O–NaBr

Salt Molality	x_1^*	y_1^*	$T(^\circ C)$
1.0	0.045	0.303	93.8
1.0	0.087	0.409	89.2
1.0	0.144	0.545	85.0
1.0	0.221	0.647	81.0
1.0	0.464	0.725	74.3
1.0	0.656	0.848	70.8
1.0	0.869	0.937	67.5
1.0	0.928	0.968	66.0
2.0	0.048	0.332	93.9
2.0	0.089	0.459	88.7
2.0	0.139	0.586	84.7
2.0	0.245	0.678	80.6
2.0	0.477	0.781	74.5
2.0	0.612	0.859	71.0
2.0	0.844	0.938	67.0
2.0	0.929	0.968	66.0
4.0	0.024	0.212	97.4
4.0	0.052	0.435	91.6
4.0	0.106	0.543	85.5
4.0	0.205	0.688	80.6
4.0	0.683	0.853	69.8
4.0	0.808	0.933	67.4
4.0	0.916	0.972	65.8

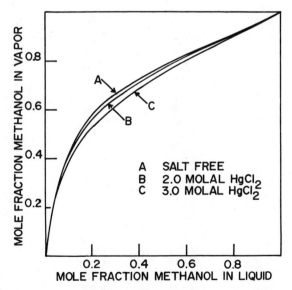

Figure 1. Smoothed equilibrium data for methanol–water–HgCl₂

Results and Discussion

The vapor–liquid equilibrium data for the five salt systems are given in Tables I–V using the pseudobinary nomenclature. Systems containing NaF, KCl, NaBr, and LiCl exhibited the expected salting-out of methanol with the degree of salting-out increasing with salt molality. Salting-in of methanol was observed with $HgCl_2$. Figure 1 shows the smoothed $x-y$ data for the $HgCl_2$ which is representative of all the data.

Using the virial equation to estimate the slight deviations from ideality exhibited by the vapor, activity coefficients were calculated from Equation 2. Infinite dilution activity coefficients for component 2* were estimated from Equation 16; infinite dilution activity coefficients for component 1* were determined graphically by extrapolating the values calculated from Equation 2

Table III.   Equilibrium Data for Methanol–Water–KCl

Salt Molality	x_1*	y_1*	$T(^\circ C)$
0.5	0.007	0.069	98.4
0.5	0.014	0.121	97.3
0.5	0.027	0.196	95.4
0.5	0.061	0.368	90.6
0.5	0.140	0.571	84.8
0.5	0.248	0.663	79.8
0.5	0.306	0.722	77.3
0.5	0.461	0.793	73.8
0.5	0.700	0.872	69.4
0.5	0.888	0.948	66.3
0.5	0.937	0.975	65.2
1.0	0.008	0.081	98.7
1.0	0.015	0.137	97.2
1.0	0.031	0.248	94.4
1.0	0.073	0.432	88.6
1.0	0.129	0.535	84.4
1.0	0.235	0.662	79.5
1.0	0.640	0.855	70.5
1.0	0.825	0.929	67.5
1.0	0.918	0.968	66.2
1.0	0.964	0.988	65.5
2.0	0.008	0.088	99.5
2.0	0.026	0.232	95.9
2.0	0.055	0.424	90.1
2.0	0.101	0.519	86.3
2.0	0.136	0.621	82.8
2.0	0.201	0.674	80.1
2.0	0.346	0.745	76.5
2.0	0.479	0.821	73.5
2.0	0.724	0.903	69.5
2.0	0.879	0.953	66.5
2.0	0.941	0.979	65.4

Table IV.   Equilibrium Data for Methanol–H$_2$O–LiCl

Salt Molality	x_1*	y_1*	$T(^\circ C)$
1.0	0.034	0.189	94.9
1.0	0.095	0.442	88.0
1.0	0.191	0.580	82.0
1.0	0.239	0.678	78.7
1.0	0.401	0.762	74.1
1.0	0.495	0.797	71.6
1.0	0.714	0.861	69.0
1.0	0.929	0.968	65.9
2.0	0.038	0.249	95.4
2.0	0.075	0.427	90.0
2.0	0.152	0.589	84.7
2.0	0.257	0.685	79.6
2.0	0.377	0.754	76.0
2.0	0.557	0.824	72.0
2.0	0.685	0.898	69.1
2.0	0.924	0.967	66.0
4.0	0.032	0.235	98.9
4.0	0.076	0.463	91.0
4.0	0.124	0.571	87.0
4.0	0.209	0.685	81.2
4.0	0.332	0.729	78.2
4.0	0.496	0.818	73.8
4.0	0.774	0.911	68.9
4.0	0.934	0.975	66.2

to obtain $\gamma_{1}*$ at $x_1* = 0$. With the infinite dilution activity coefficients determined in this manner, a general minimization algorithm was used to estimate the Wilson parameters. Both infinite dilution activity coefficients and the Wilson parameters are given in Table VI. For all but the mercuric chloride system, molar volume data of pure methanol and water were used in the Wilson equations. Volumetric data for HgCl$_2$–methanol were measured, but the variation of these properties with temperature was estimated from data on pure methanol.

Activity coefficients determined from the Wilson equations using the parameters given in Table VI are plotted in Figures 2–6. These data were then used in bubble point calculations to generate values of bubble point temperatures and vapor compsitions. Comparisons of vapor compositions calculated in this manner with those measured experimentally are shown in Figures 7–11. In almost all cases, the agreement is within 0.02 mole fraction units. There does appear, however, to be a tendency to underestimate the salt effect in the middle composition range. This may be caused by the approximations in estimating molar volumes.

By evaluating the salt parameter, k_s, for each of the salts, the utility of limited experimental data may be extended. On rearranging Equation 14:

Table V.   Equilibrium Data for Methanol–H$_2$O–NaF

Salt Molality	x_1*	y_1*	$T(°C)$
0.25	0.030	0.193	95.0
0.25	0.073	0.363	90.4
0.25	0.171	0.587	83.0
0.25	0.245	0.670	79.0
0.25	0.319	0.692	77.3
0.25	0.510	0.743	73.2
0.25	0.718	0.882	69.8
0.25	0.917	0.962	66.5
0.50	0.040	0.242	94.1
0.50	0.068	0.357	90.0
0.50	0.116	0.481	86.3
0.50	0.172	0.589	82.0
0.50	0.285	0.681	78.3
0.50	0.448	0.764	74.5
0.50	0.603	0.836	71.8
0.50	0.814	0.916	68.0
0.50	0.920	0.968	66.0
1.0	0.027	0.199	95.5
1.0	0.063	0.353	90.6
1.0	0.140	0.527	84.3
1.0	0.231	0.643	80.0
1.0	0.437	0.781	74.8
1.0	0.590	0.829	71.2
1.0	0.761	0.905	68.7
1.0	0.916	0.970	65.8

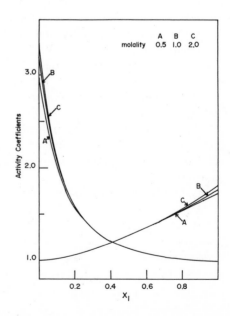

Figure 2. Activity coefficients for methanol and water in KCl system

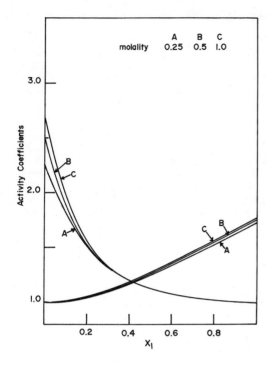

Figure 3. Activity coefficients for methanol and water in NaF system

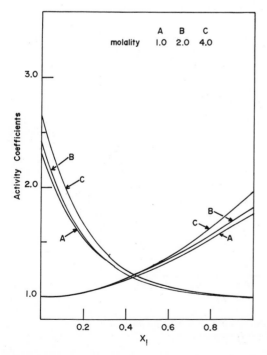

Figure 4. Activity coefficients for methanol and water in LiCl system

Table VI. Infinite Dilution Activity Coefficients and Wilson Parameters

Salt	Molality	$\gamma_1^{*\infty}$	$\gamma_2^{*\infty}$	$\lambda_{12}-\lambda_{11}$	$\lambda_{11}-\lambda_{22}$
KCl	0.5	2.97	1.74	437	410
KCl	1.0	3.20	1.77	508	400
KCl	2.0	2.25	1.78	562	386
NaF	0.25	2.31	1.75	132	509
NaF	0.5	2.50	1.75	201	495
NaF	1.0	2.68	1.78	296	461
LiCl	1.0	2.31	1.78	113	521
LiCl	2.0	2.43	1.82	156	520
LiCl	4.0	2.65	1.98	171	571
NaBr	1.0	2.62	1.78	234	498
NaBr	2.0	2.73	1.82	296	483
NaBr	4.0	3.10	1.92	292	460
$HgCl_2$	2.0	2.02	2.32	327	307
$HgCl_2$	3.0	2.26	2.35	447	196

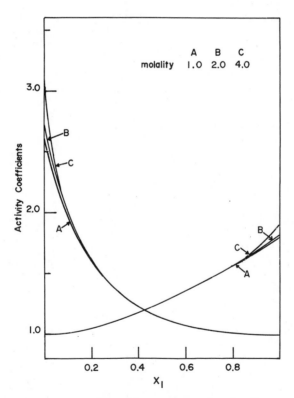

*Figure 5.   Activity coefficients for methanol and water
in NaBr system*

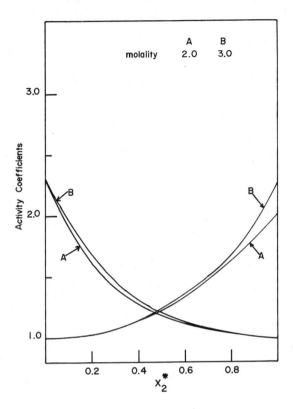

Figure 6. Activity coefficients for methanol and water in HgCl₂ system

$$k_s \, C_e = \ln \left[x_2 \frac{P^{s_1}(2^*) \, \gamma_1^{*\infty}}{P^{s_1}(2) \, \gamma_{1'}^{\infty}} \right] \tag{17}$$

Plots of the right-hand side of Equation 17 against C_e should be linear with a slope of k_s for each salt. These were constructed and are shown in Figure 12 with C_e expressed in terms of molality to avoid the need for molar volume data. Values of k_s, which were determined from a linear least squares fit through the origin, are presented for each salt in Table VII.

As shown in Figure 12, the precision of the linear fit used to estimate k_s varies among the salts tested. However, using these values of k_s to estimate the effect of a given salt on vapor–liquid equilibrium appears to give a reasonably good approximation to the vapor–liquid equilibrium behavior.

Table VII. Values of the Salt Parameter

Salt	HgCl₂	KCl	NaF	LiCl	NaBr
k_s	0.050	0.280	0.195	0.075	0.120

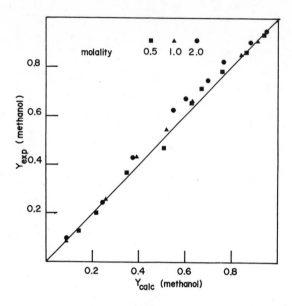

Figure 7.   Prediction of vapor phase composition for KCl system

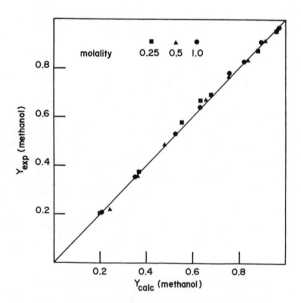

Figure 8.   Prediction of vapor phase composition for NaF system

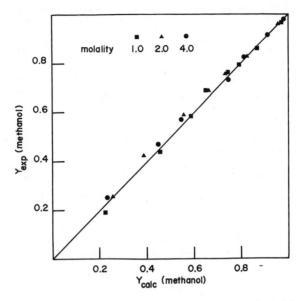

Figure 9.   Prediction of vapor phase composition for LiCl system

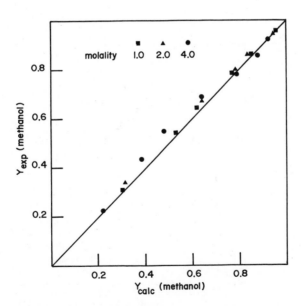

Figure 10.   Prediction of vapor-phase composition for NaBr system

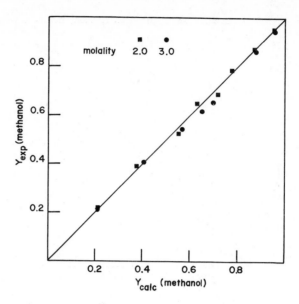

Figure 11.   *Prediction of vapor phase composition for*
HgCl$_2$

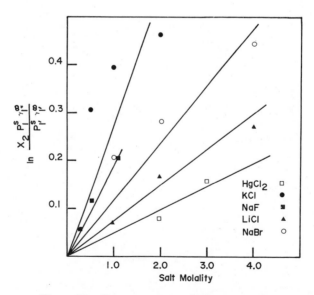

Figure 12.   *Determinations of salt parameters*

Conclusions

In this study, a thermodynamic framework has been presented for the calculation of vapor–liquid equilibria for binary solvents containing nonvolatile salts. From an appropriate definition of a pseudobinary system, infinite dilution activity coefficients for the salt-containing system may be estimated from a knowledge of vapor pressure lowering, salt-free infinite dilution activity coefficients, and a single system-dependent constant. Parameters for the Wilson equation may be determined from the infinite dilution activity coefficients.

Data have been presented for five salts in methanol–water solutions, which indicate clearly the influence of salt concentration on the observed salt effects and the Wilson parameters.

Acknowledgments

Financial support of the experimental work by the North Carolina State University Faculty Research and Professional Development Fund is gratefully acknowledged. The assistance of Gregory Miller in obtaining vapor pressure-lowering data is appreciated.

Nomenclature

C_e	= salt concentration
f_i	= fugacity of component i
$f_i{}^0$	= reference state fugacity of component i
i	= component in mixture of binary solvents and salt
i*	= component in pseudobinary mixture
i′	= component in salt-free mixture of binary solvents
k_s	= salt parameter
n_i	= moles of component i
P	= system pressure
$P^s{}_i(j)$	= vapor pressure of i determined at the normal boiling point of j
$\Delta P^s{}_i$	= vapor pressure lowering of i
R	= gas constant
S_i	= salt effect on component i
$v_i{}^L$	= molar volume of liquid i
x_i	= mole fraction of i in liquid
y_i	= mole fraction of i in vapor
γ_i	= activity coefficient of component i
$\gamma_i{}^\infty$	= infinite dilution activity coefficient of component i
Λ_{ij}	= defined by Equation 1c
$\lambda_{ij} - \lambda_{ii}$	= parameter in Wilson equation, cal/g-mole
ϕ_i	= fugacity coefficient of component i

Literature Cited

1. Johnson, A. I., Furter, W. F., *Can. J. Chem. Eng.* (1960) **38**, 78.
2. Meranda, D., Furter, W. F., *A.I.Ch.E. J.* (1971) **17**, 38.
3. Rousseau, R. W., Ashcraft, D. L., Schoenborn, E. M., *A.I.Ch.E. J.* (1972) **18**, 825.
4. Jaques, D., Furter, W. F., "Extractive and Azeotropic Distillation," ADV. CHEM. SER. (1972) **115**, 159.
5. Jaques, D., Furter, W. F., *A.I.Ch.E. J.*, (1972) **18**, 343.
6. Schreiber, L. B., Eckert, C. A., *Ind. Eng. Chem. Process Des. Dev.* (1971) **10**, 572.
7. Long, F. A., McDevitt, W. F., *Chem. Rev.* (1952) **51**, 119.
8. King, M. B., "Phase Equilibrium in Mixtures," Pergamon Press, Oxford, 1969, p 260.

RECEIVED July 13, 1975.

Prediction of Salt Effect in Vapor–Liquid Equilibrium: A Method Based on Solvation

SHUZO OHE

Ishikawajima-Harima Heavy Ind. Co. Ltd., Yokohama, Japan 235

A method of prediction of the salt effect of vapor–liquid equilibrium relationships in the methanol–ethyl acetate–calcium chloride system at atmospheric pressure is described. From the determined solubilities it is assumed that methanol forms a preferential solvate of $CaCl_2 \cdot 6CH_3OH$. The preferential solvation number was calculated from the observed values of the salt effect in 14 systems, as a result of which the solvation number showed a linear relationship with respect to the concentration of solvent. With the use of the linear relation the salt effect can be determined from the solvation number of pure solvent and the vapor–liquid equilibrium relations obtained without adding a salt.

Calcium chloride dissolves readily in methanol but less easily in ethyl acetate. Accordingly, it is assumed that the interaction between methanol and calcium chloride is dominant in the MeOH–EtOAc–CaCl₂ system. The causes of the salt effect in the system observed by the author will be discussed from the standpoint of molecular structure (Figure 1).

Causes of Salt Effect

The solubility of calcium chloride in the MeOH–EtOAc system (Figure 2) was obtained from the intersections of the x–y curves obtained at respective constant CaCl₂ concentrations and those obtained at saturated concentration. Each salt concentration at the intersection of curves of constant salt concentration to salt saturation shows the solubility of calcium chloride in the volatile binary system. The solubilities thus obtained (Figure 2) are linear. From 0 to 0.333 mole fraction of methanol, the solubility is almost zero. These solubility data indicate that if calcium chloride is dissolved by only the methanol contained in the methanol–ethyl acetate solution, both solvents exist in the form of clustered molecules comprising one methanol molecule and two ethyl acetate molecules (Figure 3).   It may be assumed that in methanol concentration over 0.333 mole fraction, free molecules forming nonclustered molecules are present in the system,

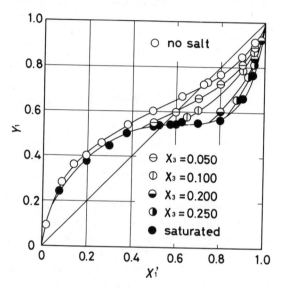

Figure 1.   MeOH–EtOAc–CaCl$_2$ system at 1 atm

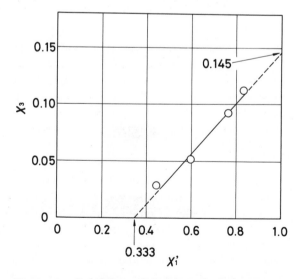

Figure 2.   Solubility of CaCl$_2$ in boiling MeOH–
EtOAc system at 1 atm

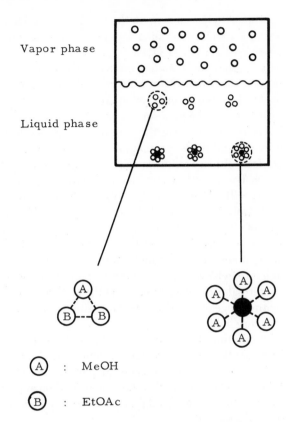

Figure 3. Preferential solvation model for
MeOH–EtOAc–CaCl₂ system at 1 atm

so that the salt is dissolved in the free molecules of methanol. From the extrapolated solubility (mole ratio of calcium chloride to methanol, $\sim$ 1:6), calcium chloride and methanol are believed to form a solvate of $CaCl_2 \cdot 6CH_3OH$. In forming this solvate, methanol is trapped by calcium chloride and cannot easily evaporate. Hence the vapor pressure of methanol drops to a corresponding level.

Prediction of Salt Effect from Preferential Solvation Number

The salt effect in the MeOH–EtOAc–CaCl₂ system can be explained by preferential solvation. As calcium chloride dissolves readily in methanol but only sparingly in ethyl acetate, it will be sufficient to consider the interaction between methanol molecules and calcium chloride molecules only in the MeOH–EtOAc solution. Referring again to Figure 2, the free methanol molecules which are not clustered with ethyl acetate increase linearly when the liquid-phase composition of methanol is above 0.333 in mole fraction. The solubility of $CaCl_2$ is proportional to the increase in the number of methanol molecules, and calcium chloride is dissolved, forming with methanol a preferential solvation which may be written $CaCl_2 \cdot 6CH_3OH$. Since solvated methanol molecules cannot be evaporated, the composition of methanol participating in the vapor–liquid equilibrium at the liquid phase is decreased. If it is assumed that the preferential solvates do not exert interaction on the volatile components in the liquid phase, the vapor–liquid equilibrium relation obtained under the addition of a salt may well be considered to be the same as the vapor–liquid equilibrium without the salt for a liquid-phase composition from which the solvents forming solvates are excluded. When a salt is expected to form a solvate with alcohol or water as in the alcohol–water–salt system, that is, when the formation of solvation is not limited to a specific component, the above-mentioned prediction cannot be made unless the solvation number of each component is calculated. For solvents consisting of ethanol and water, however, it has been found from measurements of Walden constant $(\Lambda^0\eta^0)$ of lithium chloride and lithium fluoride that preferential hydration of water molecules takes place when the ethanol concentration is less than 25 wt % (1). Therefore we can predict the salt effect in the alcohol–water–salt system, assuming that the preferential solvation takes place over the entire range of liquid-phase composition although we know that this assumption is very bold.

Method of Prediction. PREFERENTIAL SOLVATION NUMBER. The formation of preferential solvation is caused by the ionization of salt. The stability of the ion in a solution depends on the magnitude of the dielectric constant of a solvent. Debye (2) explained salting out by the formation of preferential solvation and found that the following relation exists between salting out and the dielectric constant of a solvent.

$$v_1 \ln \frac{x_2}{x_2{}^0} - v_2 \ln \frac{x_1}{x_1{}^0} = - v_2 \frac{z_i{}^2 e_i{}^2}{8\pi kT} \frac{1}{\epsilon^2 r^4} \frac{\partial \epsilon}{\partial n_1} \tag{1}$$

In Equation 1, $x_1{}^0$ and $x_2{}^0$ are x_1 and x_2 at $r = \infty$, respectively. The first component is a nonelectrolyte, while the second component is an electrolyte such as water. In other words, x_1 and x_2 represent the compositions in the neighborhood of respective salts and accordingly the solvated compositions. Assuming that changes in the dielectric constant are in the linear relationship with changes in the composition of solvent, this relation is given by $(\partial \epsilon / \partial n_1) = (\Delta \epsilon / \Delta n_1)$.

Table I. Effective Factor for Preferential Solvation in Equation 1[a]

Systems	ϵ_1	ϵ_2	$\Delta\epsilon/\Delta x_1$	$(1/\epsilon^2)$ $(\Delta\epsilon/\Delta x_1)$
Ethyl acetate (1)–methanol (2)	6.02	32.63	−26.61	−0.0712
Methanol (1)–water (2)	32.63	78.54	−45.91	−0.0088
Ethanol (1)–water (2)	24.3	78.54	−54.24	−0.0205

[a] ϵ_1, ϵ_2: value at $20°C$, $\Delta\epsilon = \epsilon_1 - \epsilon_2$, ϵ: value at $x_1{}^\circ = x_2{}^\circ = 0.5$.

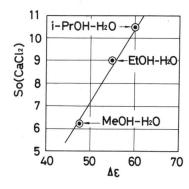

Figure 4. Difference of dielectric constants and solvation number

Therefore we can compare the values of $(1/\epsilon^2)(\Delta\epsilon/\Delta x_1)$ in several solvents (Table I).

The absolute value of $(1/\epsilon^2)(\Delta\epsilon/\Delta x_1)$ is the greatest in the ethyl acetate–methanol system but becomes smaller in the ethanol–water system and methanol–water system, in that order. In solvent systems, the greater the value of the right hand side of Equation 1, the greater the value of x_2/x_2^0 but the smaller the value of x_1/x_1^0. In other words, the preferential solvation due to methanol or water is likely to occur in such systems. Figure 4 shows the result of plotting the solvation number S_0 of pure solvent, obtained from the measurements of the salt effect, against the difference $\Delta\epsilon$ of dielectric constant of each solvent in the methanol–water, ethanol–water, and 2-propanol–water systems which are added with $CaCl_2$. As is apparent from Figure 4, the value of S_0 is greater in systems with greater value of $\Delta\epsilon$. This figure shows the same trend as Equation 1.

Then, we can obtain the preferential solvation number from the observed values of the salt effect. As the concentration of solvent is decreased by the number of solvated molecules, the actual solvent composition participating in the vapor–liquid equilibrium is changed. Assuming that a salt forms the solvate with the first component, the actual composition x_{1a} is given by

$$x_{1a} = \frac{x_1 - Sx_3}{(x_1 - Sx_3) + x_2} \tag{2}$$

Since $x_1 = x_1'(1 - x_3)$, $x_2 = x_2'(1 - x_3)$ and $x_1' + x_2' = 1$, Equation 2 is rewritten as follows:

$$x_{1a}' = \frac{x_1'(1 - x_3) - Sx_3}{(1 - x_3) - Sx_3} \tag{3}$$

Solving Equation 3 for S, we obtain

$$S = \frac{1 - x_3}{x_3} \frac{x_1' - x_{1a}'}{1 - x_{1a}'} \tag{4}$$

Therefore, the solvation number can be calculated by determining x_{1a}' from the measured values using the vapor–liquid equilibrium relation obtained without adding a salt.   When a salt forms the solvation with the second component, the following three equations can be derived in a similar manner.

$$x_{1a} = \frac{x_1}{x_1 + (x_2 - Sx_3)} \tag{5}$$

$$x_{1a}' = \frac{(1 - x_3)x_1'}{(1 - x_3) - Sx_3} \tag{6}$$

$$S = \frac{1 - x_3}{x_3} \frac{x_{1a}' - x_1'}{x_{1a}'} \tag{7}$$

The establishment of the method of prediction has been attempted by the reverse calculation of the preferential solvation number from measured values, using Equations 4 and 7 which are based on the assumption that the salt effect in the vapor–liquid equilibrium is caused by the preferential solvation formed between a volatile component and a salt.   The observed values were selected from Ciparis's data book (4), Hashitani's data (5–8), and the author's data (9–15).   S was calculated by Equation 7 when the relative volatility α_s in the vapor–liquid equilibrium with salt is increased with respect to the relative volatility α in the vapor–liquid equilibrium with salt, but by Equation 4 when α_s is decreased.   The results are shown in Figures 5–12.   From these figures, it will be seen that the following three relations exist:

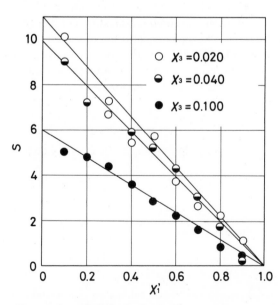

Figure 5.   MeOH–H$_2$O–CaCl$_2$ system at 1 atm

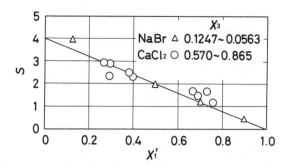

Figure 6.   MeOH–H₂O–salt system at 25°C

(a)   The preferential solvation number S shows the linear relationship with the liquid-phase composition of solvent, x_1' or x_2'.

(b)   When the concentration of salt is not saturated, S increases with increase in the mole fraction of solvent molecule which forms the solvate.

(c)   S increases with decrease in salt concentration.

These relations are established in all cases, regardless of the kinds of solvent systems, kinds of salts, isothermal equilibrium or isobaric equilibrium.   It seems that these relations are also independent of salt concentration   To be exact, however, the above-mentioned linear relation (increase or decrease with respect to solvent concentration) is assumed to exist when x_3 is constant.   If the range of salt concentration is narrow, the above-mentioned relations are assumed to be established approximately.

Let's make a comparison of individual cases.   Figures 5–7 show the cases where CaCl₂ was added to the alcohol–water system.   Figure 5 shows the values observed by the author when the salt concentration was constant in terms of mole fraction, according to which the hydration number S_0 of salt with water is 11 at $x_3 = 0.020$, 10 at $x_3 = 0.040$, and 6 at $x_3 = 0.100$.   The value of S_0 decreases as the salt concentration increases.   The reason for this may be that the activity of salt decreases and the number of molecules to be solvated decreases as the salt concentration increases.   Our calculation was made using the data obtained by Yoshida et al. (*16*).   As we obtained almost the same results as those shown in Figure 5, graphic representation is omitted.

Figure 6 shows the results at isothermal equilibrium.   The observed values in this case were taken from Ciparis's data book (*4*).   The concentration of CaCl₂, x_3, is not constant, changing from 0.1009 to 0.155, but the linear relation is shown. The salt concentration is higher than in the case of Figure 5, but S_0 is four, which is smaller.   Figure 7 is based on the values observed by Hashitani et al. (*5*), where the linear relation is also obtained although x_3 is not constant.   In particular, a similar linear relationship is obtained when a salt is dissolved to saturation at the boiling point.   In this case, S_0 is about four, which is smaller than in the case in which the salt is not saturated.   Figure 8 shows the case where CH₃COOK was

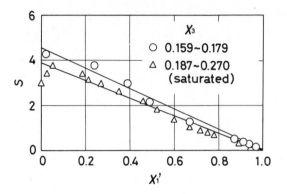

Figure 7.   EtOH–H$_2$O–CaCl$_2$ system at 1 atm

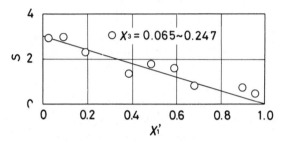

*Figure 8.   i-PrOH–H$_2$O–CH$_3$COOK system
at 1 atm*

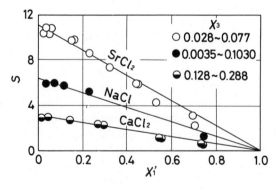

Figure 9.   CH$_3$COOH–H$_2$O–salt system at 1 atm

dissolved to saturation in the 2-propanol–water system at room temperature.   The
linear relation was also obtained, although the range of composition of x_3 was
large, 0.065–0.247.   As described above, S was calculated from the observed
values of the vapor–liquid equilibrium for such cases where a salt is assumed to
form the solvation with alcohol or water in the alcohol–water–salt system, as-

suming that the salt forms the preferential solvation with water. As a result, the linear relation was obtained. The values of S in the acetic acid–water system added with $SrCl_2$, NaCl, and $CaCl_2$, respectively, are shown in Figure 9. The same results as in the alcohol–water–salt system were obtained. Figure 10 shows the acetic acid–water system added with CH_3COONa. The addition of CH_3COONa forms the solvation with acetic acid rather than water. This may be because the affinity of CH_3COONa for acetic acid is greater than that for water. Figure 11 shows the case of the chloroform–acetone–$ZnCl_2$ system; the results obtained were the same as those in other systems. Figure 12 shows the case of the alcohol–alcohol–$CaCl_2$ system. In these three systems, the alcohol of lower boiling point forms the solvate, and the preferential solvation number decreases with increase of the carbon number in the alcohol. The effect of the alkyl group is greater and that of the hydroxyl group contributing to the formation

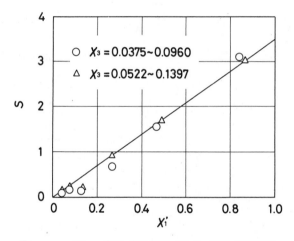

*Figure   10. $CH_3COOH–H_2O–CH_3COONa$
system at 1 atm*

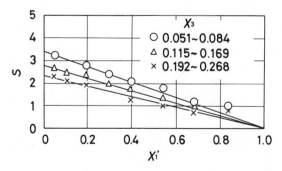

*Figure 11.   Chloroform–acetone–$ZnCl_2$ system
at 1 atm*

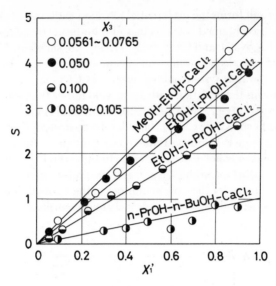

*Figure   12.   Alcohol–alcohol–CaCl$_2$ system
at 1 atm*

of solvate is smaller in systems having a larger number of carbons in the alcohol. The preferential solvation number in the ethyl acetate–ethanol–CaCl$_2$ system is shown in Figure 13.   From Figure 13, when the salt is saturated, the preferential solvation number becomes constant and is independent of the solvent concentration.   On the other hand, this relationship is not observed in the ethanol–

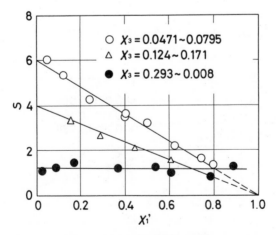

*Figure   13.   EtOAc–EtOH–CaCl$_2$ system
at 1 atm*

water–CaCl$_2$ system, as shown in Figure 7.   When a salt is dissolved to saturation, therefore, the same rule cannot be applied, thus making prediction impossible.

We can now discuss the solvation number.   In systems such as the methanol–water–CaCl$_2$ system shown in Figure 5, the hydration number is the greatest, that is, 11 at $x_3 = 0.020$.   If the hydration number of ions is calculated from the hydration entropy, Ca^{2+} is seven and Cl^- is two (3).   If it is assumed that CaCl$_2$ is completely dissociated and both the cation and anion forms hydrate, the hydration number becomes: $7 + 2 \times 2 = 11$, which agrees with the value obtained from the salt effect.

PREDICTION OF SALT EFFECT.   The procedure for calculation of the preferential solvation number S has been described above.   By reversing this procedure, that is, by determining x_{1a}' from S, we can estimate the salt effect using the vapor–liquid equilibrium without a salt.   When the salt concentration is below saturation, the preferential solvation number S can be expressed as follows in cases where the solvation is formed with the first component.

$$S = S_0 x_1' \tag{8}$$

where S_0 is the solvation number of the pure solvent with a salt.

When the solvation is formed with the second component, the solvation number is given by

$$S = S_0(1 - x_1') \tag{9}$$

Accordingly, Equations 3 and 6 are rewritten as follows:

$$x_{1a}' = \frac{(1 - x_3)x_1' - S_0 x_1' x_3}{(1 - x_3) - S_0 x_1' x_3} \tag{10}$$

$$x_{1a}' = \frac{(1 - x_3)x_1'}{(1 - x_3) - S_0(1 - x_1')x_3} \tag{11}$$

The procedure for predicting the salt effect from the solvation number of pure solvent is described below.

(a)   Decide with what component the salt forms the preferential solvate in the system being predicted.

(b)   Obtain the solvation number S_0 to be formed by a pure solvent.

(c)   Calculate x_{1a}' by Equation 10 or 11.

(d)   Set the vapor-phase composition at x_{1a}' in the absence of salt, at x_1' in the presence of salt.

Example of prediction:

The following is the prediction of vapor–liquid equilibrium composition when CaCl$_2$ is added in 7.24 mol % to the ethyl acetate–ethanol system in which the liquid-phase composition of ethyl acetate is 0.502 in terms of mole fraction.

First, the component that forms the preferential solvation is determined according to step (a) as described above.   Since ethanol is stronger in polarity and greater in dielectric constant ϵ than ethyl acetate, we can assume that ethanol

Table II. Systems of Salt Effect

No.	Systems	Condition
1	Methanol + water + $CaCl_2$[a]	1 atm
2	Methanol + water + $CaCl_2$ (No. 33)[c]	25° C
3	Methanol + water + NaBr (No. 29)[c]	25° C
4	Ethanol + water + $CaCl_2$[b]	1 atm
5	2-Propanol + water + CH_3COOK[a]	1 atm
6	Acetic acid + water + $CaCl_2$ (No. 61)[c]	1 atm
7	Acetic acid + water + NaCl (No. 41)[c]	1 atm
8	Acetic acid + water + $SrCl_2$ (No. 65)[c]	1 atm
9	Acetic acid + water + CH_3COONa(No. 44)[c]	1 atm
10	Ethyl acetate + ethanol + $CaCl_2$[b]	1 atm
11	Chloroform + acetone + $ZnCl_2$[b]	1 atm
12	Methanol + ethanol + $CaCl_2$[a]	1 atm
13	Ethanol + i-propanol + $CaCl_2$[a]	1 atm
14	1-Propanol + 1-butanol + $CaCl_2$[a]	1 atm

[a] Author's data.
[b] Hashitani's data.

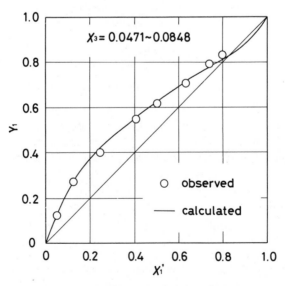

Figure 14. Result of prediction for EtOAc–EtOH–H_2O–$CaCl_2$ system at 1 atm

Predicted by the Solvation

x_3	S_0	Error (%)	Figure	Table
0.020	11.0	1.2	5	III
0.040	10.0	0.5	5	III
0.0887–0.175	4.0	1.4	6	IV
0.0563–0.1247	4.0	8.6	6	V
0.159–0.179	4.5	5.2	7	VI
0.0636–0.1052	8.0	7.1	7	VI
0.066–0.247	3.0	6.3	8	VII
0.128–0.228	3.0	10.2	9	VIII
0.0035–0.1025	6.4	2.9	9	IX
0.028–0.077	11.0	6.4	9	X
0.0552–0.1397	3.5	1.2	10	XI
0.0375–0.0960	3.5	9.8	10	XI
0.0471–0.0848	6.0	1.1	13	XII
0.124–0.171	4.0	3.6	13	XII
0.051–0.084	3.3	3.3	11	XIII
0.115–0.169	2.7	2.7	11	XIII
0.192–0.268	2.3	2.3	11	XIII
0.0561–0.0765	5.0	2.5	12	XIV
0.050	4.0	3.8	12	XV
0.110	3.0	3.6	12	XV
0.089–0.105	1.0	2.8	12	XVI

[c] Table No. in Ciparis' data book.

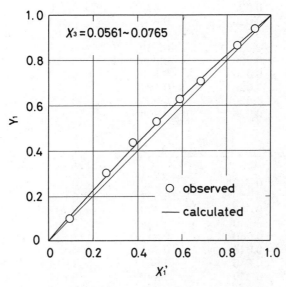

*Figure 15. Result of prediction for MeOH–
EtOH–$CaCl_2$ system at 1 atm*

and $CaCl_2$ form the solvate. Then the solvation number is determined as described in step (b). As the solvation number of ethanol and $CaCl_2$ is not available in the literature, we can use $S_0 = 6$ obtained from the observed values shown in Figure 13. For step (c), we can obtain $x_{1a}' = 0.655$ by substituting $x_1' = 0.502$, $x_3 = 0.8724$ and $S_0 = 6$ in Equation 11. Finally we can determine the vapor-phase composition as described in step (d). When the liquid-phase composition of ethyl acetate is 0.655 mole fraction, the vapor-phase composition in the absence of $CaCl_2$ is 0.613 mole fraction. Accordingly, when the liquid-phase composition of ethyl acetate is 0.502 mole fraction, the vapor-phase composition in the presence of $CaCl_2$ in 7.24 mol % will be 0.655 mole fraction. Since the observed value is 0.620 mole fraction, the error is 1.1%.

The results of prediction over the entire range of liquid-phase composition are given in Table XII. The comparison of the prediction results with the observed values shown in the x-y diagram is shown in Figure 14.

RESULTS OF PREDICTION AND DISCUSSION. The systems predicted are given in Table II together with error. The results of prediction of each system are given in Tables III–XVI.

The values of S_0 obtained from the observed values shown in Figures 4–12 were used as the preferential solvation number for prediction. As S_0 is the solvation number between a pure solvent and a salt, it should not be obtained from

Table III.   Salt Effect Predicted from the Preferential Solvation for
Methanol–Water–CaCl$_2$ System at 1 atm

$S_0 = 11.0$

x_1'	x_3	y_1 calc.	y_1 obs.	Error (%)
0.100	0.020	0.463	0.478	3.1
0.200	0.020	0.616	0.594	3.7
0.300	0.020	0.700	0.702	0.3
0.400	0.020	0.760	0.757	0.4
0.500	0.020	0.806	0.810	0.5
0.600	0.020	0.850	0.849	0.1
0.700	0.020	0.892	0.890	0.2
0.800	0.020	0.931	0.920	1.2
0.900	0.020	0.968	0.955	1.4
				av. 1.2%

$S_0 = 10.0$

x_1'	x_3	y_1 calc.	y_1 obs.	Error (%)
0.100	0.040	0.529	0.530	0.2
0.200	0.040	0.667	0.655	1.8
0.300	0.040	0.743	0.737	0.8
0.400	0.040	0.794	0.794	0
0.500	0.040	0.841	0.843	0.2
0.600	0.040	0.881	0.884	0.3
0.700	0.040	0.915	0.914	0.1
0.800	0.040	0.948	0.944	0.4
0.900	0.040	0.975	0.966	0.9
				av. 0.5%

Table IV.   Salt Effect Predicted from the Preferential Solvation for
Methanol–Water–$CaCl_2$ System at 25° C

$S_0 = 4.0$

x_1'	x_3	y_1 calc.	y_1 obs.	Error (%)
0.2633	0.155	0.867	0.866	0.1
0.2909	0.175	0.920	0.866	6.2
0.2941	0.150	0.872	0.877	0.6
0.3791	0.1464	0.896	0.893	0.3
0.3977	0.1532	0.910	0.901	1.0
0.6626	0.1171	0.942	0.956	1.5
0.6895	0.1242	0.953	0.960	0.7
0.7257	0.0887	0.945	0.962	1.8
0.7542	0.1009	0.956	0.962	0.6

av. 1.4%

Table V.   Salt Effect Predicted from the Preferential Solvation for
Methanol–Water–NaBr System at 25° C

$S_0 = 4.0$

x_1'	x_3	y_1 calc.	y_1 obs.	Error (%)
0.148	0.1247	0.575	0.756	23.9
0.292	0.1118	0.725	0.820	11.5
0.500	0.0909	0.840	0.884	5.0
0.700	0.0722	0.915	0.932	1.8
0.900	0.0563	0.970	0.979	0.9

av. 8.6%

the vapor–liquid equilibrium relations.   At present, however, data are not available.   In each table, the comparison of prediction results of each system and observed values is shown as the error.

The mean error was 0.5% minimum and 10.2% maximum.   A general trend is that the error is great when the value of S is large.   For example, in the methanol–water–NaBr system shown in Table V, S decreases with the increase in x_1' as shown in Figure 6.   While the error is 23.9% at $x_1' = 0.148$, it decreases to 0.9% at $x_1' = 0.900$.   The same trend is observed in all systems given in Tables VI–XI, XIV–XVI.   The cause of this may be that the difference between the calculated value of S and the measured value of S is large where the value of S is great.   An example of prediction results is shown in Figure 15 in the form of an *x–y* diagram.

The errors in this method of prediction are assumed to result from:

(a)   the degree of linearity of values of S to x_1', and
(b)   the amount of scatter of values of S from the above-mentioned line.

The amount of scatter must be smaller if observed values are accurate.   The linearity of values of S affects the validity of this method of prediction.   To discuss

Table VI. Salt Effect Predicted from the Preferential Solvation for
Ethanol–Water–CaCl$_2$ System at 1 atm

$$S_0 = 4.5$$

x_1'	x_3	y_1 calc.	y_1 obs.	Error (%)
0.106	0.168	0.678	0.785	13.6
0.233	0.161	0.745	0.830	10.2
0.385	0.159	0.814	0.860	5.4
0.482	0.170	0.819	0.868	5.9
0.666	0.179	0.988	0.908	8.8
0.860	0.164	0.977	0.952	2.6
0.907	0.171	0.990	0.960	3.1
0.917	0.167	0.988	0.973	1.5
0.929	0.160	0.986	0.973	1.3
0.962	0.165	0.995	0.989	0.6
				av. 5.2%

$$S_0 = 8.0$$

x_1'	x_3	y_1 calc.	y_1 obs.	Error (%)
0.105	0.0636	0.525	0.655	19.9
0.426	0.0952	0.805	0.836	3.7
0.457	0.0970	0.863	0.840	2.7
0.583	0.1052	0.954	0.880	8.4
0.791	0.0919	0.938	0.931	0.8
				av. 7.1%

Table VII. Salt Effect Predicted from the Preferential Solvation for
1-Propanol–Water–CH$_3$COOK System at 1 atm

$$S_0 = 3.0$$

x_1	x_3	y_1 calc.	y_1 obs.	Error (%)
0.024	0.247	0.641	0.672	4.6
0.090	0.240	0.662	0.801	17.4
0.190	0.239	0.762	0.853	10.7
0.390	0.237	0.860	0.891	3.5
0.485	0.235	0.880	0.889	1.0
0.592	0.222	0.865	0.894	3.2
0.682	0.200	0.850	0.906	6.2
0.896	0.120	0.900	0.959	6.2
0.952	0.066	0.936	0.975	4.0
				av. 6.3%

Table VIII. Salt Effect Predicted from the Preferential Solvation for
Acetic Acid–Water–CaCl$_2$ System at 1 atm
$S_o = 3.0$

x_1'	x_3	y_1 calc.	y_1 obs.	Error (%)
0.016	0.228	0.085	0.086	1.2
0.044	0.227	0.193	0.250	22.8
0.146	0.212	0.346	0.414	16.4
0.274	0.199	0.460	0.536	14.2
0.297	0.197	0.480	0.547	12.2
0.544	0.195	0.715	0.655	9.2
0.558	0.186	0.695	0.646	7.6
0.732	0.136	0.745	0.719	3.6
0.739	0.128	0.740	0.708	4.5
				av. 10.2%

Table IX. Salt Effect Predicted from the Preferential Solvation for
Acetic Acid–Water–NaCl System at 1 atm
$S_o = 6.4$

x_1'	x_3	y_1 calc.	y_1 obs.	Error (%)
0.033	0.1025	0.075	0.068	10.3
0.071	0.0934	0.123	0.121	1.6
0.114	0.0863	0.165	0.167	1.2
0.232	0.0642	0.250	0.252	0.8
0.412	0.0384	0.360	0.382	5.8
0.745	0.0129	0.645	0.641	0.6
0.994	0.0035	0.986	0.985	0.1
				av. 2.9%

Table X. Salt Effect Predicted from the Preferential Solvation for
Acetic Acid–Water–SrCl$_2$ System at 1 atm
$S_o = 11.0$

x_1'	x_3	y_1 calc.	y_1 obs.	Error (%)
0.025	0.077	0.162	0.128	26.6
0.031	0.073	0.130	0.143	9.1
0.051	0.075	0.235	0.211	11.4
0.059	0.070	0.183	0.220	16.8
0.150	0.061	0.273	0.296	7.8
0.161	0.058	0.267	0.295	9.5
0.229	0.056	0.351	0.343	2.3
0.319	0.049	0.389	0.386	0.8
0.445	0.039	0.466	0.449	3.8
0.449	0.038	0.465	0.448	3.8
0.527	0.033	0.504	0.502	0.4
0.529	0.034	0.512	0.483	6.0
0.696	0.024	0.643	0.633	1.6
0.707	0.028	0.667	0.638	4.5
				av. 6.4%

Table XI.   Salt Effect Predicted from the Preferential Solvation for
Acetic Acid–Water–CH_3COONa System at 1 atm

$S_0 = 3.5$

$x_1{'}$	x_3	y_1 calc.	y_1 obs.	Error (%)
0.035	0.0552	0.020	0.020	0
0.074	0.0597	0.040	0.041	2.4
0.130	0.0657	0.070	0.070	0
0.268	0.0809	0.137	0.140	2.1
0.488	0.1037	0.257	0.262	1.9
0.863	0.1397	0.606	0.610	0.7
				av. 1.2%

$S_0 = 3.5$

$x_1{'}$	x_3	y_1 calc.	y_1 obs.	Error (%)
0.034	0.0375	0.020	0.022	9.1
0.073	0.0405	0.043	0.046	6.5
0.122	0.0447	0.052	0.076	31.6
0.264	0.0551	0.143	0.158	9.5
0.468	0.0702	0.284	0.285	0.4
0.835	0.0960	0.644	0.632	1.9
				av. 9.8%

Table XII.   Salt Effect Predicted from the Preferential Solvation for
Ethyl Acetate–Ethanol–$CaCl_2$ System at 1 atm

$S_0 = 6.0$

$x_1{'}$	x_3	y_1 calc.	y_1 obs.	Error (%)
0.049	0.0471	0.124	0.123	0.8
0.123	0.0535	0.265	0.266	0.4
0.240	0.0631	0.417	0.407	2.5
0.401	0.0743	0.560	0.558	0.4
0.403	0.0720	0.558	0.550	1.5
0.502	0.0724	0.613	0.620	1.1
0.629	0.0848	0.713	0.710	0.4
0.741	0.0824	0.781	0.789	1.0
0.797	0.0795	0.815	0.830	1.8
				av. 1.1%

$S_0 = 4.0$

$x_1{'}$	x_3	y_1 calc.	y_1 obs.	Error (%)
0.149	0.124	0.373	0.363	2.8
0.290	0.148	0.563	0.543	3.7
0.442	0.161	0.696	0.673	3.4
0.604	0.171	0.824	0.789	4.4
				av. 3.6%

Table XIII.   Salt Effect Predicted from the Preferential Solvation for Chloroform–Acetone–$ZnCl_2$ System at 1 atm

$S_0 = 3.3$

x_1'	x_3	y_1 calc.	y_1 obs.	Error (%)
0.0497	0.0511	0.0235	0.0325	0
0.102	0.0562	0.108	0.0725	50.0
0.198	0.0622	0.169	0.167	1.2
0.295	0.0628	0.267	0.267	0
0.398	0.0683	0.406	0.408	0.5
0.549	0.0757	0.615	0.640	3.9
0.689	0.0803	0.804	0.812	1.0
0.838	0.0836	0.922	0.948	2.7
0.951	0.0254	0.972	0.981	0.9
				av. 3.3%

$S_0 = 2.7$

x_1'	x_3	y_1 calc.	y_1 obs.	Error (%)
0.0497	0.1147	0.0439	0.0439	0
0.102	0.1187	0.097	0.0963	0.7
0.198	0.1303	0.212	0.219	3.2
0.295	0.1519	0.385	0.383	0.5
0.398	0.1465	0.515	0.530	2.8
0.549	0.1619	0.755	0.765	1.3
0.689	0.1694	0.870	0.904	3.8
0.838	0.1088	0.925	0.971	4.7
				av. 2.7%

$S_0 = 2.3$

x_1'	x_3	y_1 calc.	y_1 obs.	Error (%)
0.0497	0.1018	0.0620	0.0663	6.5
0.102	0.1914	0.135	0.137	1.5
0.198	0.2098	0.309	0.313	1.3
0.398	0.2345	0.715	0.672	6.4
0.549	0.2541	0.896	0.883	1.5
0.689	0.2679	0.958	0.946	1.3
				av. 2.3%

Table XIV.   Salt Effect Predicted from the Preferential Solvation for
Methanol–Ethanol–CaCl$_2$ System at 1 atm

$$S_0 = 5.0$$

x_1'	x_3	$y_{1\ calc.}$	$y_{1\ obs.}$	Error (%)
0.098	0.077	0.099	0.100	1.0
0.265	0.071	0.285	0.310	8.1
0.382	0.069	0.407	0.437	6.9
0.490	0.066	0.522	0.530	1.5
0.590	0.064	0.620	0.627	1.1
0.684	0.062	0.715	0.705	0.4
0.852	0.058	0.864	0.864	0
0.928	0.056	0.934	0.934	0

av.   2.5%

Table XV.   Salt Effect Predicted from the Preferential Solvation for
Ethanol–1-Propanol–CaCl$_2$ System at 1 atm

$$S_0 = 3.0$$

x_1'	x_3	$y_{1\ calc.}$	$y_{1\ obs.}$	Error (%)
0.056	0.110	0.040	0.046	13.0
0.112	0.110	0.082	0.082	0
0.225	0.110	0.175	0.171	2.3
0.339	0.110	0.274	0.269	1.9
0.449	0.110	0.377	0.401	6.0
0.562	0.110	0.485	0.509	4.7
0.674	0.110	0.602	0.607	0.8
0.786	0.110	0.732	0.749	2.3
0.899	0.110	0.873	0.881	0.9

av.   3.6%

$$S_0 = 4.0$$

x_1'	x_3	$y_{1\ calc.}$	$y_{1\ obs.}$	Error (%)
0.053	0.050	0.046	0.044	4.6
0.105	0.050	0.095	0.090	5.6
0.210	0.050	0.198	0.182	8.8
0.316	0.050	0.298	0.274	8.8
0.421	0.050	0.403	0.385	4.7
0.526	0.050	0.505	0.508	0.6
0.632	0.050	0.613	0.624	1.8
0.737	0.050	0.724	0.741	2.3
0.842	0.050	0.837	0.844	0.8
0.947	0.050	0.948	0.948	0

av.   3.8%

this point in detail, it is necessary to establish an elaborate theory from the standpoint of physical chemistry. It was possible for various systems, however, to obtain the linear relation empirically from the measured values as described above. Accordingly, we are of the opinion that the faithful utilization of such results is meaningful as the first step toward the establishment of this type of prediction method since there is no prediction method of salt effect available at present.

Table XVI.   Salt Effect Predicted from the Preferential Solvation for 1-Propanol–1-Butanol–CaCl$_2$ System at 1 atm

$$S_0 = 1.0$$

x_1'	x_3	y_1 calc.	y_1 obs.	Error (%)
0.100	0.136	0.168	0.183	8.2
0.300	0.141	0.458	0.451	1.6
0.400	0.156	0.553	0.566	2.3
0.500	0.187	0.627	0.662	5.3
0.600	0.176	0.715	0.736	2.9
0.700	0.180	0.795	0.807	1.5
0.800	0.177	0.878	0.873	0.6
0.900	0.134	0.945	0.944	0.1
				av. 2.8%

Conclusion

The salt effect is attributable to the formation of preferential solvation from the standpoint of molecular structure. In other words, when calcium chloride, which dissolves readily in methanol but very little in ethyl acetate, was added to the methanol–ethyl acetate system to saturation, calcium chloride formed with methanol the preferential solvate which may be written CaCl$_2$·6CH$_3$OH. It was also shown from the observation of solubility that the solvated methanol molecules did not participate in the vapor–liquid equilibrium.

The preferential solvation number was calculated from the observed values of salt effect in 14 systems, as a result of which the solvation number showed the linear relationship with respect to the concentration of solvent. It has been made clear that the solvation number increases with increase in the concentration of a solvent forming the solvation when the salt concentration is not saturated, but is kept constant when the salt concentration is saturated. Thus the salt effect was predicted with the use of the above-mentioned relationship. According to this prediction method, the salt effect can be determined from the solvation number of pure solvent and the vapor–liquid equilibrium relations without a salt.

Nomenclature

e	= electric charge of electron	$[1.6018 \times 10^{-19}$ coulomb]
k	= Boltzmann constant	$[(1.38044 \pm 0.00007) \times 10^{-16}$ erg/deg]
n	= number of nonelectrolyte molecule	$[-]$
r	= distance between ions	$[10^{-8}$ cm]
S	= preferential solvation number	$[-]$
T	= Absolute temperature	$[^\circ K]$
t	= temperature	$[^\circ C]$
v	= molar volume	[cc/mol]
x	= liquid phase composition	[mole fraction]
y	= vapor phase composition	[mole fraction]
z	= electric charge number	$[-]$
ϵ	= dielectric constant	$[-]$
π	= ratio of the circumference to its diameter	$[-]$

Superscript

$'$ = salt free

Subscripts

$_1$ = first component
$_2$ = second component
$_3$ = third component
$_a$ = free solvent molecule not solvated
$_i$ = ion

Literature Cited

1. Wada, G., Itoh, C., *J. Chem. Soc. Jpn.* Pure Chemistry Section (1956) **77**, 391.
2. Debye, P., *Z. Phys. Chem. (Leipzig)* (*1927*) **130**, 55.
3. Harned, H. S., Owen, B. B., "Physical Chemistry of Electrolytic Solutions," 3rd ed., Reinhold, N.Y., 1957, p 546.
4. Ciparis, J. N., "Data of Salt Effect in Vapour–Liquid Equilibrium," Lithuanian Agricultural Academy, Kaunas, USSR, 1966.
5. Hashitani, M., Hirose, Y., Hirata, M., *Kagaku Kogaku* (1968) **32**, 182.
6. Hashitani, M., Hirata, M., *J. Chem. Eng. Jpn* (1968) **1**, 116.
7. Hashitani, M., Hirata, M., *J. Chem. Eng. Jpn* (1969) **2**, 149.
8. Hashitani, M., Ph.D. Thesis, Tokyo Metropolitan University, Tokyo, Japan (1970).
9. Ohe, S., Yokoyama, K., Nakamura, S., *Kogyo Kagaku Zasshi* (1969) **72**, 313.
10. Ohe, S., Yokoyama, K., Nakamura, S., *J. Chem. Eng. Jpn* (1969) **2**, 1.
11. Ohe, S., Yokoyama, K., Nakamura, S., *J. Chem. Eng. Data* (1971) **16**, 70.
12. Ohe, S., Yokoyama, K., Nakamura, S., *Kogyo Kagaku Zassni* (1970) **73**, 1647.
13. Ohe, S., Yokoyama, K., Nakamura, S., *Kagaku Kogaku* (1970) **34**, 325.
14. Ohe, S., Yokoyama, K., Nakamura, S., *Kagaku Kogaku* (1970) **34**, 1112.
15. Ohe, S., Yokoyama, K., Nakamura, S., *Kagaku Kogaku* (1971) **35**, 104.
16. Yoshida, F., Yasunishi, A., Hamada, Y., *Kagaku Kogaku* (1964) **28**, 133.

RECEIVED July 6, 1975.

Salt Effect on Isothermal Vapor–Liquid Equilibrium of 2-Propanol–Water Systems

EIZO SADA, TETSUO MORISUE, and NORIO TSUBOI

Department of Chemical Engineering, Nagoya University, Furo-cho, Chikusa-ku, Nagoya, 464 Japan

Isothermal vapor–liquid equilibrium data at 75°, 50° and 25°C for the system of 2-propanol–water–lithium perchlorate were obtained by using a modified Othmer still. In the 2-propanol-rich region 2-propanol was salted out from the aqueous solution by addition of lithium perchlorate, but in the water-rich region 2-propanol was salted in. It is suggested from the experimental data that the simple electrostatic theory cannot account for the salt effect parameter of this system.

A dissolved salt in a mixed solvent consisting of water and a nonelectrolyte changes the phase equilibrium by causing preferential solvation in the liquid phase. According to Johnson and Furter (1) the salt effect on the vapor–liquid equilibrium is defined by the relative change of the chemical potentials of both solvents by dissolution of salt in the mixed solvent. If temperature, pressure, and a mixed solvent composition are fixed, the salt effect is a function of salt concentration only. Johnson and Furter assume a linear relation between the relative change of each chemical potential and the salt concentration as a first approximation. No one so far has reported the results of testing the linearity of the relationship under the originally derived conditions.

The thermodynamic excess functions for the 2-propanol–water mixture and the effects of lithium chloride, lithium bromide, and calcium chloride on the phase equilibrium for this binary system have been studied in previous papers (2, 3). In this paper, the effects of lithium perchlorate on the vapor–liquid equilibrium at 75°, 50°, and 25°C for the 2-propanol–water system have been obtained by using a dynamic method with a modified Othmer still. This system was selected because lithium perchlorate may be more soluble in alcohol than in water (4).

Experimental

Materials. Distilled water was used; 2-propanol and trihydrous lithium perchlorate, of guaranteed reagent quality from Wakô Pure Chemicals Co., were used without further purification. The purity of the 2-propanol was checked by gas chromatography, with Porapak-Q as the column packing, and found to be more than 99.9 mol %. The physical properties of pure solvents were compared with the literature values in a previous paper (2), and the agreement was satisfactory.

Apparatus. All vapor–liquid equilibrium measurements were made by using a modified Othmer still provided with an external electric heater. Total volume of the still was about 500 cm³, of which about 300 cm³ was occupied by liquid. The liquid loaded in the condensate receiver was about 7 cm³. Details of the still are described in a previous paper (5).

The boiling point temperature was maintained within ±0.02°C of the selected temperature, and measured by using a mercury-in-glass thermometer. The equilibrium pressure was measured by means of a mercury-in-glass manometer, and was readable within an accuracy of ±0.1 mm.

Equilibrium vapor condensate was analyzed by means of density measurement at 25.00° ±0.02°C. An Ostwald pycnometer (capacity ca. 5 cm³) was used. Liquid phase composition was calculated by taking a material balance. In this case, the three moles of water present in trihydrous lithium perchlorate were considered water component. The accuracies of both compositions were ±0.001 mole fraction.

Thermodynamic Consideration

To 2-propanol and water (designated by the subscripts 1 and 2 respectively) a nonvolatile salt (designated by subscript 3) is added. At infinite dilution this salt dissociates into an ion couple.

The problem is to define the extent of the salt effect on a two-component solvent. If the salt has an equal effect on both components of the solvents, then it can be assumed

$$(\partial\mu_1/\partial n_3)_{n_1,n_2} = (\partial\mu_2/\partial n_3)_{n_1,n_2} \tag{1}$$

where n_3 is the formulary moles of salt. However, the salt effect on each solvent is not equal, that is,

$$\left[\frac{\partial(\mu_1 - \mu_2)}{\partial n_3}\right]_{n_1,n_2} \gtrless 0 \tag{2}$$

If the inequality is positive, solvent 1 (nonelectrolyte) is salted out; on the other hand, if it is negative, solvent 1 is salted in. Thus the extent of the salt effect on the mixed solvent should be measured by the relative change of the chemical potential of each solvent in the addition of salt. Such an approach has been adopted by Grunwald and Bacarella (*6, 7, 8*). It is convenient to express μ_1 and μ_2 as functions of the composition variables Z_1 and N_3,

$$Z_1 = n_1/(n_1 + n_2) \text{ and } Z_1 + Z_2 = 1 \tag{3}$$

$$N_3 = n_3/(n_1 + n_2) \tag{4}$$

Since the cross differential relation

$$\left(\frac{\partial \mu_1}{\partial n_3}\right)_{n_1,n_2} = \left(\frac{\partial \mu_3}{\partial n_1}\right)_{n_2,n_3} \text{ and } \left(\frac{\partial \mu_2}{\partial n_3}\right)_{n_1,n_2} = \left(\frac{\partial \mu_3}{\partial n_2}\right)_{n_1,n_3} \tag{5}$$

an analogous equation to Equation 2 can be obtained.

$$\left[\frac{\partial(\mu_1 - \mu_2)}{\partial N_3}\right]_{Z_1} = \left(\frac{\partial \mu_3}{\partial Z_1}\right)_{N_3} \geq 0 \tag{6}$$

Here μ_3 is completely defined by the variables Z_1 and N_3. For any given values of Z_1 and N_3, the reference of the activity coefficient will be chosen as the extremely dilute state ($N_3 = 0$) of the given solute in a binary mixed solvent of the same composition Z_1. By the definition, the chemical potential of the reference state varies with Z_1. Hence, one obtains for the 1-1 salts

$$\left(\frac{\partial \mu_3}{\partial Z_1}\right)_{N_3} = 2 \left(\frac{d\mu_\pm^*}{dZ_1}\right) + 2RT \left(\frac{\partial \ln \gamma_\pm}{\partial Z_1}\right)_{N_3} \tag{7}$$

where the superscript * denotes the infinite dilution of salt in the solvent, and the subscript $\pm$ denotes the mean value for the ions of a salt. Substituting Equation 7 into Equation 6 and integrating, we have

$$\Delta \mu = k_0 N_3 + 2 \int_0^{N_3} \left(\frac{\partial \ln \gamma_\pm}{\partial Z_1}\right)_{N_3} dN_3 \tag{8}$$

where

$$\Delta \mu = \left(\frac{\mu_1 - \mu_2}{RT}\right) - \left(\frac{\mu_1 - \mu_2}{RT}\right)_0 \tag{9}$$

$$k_0 = \frac{2}{RT} \left(\frac{d\mu_\pm^*}{dZ_1}\right) \tag{10}$$

and the subscript 0 denotes the system without salt. If the salt is very dilute, the second term on the right side in Equation 8 is proportional to $N_3^{3/2}$ (*6, 7, 8*), so this term may be neglected in comparison with the first term. Thus

$$\Delta \mu = k_0 N_3 \tag{11}$$

Equation 11 is similar to the linear approximation by Johnson and Furter (*1*).

Results and Discussion

Vapor–liquid equilibrium data obtained for the 2-propanol–water binary system at 75°C agreed well with the values calculated from the total pressure data used in the numerical method of Mixon et al. (9). Thus, the apparatus used in this work gives consistent data.

The isothermal vapor–liquid equilibrium data at 75°, 50°, and 25°C for the 2-propanol–water–lithium perchlorate system are listed in Tables I and II. It can be seen from these tables that in the alcohol-rich region the vapor phase composition of 2-propanol increases with an increase in salt concentration. However, in a mixed solvent of 10 mol % alcohol the change of the vapor phase composition is small, and at a temperature of 50° and 25°C it even decreases a

Table I.   Vapor–Liquid Equilibrium Data for
2-Propanol–Water–LiClO$_4$ System at 75.08 ± 0.02°C

Z_1	$N_3 \times 100$	y_1	P (mmHg)	a_1	a_2
0.099	0.0	0.498	552.9	0.487	0.950
0.100	0.51	0.498	546.6	0.482	0.939
0.100	1.04	0.502	542.4	0.482	0.925
0.096	1.57	0.501	537.0	0.477	0.917
0.095	2.11	0.501	535.3	0.475	0.915
0.095	3.23	0.499	531.4	0.470	0.912
0.300	0.0	0.542	587.2	0.560	0.926
0.296	0.51	0.550	584.8	0.568	0.899
0.300	1.03	0.554	580.6	0.567	0.886
0.296	1.58	0.563	578.0	0.575	0.863
0.296	2.11	0.565	573.3	0.572	0.852
0.296	3.25	0.565	563.7	0.563	0.839
0.500	0.0	0.596	604.8	0.634	0.826
0.498	0.52	0.613	600.8	0.649	0.795
0.500	1.06	0.621	600.1	0.656	0.778
0.498	1.59	0.625	595.4	0.656	0.764
0.498	2.13	0.633	593.3	0.662	0.745
0.497	3.27	0.647	582.3	0.665	0.703
0.703	0.0	0.698	613.6	0.754	0.633
0.700	0.52	0.703	608.8	0.754	0.618
0.700	1.02	0.712	607.9	0.762	0.598
0.701	1.58	0.726	601.7	0.769	0.564
0.699	2.13	0.735	597.6	0.773	0.540
0.699	3.26	0.750	588.1	0.778	0.502
0.900	0.0	0.869	591.1	0.906	0.266
0.901	0.52	0.873	588.4	0.906	0.256
0.903	1.02	0.879	583.8	0.906	0.240
0.900	1.59	0.889	581.4	0.912	0.221
0.900	2.13	0.901	577.1	0.918	0.195
0.900	3.26	0.913	571.3	0.921	0.169

Table II. Vapor–Liquid Equilibrium Data for
2-Propanol–Water–LiClO₄ System at 50° and 25 °C

Z_1	$N_3 \times 100$	y_1	P (mmHg)	a_1	a_2
			$50.01 \pm 0.02°C$		
0.097	0.0	0.489	175.7	0.477	0.970
0.096	0.51	0.486	172.4	0.465	0.956
0.099	1.01	0.480	172.5	0.460	0.969
0.095	1.51	0.485	174.0	0.469	0.968
0.096	2.03	0.484	169.3	0.455	0.943
0.095	3.04	0.486	166.5	0.450	0.924
			$25.13 \pm 0.02°C$		
0.098	0.0	0.456	43.0	0.430	0.977
0.096	0.51	0.444	42.5	0.415	0.987
0.096	1.01	0.441	41.8	0.405	0.976
0.096	1.52	0.448	41.4	0.407	0.955
0.097	2.03	0.443	40.9	0.398	0.951
0.096	3.04	0.444	40.0	0.390	0.928

little. On the other hand, the total pressure decreases with an increase in salt concentration over the whole region of the solvent composition.

The activities of each solvent are shown in columns five and six of Tables I and II and are plotted in Figure 1 for 2-propanol. The standard state for the activities of each solvent was taken as the state for the pure component at the same temperature and pressure as that of the mixture. Assuming the vapor mixture obeys the ideal solution law, that is $\beta_{12} = (\beta_{11} + \beta_{22})/2$, the activities were calculated by the following equation

$$a_i = (P_i/P_i°) \exp[(\beta_{ii} - v_i)(P - P_i°)/RT] \tag{12}$$

where the superscript o denotes the pure component. The second virial coefficients and the molar volumes in the liquid phase of each pure solvent used in these calculations are described in a previous paper (3). The activity of water decreases gradually, and then abruptly decreases with the salt concentration at a particular 2-propanol concentration. However, in the alcohol-rich region the activity of 2-propanol increases with the salt concentration, but in the water-rich region it decreases.

The extent of the salt effect on a mixed solvent, however, should be measured not by the activity change of nonelectrolyte caused by the salt addition, but by the relative change of the chemical potentials of each solvent. As discussed in the previous paper (10), $\Delta\mu$ may be evaluated at a constant temperature by the following equation,

$$\Delta\mu = \ln\frac{\alpha}{\alpha_0} + \int_{P_0}^{P} \frac{(v_1 - v_2)}{RT} dP \tag{13}$$

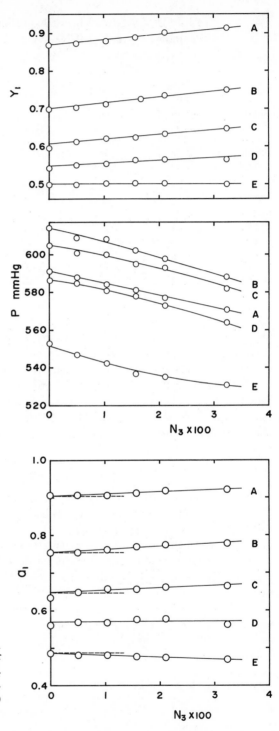

Figure 1. Activity of 2-propanol as a function of salt concentration at the mixed solvent compositions: A = 0.9; B = 0.7; C = 0.5; D = 0.3; E = 0.1.

Table III. Salt Effect Parameter of $LiClO_4$ at 75°C

Z_1	k_0	$y_{10,cal}$	σ
0.1	0.17	0.499	0.291
0.3	2.95	0.546	0.904
0.5	6.16	0.602	0.979
0.7	8.46	0.696	0.994
0.9	15.0	0.865	0.990

Table IV. Salt Effect Parameter of $LiClO_4$ at $Z_1 = 0.1$

T (°C)	k_0	$y_{10,cal}$	σ
75	0.17	0.499	0.291
50	−0.24	0.486	−0.217
25	−1.02	0.450	−0.513

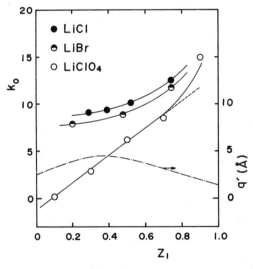

Figure 2. Salt effect parameters k_0 and q' at 75°C as a function of mixed solvent composition

where

$$\alpha = a_1/a_2 \tag{14}$$

The second term on the right side in Equation 13 is entirely negligible at low pressures in comparison with the activity term. Therefore from Equations 11 and 13, we can write

$$\ln \alpha = k_0 N_3 + \text{const.} \tag{15}$$

The values of the two constants in Equation 15 were obtained by the least-squares method, the results of which are shown in Tables III and IV. In these tables σ is the correlation coefficient in the experimental values of $\ln\alpha$ from the fitted curve between $\ln\alpha$ and N_3.

The salt effect parameter k_0 is plotted in Figure 2, and the data for lithium chloride and lithium bromide reported in the previous paper (3) are also plotted for the purposes of comparison. It can be seen from Figure 2 that k_0 depends markedly on the solvent composition. The values of k_0 decrease and in the extremely water-rich region k_0 is negative at 50° and 25°C. In other words, 2-propanol is salted in by the addition of lithium perchlorate. The salting-in effect of 2-propanol increases with reduction in temperature.

The chemical potential of ion at the reference state μ_i^* is conveniently separated into an electrostatic part and other nonelectrostatic parts. If the Born approximation (11) is taken for the electrostatic part, the salt effect parameter for 1–1 salt may be given by the following equation:

$$k_0 = \frac{2}{RT}\left(\frac{d\mu_{\pm,\text{ne}}^*}{dZ_1}\right) + \frac{q'}{r_\pm} \tag{16}$$

where the subscript ne denotes the nonelectrostatic part, $r_\pm$ is the harmonic mean of the two ionic radii, and

$$q' = \frac{N_A e^2}{2RT}\frac{d(1/\epsilon_0)}{dZ_1} \tag{17}$$

If the first term on the right side in Equation 16 is negligible in comparison with the second term, the salt effect parameter may be evaluated from the dielectric constant of the solvent. This equation is similar to the Debye–McAulay equation (12). The values of q' at 75°C are plotted in Figure 2. The values of ϵ_0 used in the calculation were interpolated from the experimental data by Åkerlöf (13). The values calculated using the harmonic mean of the crystal radii of lithium ion and perchlorate ion do not agree well with the measured values, especially in the negative region of k_0 and the isopropanol-rich region. The large discrepancy indicates that the ions are solvated and form a short-range ion pair.

Conclusion

Isothermal vapor–liquid equilibrium data at 75°, 50°, and 25°C for the system of 2-propanol–water–lithium perchlorate were obtained by using a modified Othmer still. In the 2-propanol-rich region 2-propanol was salted out from the aqueous solution by addition of lithium perchlorate, but in the water-rich region 2-propanol was salted in. The salt effect parameter of lithium perchlorate was evaluated.

Nomenclature

a_i	=	activity of component i
e	=	electronic charge
k_0	=	salt effect parameter defined by Equation 10
N_A	=	Avogadro number
N_3	=	moles of salt per one mole of solvent
n_i	=	moles of component i
P	=	pressure
P_i	=	partial pressure of solvent i
q'	=	defined by Equation 17
R	=	gas constant
r	=	ionic radius
T	=	temperature
v_i	=	molar volume of pure solvent i in liquid phase
y_i	=	mole fraction of solvent i in vapor phase
Z_i	=	mole fraction of solvent i in mixed solvent
α	=	defined by Equation 14
β	=	second virial coefficient
γ	=	activity coefficient
$\Delta\mu$	=	defined by Equation 9
ϵ_0	=	dielectric constant of mixed solvent
μ_i	=	chemical potential of component i

Superscript

o = pure state
* = infinite dilution

Subscript

0 = without salt
1 = 2-propanol
2 = water
3 = salt
± = ionic mean value

Literature Cited

1. Johnson, A. I., Furter, W. F., *Can. J. Chem. Eng.* (1960) **38**, 78.
2. Sada, E., Morisue, T., *J. Chem. Eng. Jpn.* (1975) **8**, 191.
3. *Ibid.*, 196.
4. Seidell, A., Linke, W. F., "Solubilities of Inorganic and Metal-organic Compounds," American Chemical Society, Washington D.C., 1965.
5. Sada, E., Morisue, T., Miyahara, K., *J. Chem. Eng. Data* (1975) **20**, 283.
6. Grunwald, E., Bacarella, A. L., *J. Am. Chem. Soc.* (1958) **80**, 3840, (1960) **82**, 5801, (1974) **96**, 423.
7. *Ibid.* (1960) **82**, 5801.

8. *Ibid.* (1974) **96**, 423.
9. Mixon, F. O., Gumovski, B., Carpenter, B. H., *Ind. Eng. Chem. Fundam.* (1965) **4**, 455.
10. Sada, E., Morisue, T., *J. Chem. Eng. Jpn.* (1973) **6**, 385.
11. Moelwyn-Hughes, E. A., "Physical Chemistry," 2nd Ed., Pergamon, England, 1961, Chap. 18.
12. Debye, P., McAulay, L., *Phys. Z.* (1925) **26**, 22.
13. Akerlöf, G., *J. Am. Chem. Soc.*, (1932) **54**, 4125.

RECEIVED June 30, 1975.

Vapor–Liquid Equilibrium in the Ethanol–Water System Saturated with Chloride Salts

M. A. GALAN, M. D. LABRADOR, and J. R. ALVAREZ

Department of Chemistry (Technical), Faculty of Science, University of Salamanca, Salamanca, Spain

Vapor–liquid equilibrium data at atmospheric pressure (690–700 mmHg) for the systems consisting of ethyl alcohol–water saturated with copper(II) chloride, strontium chloride, and nickel(II) chloride are presented. Also provided are the solubilities of each of these salts in the liquid binary mixture at the boiling point. Copper(II) chloride and nickel(II) chloride completely break the azeotrope, while strontium chloride moves the azeotrope up to richer compositions in ethyl alcohol. The equilibrium data are correlated by two separate methods, one based on modified mole fractions, and the other on deviations from Raoult's Law.

Many papers concerning salt effect on vapor–liquid equilibrium have been published. The systems formed by alcohol–water mixtures saturated with various salts have been the most widely studied, with those based on the ethyl alcohol–water binary being of special interest (*1–6,8,10,11*). However, other alcohol mixtures have also been studied: methanol (*10,16,17,20,21,22*), 1-propanol (*10,12,23,24*), 2-propanol (*12,23,25,26*), butanol (*27*), phenol (*28*), and ethylene glycol (*29,30*). Other binary solvents studied have included acetic acid–water (*22*), propionic acid–water (*31*), nitric acid–water (*32*), acetone–methanol (*33*), ethanol–benzene (*27*), pyridine–water (*25*), and dioxane–water (*26*).

Although the third component of these systems is usually a single inorganic salt, mixtures of two or more salts have been studied, and some research has been done with third components of low vapor pressure (*18,19*). Some qualitative studies have been done on salt effect in vapor–liquid equilibrium with salts which are either: soluble in only one or both components, hygroscopic or non-hygroscopic, etc.

The present work studies the vapor–liquid equilibrium of the ethanol–water system saturated with copper(II) chloride, strontium chloride, and nickel(II) chloride.

Salts soluble in ethanol as well as water have been found to break the azeotrope, while salts which are very soluble in water and only slightly soluble in the alcohol move the azeotrope to richer ethanol regions without breaking it. Furthermore, the salts or compounds which dissolve more in one component are found to raise the volatility of the other component. This finding is in conformity with that of previous workers in the field (8,11,12,13,18,19,23,24,27). In this work the equilibrium diagrams were obtained at atmospheric pressure (690–700 mmHg), and under saturation conditions.

From experimental data for the ethanol–water system without salt, obtained at 700 and 760 mmHg, it can be seen that within this pressure range the effects of pressure on the equilibrium data are small enough to be within the experimental scatter. In fact, in previous works (8,11,12,13,18,19,23,24,27) there seems to be no clear difference between the equilibrium data at 700 and at 760 mmHg. Errors obtained in the determination of liquid and vapor compositions are approximately ±0.05 wt % for the systems without salt. For salt-saturated systems, the same error prevails for the vapor phase, while the error is between 0.1 and 0.2 wt % for liquid phase compositions. The error for the boiling temperature is less than 0.1°C for the systems without salt, but for saturated solutions the error is much greater: from 0.2°C for nonconcentrated solutions to 3°C or more for highly concentrated solutions.

Experimental Procedure

The apparatus and experimental procedure used here are the same as those used in previous work (8,11,12,13,18,19,23,24,27). An improved Othmer still is used. In the lower part of the still, at five cm from the base, there is a stopcock through which liquid samples can be taken. The cock bore is larger than the standard bore, thus preventing sticking by the salt. A thermometer which measures the boiling temperature is introduced into the vessel through an inlet on either side of the still, with the bulb submerged in the solution. This solution is vigorously stirred magnetically. The still neck is covered with nichrome thread through which an electric current is passed. In this way, the temperature of the vapor can be kept higher than the boiling point, thus avoiding any condensation. The recycling tube has a narrow bore section, which prevents mixing of solution and vapor.

The vapor composition was determined by picnometry. The liquid phase was evaporated to dryness; the solubility of the salt and the composition of the liquid on a salt-free basis were determined from the salt and the liquid obtained by weighing and picnometry.

Results and Discussion

Figure 1 shows the equilibrium data for the ethanol–water systems saturated with copper(II) chloride, strontium chloride, or nickel(II) chloride. Figure 2 shows the temperature-compositions diagrams corrected to 700 mmHg.

For the copper(II) chloride system, a maximum and minimum were observed. Maximum and minimum points have also been observed for the 1-propanol–water system saturated with copper(II) chloride. The explanation for these singular points will be made after demonstration of the solubility data.

The temperature data for the ethanol–water–strontium chloride system show that the curve for the liquid phase cuts the curve corresponding to the salt-free system. This tends to happen with salts that are very soluble in water and only slightly soluble or insoluble in ethanol (8,12). This is because, for a binary

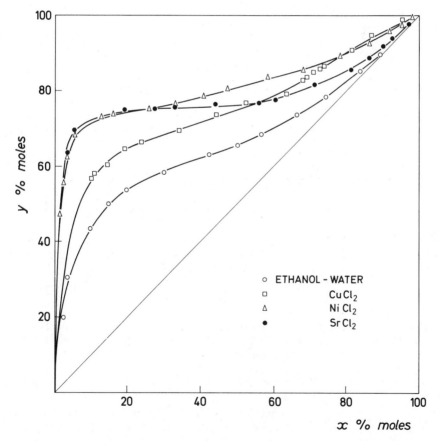

Figure 1. Vapor–liquid equilibrium data for the ethanol–water systems saturated with copper(II) chloride, nickel(II) chloride, and strontium chloride

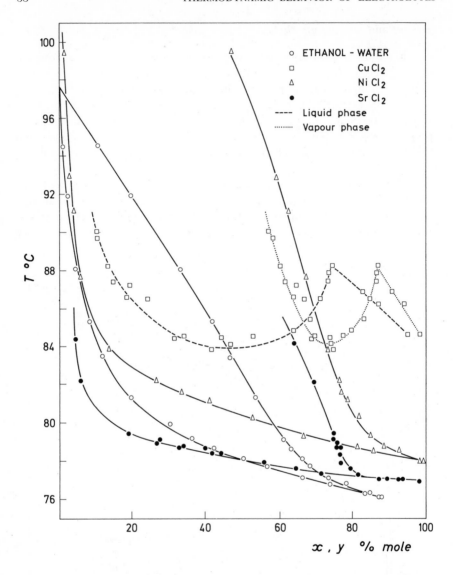

Figure 2.   Temperature–composition diagrams, corrected to 700 mmHg, for the ethanol–water systems saturated with copper(II) chloride, nickel(II) chloride, and strontium chloride

mixture of a determined composition and at a determined pressure, the boiling point is a fixed temperature.

   If a salt which is somewhat soluble in both water and ethanol is added to a binary mixture, the composition of the liquid phase is modified such that the solution can be considered to consist of a binary mixture formed by free water and ethanol with a composition richer in ethanol than the initial binary mixture (without salt), and the salt with fixed water; hence the salt effect is to fix the water, thus raising the alcohol volatility. The boiling temperature of the solution with salt will be the sum of: the boiling temperature of free water and ethanol (lower than the boiling temperature of the initial solution), and the temperature increment due to the salt. If this increment is small (as with ions of small activity such as Na^+, K^+, Sr^{2+}), the solution with salt may have a lower boiling point than the same solution without salt.

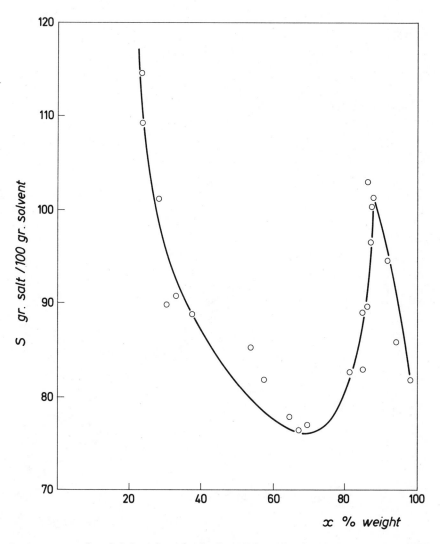

Figure 3. Salt solubility data for boiling ethanol–water mixtures saturated with copper(II) chloride

From the equilibrium composition and temperature, it can be deduced that copper(II) chloride and nickel(II) chloride break the azeotrope, while strontium chloride moves it toward a richer ethanol concentration. Figures 3 and 4 show solubility data for these systems, expressed in grams of salt per hundred grams of solvent, plotted against liquid composition in wt % on a salt-free basis. Solubility data for the copper(II) chloride system (*see* Figure 3) show a maximum

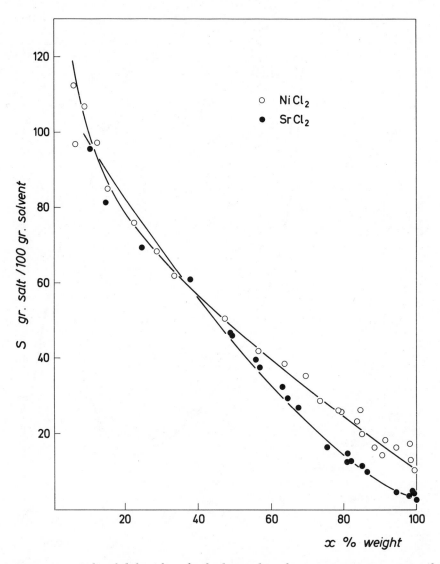

Figure 4. Salt solubility data for boiling ethanol–water mixtures saturated with nickel(II) chloride and strontium chloride

and minimum for two different compositions of the binary mixture. This was also observed in the 1-propanol–water–copper(II) chloride system (*24*). These singular points were also found in other systems: a minimum in the solubility data was found in the 1-propanol–water–cobalt(II) chloride system (*24*), and a maximum was found for the ethyl alcohol–water system saturated with phenolphthalein (*19*). The solubility maximum and minimum for the ethanol–water–copper(II) chloride system can account for the temperature maximum and minimum (*see* Figure 2).

On the other hand, the system shown in Figure 1 breaks the azeotrope completely. This means that the vapor phase composition must always be larger than the liquid phase compositions over the entire range of compositions. From this, the boiling point of the vapor must be lower and the composition of vapor must be higher than those of the liquid. Therefore, the lines for the liquid phase and the vapor phase in this diagram go through the maximum and minimum without crossing (cutting) over the entire range of concentrations. These singular points are not observed in the case of nickel(II) and strontium chlorides.

Correlation

Equations studied previously (*19,27*) have been used to correlate the equilibrium data. Alvarez and Vega (*27*) correlate the equilibrium data as a function of a modified mole fraction X'_i, and the liquid phase composition on a salt-free basis X^+_i.

$$\log X'_i = a \log X_i^+ + \log b \qquad (1)$$

In this equation, a and b are constants characteristic of the system. The modified mole fraction is the one defined by Lu (*34*) from the compositions on a salt-free basis and from the vapor pressure of the pure components and of the salt plus pure liquid solutions. Figures 5 and 6 show the values of X'_i and X^+_i corresponding, respectively, to ethanol and water for each of the three systems. For nickel(II) chloride and strontium chloride, the experimental data follow a straight line, while for copper(II) chloride the data form three straight lines, as was expected (*24*) from the maximum and minimum in the temperature diagram.

For nickel(II) and strontium chlorides the experimental data are in good agreement with data obtained by Alvarez Gonzalez and Vega Zea (*27*); that is, when the solubility of the salt in one component increases, the value of X'_i decreases for the pure component. Moreover, the decrease is proportional to the solubility of the salt in this component. Accordingly, a salt that is insoluble in one component must pass through the point $X'_i = 100$, $X^+_i = 100$. The values obtained for constants a and b of Equation 1 are shown in Table I.

The equilibrium data were also correlated by studying the deviations from Raoult's Law. Since the salt-free system obeys this law when the concentration of a specified component is substituted by its activity, it follows that for the salt-free system:

$$\pi y_i = P_i^0(X_i^0\gamma_i^0)$$

and for the system with salt,

$$\pi y_i = P_i^0(X_{si}\gamma_i)^n$$

where X_{si} is the real concentration of component i in the salt solution, and the salt is considered a third component with no vapor pressure. X_i^0 is the concentration of the salt-free system, and γ_i^0 the activity coefficient of the salt-free system for the concentration X_i^0 corresponding to X_{si}. It is interesting to remember that $X_i^0 = X_i^+$.

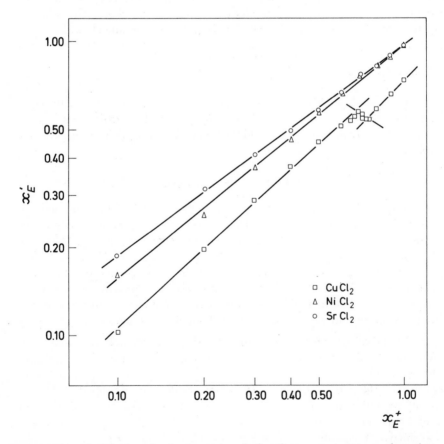

Figure 5.   Modified molar fraction vs. salt-free base composition referred to ethanol for the systems saturated with copper(II) chloride, nickel(II) chloride, and strontium chloride

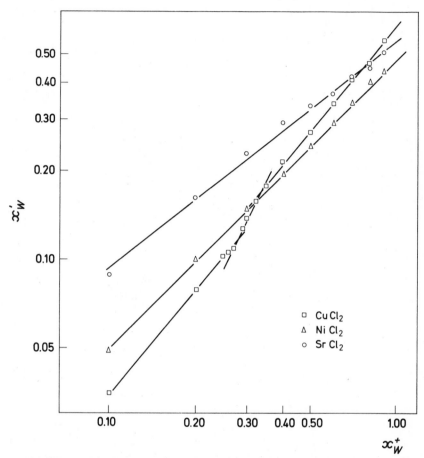

Figure 6. Modified molar fraction vs. salt-free base composition referred to water for the systems saturated with copper(II) chloride, nickel(II) chloride, and strontium chloride

Table I. Experimental Values Obtained for Constants a and b in Equation 1

Reference Component	Second Component	a	b
Ethanol–strontium chloride	water	0.717	3.553
Water–strontium chloride	ethanol	0.768	1.613
Ethanol–nickel(II) chloride	water	0.794	2.504
Water–nickel(II) chloride	ethanol	0.998	0.497
Ethanol–copper(II) chloride	water	0.877	1.407
Ethanol–copper(II) chloride	water	−0.537	547.4
Ethanol–copper(II) chloride	water	1.012	0.695
Water–copper(II) chloride	ethanol	1.164	0.240
Water–copper(II) chloride	ethanol	1.967	0.017
Water–copper(II) chloride	ethanol	1.193	0.256

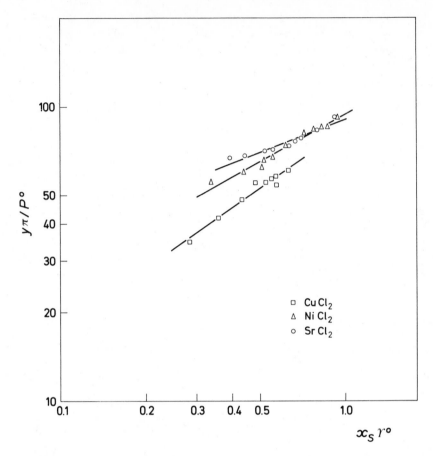

Figure 7.   Alvarez–Bueno–Galan correlation for the ethanol–water systems saturated with copper(II) chloride, nickel(II) chloride, and strontium chloride

The exponent n indicates the deviation of the system with salt from the salt-free system.   Figure 7 shows the values of log $\pi y_i/P_i^0$ vs. log $(X_{si}\gamma_i^0)$ for the ethanol–water system saturated with copper(II) chloride, nickel(II) chloride, and strontium chloride respectively.   The values for n obtained from the above system are shown in Table II.

Table II.   Experimental Values Obtained for Constant n in Equation 2

System	n
Ethanol–water–strontium chloride	0.38
Ethanol–water–nickel(II) chloride	0.53
Ethanol–water–copper(II) chloride	0.67

Alvarez Gonzalez et al. (*19*) stated that those systems in which the value of n is less than unity move the azeotrope to richer ethanol compositions, even to breaking the azeotrope. This was also observed for the systems presently under consideration. Reference *19* proposes a correlation of the equilibrium data by the empirical equation:

$$\frac{\pi y_i}{P_R} = K(X_i^0 \gamma_i^0)^n \qquad (3)$$

where P_R is the vapor pressure of an ideal ethanol–water system which obeys Raoult's law, and K is a constant. The exponent n gives the deviation of the system, salt-free or not, with respect to the ideal system, which follows Raoult's law. The log $\pi y_i/P_R$ vs. log $(X_i^0 \gamma_i^0)$ is presented in Figure 8 for the three systems studied. The values obtained for n are presented in Table III.

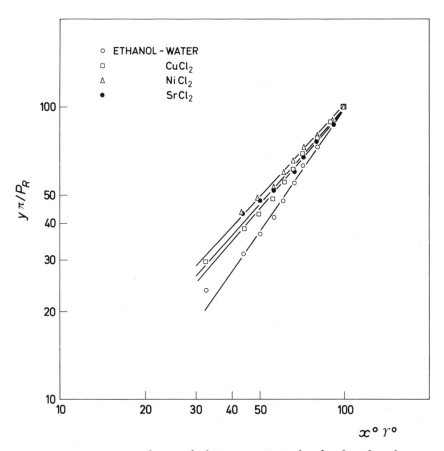

Figure 8.   Empiric correlation of Alvarez–Bueno–Galan for the ethanol–water system and ethanol–water saturated with copper(II) chloride, nickel(II) chloride, and strontium chloride

Table III. Experimental Values Obtained for Constant n
in Equation 3

System	n
Ethanol–water	1.39
Ethanol–water–strontium chloride	1.03
Ethanol–water–nickel(II) chloride	1.05
Ethanol–water–copper(II) chloride	1.13

From these it can be deduced that the system with a slope of less than 1.39 moves the azeotrope toward richer ethanol composition, as was also deduced in Reference *19*.

To correlate the temperatures, these same authors (*19*) proposed the representation of $\log T_{Ri}\gamma_i^0/T_i$ vs. $\log (T_{Ri}\gamma_i^0)$, where T_{Ri} is the boiling temperature

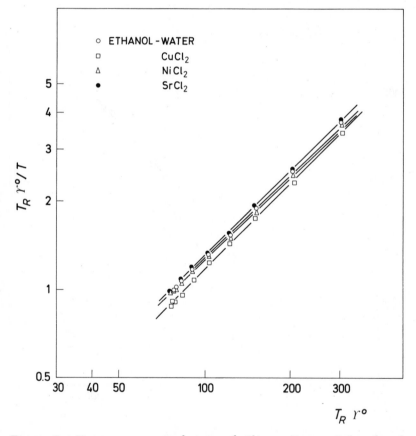

Figure 9.   Temperature correlations of Alvarez–Bueno–Galan for the ethanol–water system and ethanol–water saturated with copper(II) chloride, nickel(II) chloride, and strontium chloride

of the ideal system and T_i is the boiling temperature of the real system without salt.   Figure 9 shows that a straight line was obtained for each system.

These correlations allow the prediction of equilibrium data for systems saturated with salt from only one or two experimental points.   Nevertheless, in all work done on salt effect it seems that what was stated at the beginning of this paper is true, i.e., the effects on the volatilities of components and hence the variation in relative volatility depend on the solubility of the salt in both components.

Literature Cited

1. Kyrides, L. P., Carwell, T. S., Pfeifer, C. E., Wobus, R. S., *Ind. Eng. Chem.* (1932) **24**, 795.
2. Rieder, R. M., Thompson, A. R., *Ind. Eng. Chem.* (1950) **42**, 379.
3. Tursi, R. R., Thompson, A. R., *Chem. Ing. Prog.* (1951) **47**, 304.
4. Jost, W., *Chem. Eng. Tech.* (1951) **23** No. 3, 64.
5. Costa, E., Moragues, J., *An. R. Soc. Esp. Fis. Quim.* (1952) **48B**, 397–408; (1952) **48B**, 441.
6. Yamamoto, Y., Maruyano, T., *Chem. Eng. Jpn* (1952) **16**, 166.
7. Dobroserdov, L. L., Il'Yina, V. P., *Tr. Leningr. Tekhnol. Inst. Pishch. Prom.* (1956) **13**, 92.
8. Rius Miro, A., Otero de la Gandara, J. L., Alvarez Gonzalez, J. R., *An. R. Soc. Esp. Fis. Quim.* (1957) **53B**, 171; (1957) **53B**, 185.
9. Dobroserdov, L. L., Il'Yina, V. P., *Tr. Leningr. Tekhnol. Inst. Pishch. Prom.* (1958) **14**, 147.
10. Johnson, A. I., Furter, W. F., *Can. J. Chem. Eng.* (1960) **38**, 78.
11. Rius Miro, A., Alvarez Gonzalez, J. R., Uriarte, A., *An. R. Soc. Esp. Fis. Quim.* (1960) **56B**, 629.
12. Alvarez Gonzalez, J. R., Artacho, E., *An. R. Soc. Esp. Fis. Quim.* (1961) **57B**, 219.
13. Alvarez Gonzalez, J. R., Uriarte, A., *An. R. Soc. Esp. Fis. Quim.* (1962), **58B**, 145.
14. Proinova, Z. A., Toncheva, G., *Khim. Ind. Sofia* (1966) **38** (3), 130.
15. Meranda, D., Furter, W. F., *Can. J. Chem. Eng.* (1966) **44**, 298.
16. Meranda, D., Furter, W. F., *A.I.Ch.E. J.* (1971) **17**, 38.
17. Meranda, D., Furter, W. F., *A.I.Ch.E. J.* (1972) **18**, 111.
18. Alvarez Gonzalez, J. R., Galan Serrano, M. A., *An. R. Soc. Esp. Fis. Quim.* (1973) **69**, 545.
19. Alvarez Gonzalez, J. R., Bueno Cordero, J., Galan Serrano, M. A., *An. R. Soc. Esp. Fis. Quim.* (1974) **70**, 262.
20. Jaques, D., Furter, W. F., *Ind. Eng. Chem. Fundam.* (1974) **13** (3), 238.
21. Meranda, D., Furter, W. F., *A.I.Ch.E. J.* (1974) **20** (1), 103.
22. Yoshida, F., Yasunishi, A., Hamada, Y., *Chem. Eng. (Tokyo)* (1964) **28** (2), 133.
23. Rius Miro, A., Alvarez Gonzalez, J. R., *An. R. Soc. Esp. Fis. Quim.* (1958) **54B**, 797.
24. Alvarez Gonzalez, J. R., Galan Serrano, M. A., *An. R. Soc. Esp. Fis. Quim.* (1974) **70**, 271.
25. Ohe, S., Yokayama, L., Nakamura, S., *Ishikawajima Harima Giho* (1971) **11** (5), 368.
26. Prausnitz, J. M., Targovnik, *J. Chem. Eng. Data* (1958) **3**, 234.
27. Alvarez Gonzalez, J. R., Vega Zea, J., *An. R. Soc. Esp. Fis. Quim.* (1965) **61B**, 831; (1967) **63B**, 749.
28. Bogart, M. J. P., Brunjes, A. S., *Chem. Eng. Progress* (1948) **44**, 95.
29. Fogg, E. T., Diss. Abstr. (1953), **13**, 739–740. Univ. Microfilm Pub. No. 5589. Univ. of Pennsylvania. EE. UU.
30. Fox, P. M., M.S. Thesis., Univ. of Pennsylvania, 1949.

31. Ramalho, R. S., Edgett, N. S., *J. Chem. Eng. Data* (1964) **9,** 324.
32. Baranov, A. V., Karev, V. G., *Tr. Sib. Tekh. Inst.* (1963) **36,** 61.
33. Proszt, J., Kollar, G., *Rocz. Chem.* (1958) **32,** 611.
34. Benjamin, C., Lu, Y., *Ind. Eng. Chem.* (1960) **52,** 871.

RECEIVED July 15, 1975.

Effects of Salts Having Large Organic Ions on Vapor–Liquid Equilibrium

JOHN A. BURNS and WILLIAM F. FURTER[1]

Department of Chemical Engineering, Royal Military College of Canada, Kingston, Ontario, Canada, K7L 2W3

The salt effects of potassium bromide and a series of five symmetrical tetraalkylammonium bromides on vapor-liquid equilibrium at constant pressure in various ethanol-water mixtures were determined. For these systems, the composition of the binary solvent was held constant while the dependence of the equilibrium vapor composition on salt concentration was investigated; these studies were done at various fixed compositions of the mixed solvent. Good agreement with the equation of Furter and Johnson was observed for the salts exhibiting either mainly electrostrictive or mainly hydrophobic behavior; however, the correlation was unsatisfactory in the case of the one salt (tetraethylammonium bromide) where these two types of solute-solvent interactions were in close competition. The transition from salting out of the ethanol to salting in, observed as the tetraalkylammonium salt series is ascended, was interpreted in terms of the solute-solvent interactions as related to physical properties of the system components, particularly solubilities and surface tensions.

The addition of salts to a liquid mixture to aid in the separation of the components of that mixture by fractional distillation has important implications in terms of theoretical studies and practical applications. The complexity of the salt effect in vapor–liquid equilibria and the sparse conclusive work in this field are largely responsible for the limited applications it has received industrially, despite a potential for dramatically improved separation performance in certain systems. Not only is the effect of the salt on a system complex, but variations occur from system to system; that is, each system is unique. In addition, the nature of the effects, as well as their magnitudes, tends to be com-

---

[1] To whom correspondence should be addressed.

position dependent, even within a given system. Consequently, the scope of the salt effect even on binary, let alone multicomponent, systems becomes astronomical. Despite the many proposals submitted, no one theory has yet been able to predict satisfactorily the effects of salts on vapor–liquid equilibria from pure-component properties alone. If a treatment applies for a certain type of salt, e.g., inorganic, then it fails in the case of others, e.g., the tetraalkylammonium (TAA) salts of the present investigation. Regardless, compensating factors aid in producing convenient behavioral trends for which plausible explanations can be tendered and empirical generalizations noted. In this way it is possible to simplify an otherwise unmanageable situation.

Applications of salt effects to both vapor–liquid and liquid–liquid phase equilibria were reviewed in 1958 by Prausnitz and Targovnik (1). More recently several authors (2, 3, 4) have offered comprehensive treatises of salt effects on nonelectrolytes in mixed-solvent solution; however, most studies have been confined to effects of salts on activity coefficients and solubilities of the nonelectrolyte in aqueous solution. In some cases, a detailed mathematical treatment has been applied to those salt effects (4, 5, 6). An extensive review of the literature of salt effect in vapor–liquid equilibrium and distillation processes prior to 1966 has been compiled by Furter and Cook (8) who list over two hundred references. Previously, studies have concentrated on saturated solutions of mainly inorganic salts in binary solvent mixtures (9–15). Similar experiments using a fixed salt concentration below saturation have been reported (16, 17). Long and McDevit (7), and more recently Gubbins and Tiepel (5), reviewed the salt effects of aqueous solutions of tetraalkylammonium (TAA) halides. These authors report that the TAA cations salt in most nonelectrolytes, and this effect becomes more pronounced as the size of the TAA cation increases, as opposed to salting-out of these solutes by most other electrolytes. This observed behavior has been accounted for by either an influence on water structure (18) or association between the ion and the nonelectrolyte (18, 19, 20). This latter tendency of nonpolar groups to adhere to one another in an aqueous environment has become known as hydrophobic bonding (21).

From thermodynamic considerations and after a sequence of simplifying assumptions has been applied, including those of constant temperature and pressure, an equation for the salt effect in vapor–liquid equilibria under conditions of constant mixed-solvent composition has been derived (22, 23). The equation, in its simplest form, reduces to

$$\ln \frac{\alpha_s}{\alpha} = kz. \tag{1}$$

This equation relates a so-called "improvement factor," the logarithm of the ratio of relative volatility with and without salt present, to the salt concentration in the liquid phase under the condition of fixed mixed-solvent composition, by a salt effect parameter k. Usually, the added salt lowers the volatility of both components in the liquid phase. If the extent of this lowering is different for the two volatile components, as is usually the case, there will be an enhancement of one component and depletion of the other in the equilibrium vapor phase, that is, an alteration in vapor composition caused by the addition of the salt. Equation 1 also appears to apply (8, 12, 24) to some ternary systems even if the liquid composition is varied. (However, under such conditions Equation 1 would become primarily empirical and lose much of its theoretical significance.) Also, k has been found to vary with the system and the temperature (23, 25). Of the many systems investigated, few studies involved the simpler conditions of constant liquid composition while varying the salt concentration. These are precisely the conditions to which Equation 1 should be applied, but there are other factors involved (for example, temperature, pressure, solute–solute interactions). Nevertheless, it is possible to explain certain salt effect trends in general terms by the physical nature of the components involved in the mixture, although inconsistencies with such generalities exist (15). Other relations have been devised to correlate more adequately the salt effects on vapor–liquid equilibria under conditions of varying liquid composition (26, 27), but either the complexity or the empirical nature of the resulting equations limits their usefulness. Recently Jaques and Furter (28) reported results under suitable conditions to which Equation 1 should apply rigorously, and they observed that the equation was indeed satisfactory for additions of sodium bromide, ammonium chloride, and sodium chloride to ethanol–water mixtures, and for the calcium chloride–methanol–water system. However, for these systems the solubilities and boiling point elevations at the liquid composition values used were not large. Burns and Furter (29) found similar results for the potassium bromide–, ammonium bromide–, and potassium iodide– ethanol–water systems, but the correlation was unsatisfactory for potassium and sodium acetates in ethanol–water mixtures.

Alcohols exhibit a bifunctional nature in aqueous solution. On the one hand, there exists a hydrophobic hydrocarbon group which resists aqueous solvation; on the other, there is the hydrophilic hydroxyl group which interacts intimately with the water molecules. Franks and Ives (30, 31) have reviewed experimentation and theoretical treatises on the structure of water, the structure of liquid alcohols, and the thermodynamic, spectroscopic, dielectric, and solvent properties and P–V–T relationships of alcohol–water mixtures. Sada et al. (27) reviewed, in particular, the salt effects of electrolytes in alcohol–water systems and discussed the various correlations of the salt effect applied to these systems. Inorganic salts were used almost universally in these salt effect studies.

The present study investigates the effects of a series of salts while maintaining certain fixed factors. The contribution of the cation to the salt effect by tetraalkylammonium bromides (R_4NBr) on the ethanol–water system with the liquid composition held constant is studied. These TAA cations exhibit an ambivalent nature in aqueous solution; the smaller cations display predominantly electrostrictive interactions (32), i.e., hydrophilic bonding or net breaking of hydrogen bonds in water, while the larger ones display hydrophobic bonding (33), or the net making of hydrogen bonds in water when the salt is added. The salts of the present investigation were chosen to complement and extend studies by Jaques and Furter (28) and further work done by the present authors (29). They include potassium, ammonium, tetramethylammonium, tetraethylammonium, tetra-n-propylammonium, and tetra-n-butylammonium bromides in ethanol–water mixtures at fixed liquid composition.

Experimental

The data were obtained by means of an improved Othmer recirculation still (34), as modified for salt effect studies by Johnson and Furter (22, 35). Thermal energy was applied to the still by means of a heating mantle controlled by a rheostat in order that the boiling rate could be adjusted effectively. Suppression of bumping of the solution and both thermal and physical homogeneity were maintained by means of a magnetic stirring mechanism.

Analyses of salt-free aqueous ethanol mixtures were obtained using 10-ml Weld-type pycnometers. These calibrated specific gravity bottles were immersed in a 20-l. water bath thermostatted to 25.00°C, for two hr before being weighed. The mass of the bottles was obtained using a Becker Chainomatic Balance accurate to four decimal places. Corrections for buoyancy were applied to the weights, and a tare was utilized. Alcohol concentrations were then obtained from the experimental specific gravities by interpolation of literature data (36) with the use of a five-coefficient nonlinear least squares program adapted to an XDS Sigma 3 computer.

Salt concentrations were obtained by the addition of known weights of dried salt to the solvent mixture. The compositions of the condensed equilibrium vapor samples and the previously prepared ethanol–water charges to the still were determined as previously outlined. The mole fractions of the salt, ethanol, and water charged to the Othmer still were thus accurately determined by mass balance calculations.

Anhydrous ethanol (99.9+ % purity), conforming to the specifications of the British Pharmacopoeia, was obtained from Gooderham and Worts Ltd., Toronto, Ontario. The potassium and ammonium bromides were British Drug Houses Analar analytical reagent grade. The tetramethylammonium and tetraethylammonium bromides were reagent grade from J. T. Baker Chemical Co., Phillipsburg, N. J. The tetra-n-propylammonium and tetra-n-butylammonium bromides were purchased from Eastman Kodak Company, Rochester, N. Y. All salts were dried for at least 72 hr immediately prior to use in a vacuum drying oven; the temperature during drying depended on the thermal stability and fusion point of the salt. The drying temperature was adjusted to 120°C for the ammonium and potassium bromides, 100°C for tetramethylammonium bromide, 90°C for tetraethylammonium bromide, 75°C for tetra-n-propylammonium bromide and 65°C for tetra-n-butylammonium bromide. The dried salts were stored under vacuum over P_2O_5. Laboratory distilled water was further purified by distillation over basic permanganate followed by an ion exchange treatment through a 38-cm Corning column.

The surface tension measurements were obtained with a Precision Cenco-du-Nouy Tensiometer, No. 70540. The solvents and saturated solutions were prethermostatted to 25.0°C prior to each measurement. The temperature of the solutions was determined subsequent to the measurements, and adjustments were made to coincide with a temperature deviation from 25.0°C. Only reproducible values were retained. R/r and L for the platinum ring used for the surface tension measurements were 53.6 and 5.997 cm, respectively. The factor F, which corrects for liquid elevated above the free surface of the liquid by the ring in the relationship $\gamma = Mg/2L \times F$, was determined (37) for each particular solvent.

The solubility studies were conducted gravimetrically after the solutions were frequently shaken vigorously and maintained at 25°C for 24 hr.

Results

The data in Tables I–XVI (*see* Appendix *for all tables*) show the isobaric vapor–liquid equilibrium results at the boiling point for potassium, ammonium, tetramethylammonium, tetraethylammonium, tetra-n-propylammonium, and tetra-n-butylammonium bromides in various ethanol–water mixtures at fixed liquid composition ratios. The temperature, t, is the boiling temperature for all solutions in these tables. In all cases, the ethanol–water composition was held constant between 0.20 and 0.35 mole fraction ethanol since it is in this range that the most dramatic salt effects on vapor–liquid equilibrium in this particular system should be observed. That is, previous data (*12–15, 38*) have demonstrated that a maximum displacement of the vapor–liquid equilibrium curve by salts frequently occurs in this region. In the results presented here, it should be noted that Equation 1 has been modified to

$$\log_{10} \frac{\alpha_s}{\alpha} = k'z \qquad (2)$$

where $k' = k/2.303$.

Tables IV–XVI show that the tetraalkylammonium salts have a large effect on both solvents in the binary solvent mixture, especially the larger tetraalkylammonium bromides, i.e., $(n\text{-}C_3H_7)_4NBr$ and $(n\text{-}C_4H_9)_4NBr$. This can be seen from consideration of the boiling temperature alone. This observation is also borne out by the surface tensions and solubilities at 25°C of the individual salts studied, the results of which are tabulated in Table XVII in water, in ethanol, and in an ethanol–water mixture at $x = 0.206$. For the higher homologs of the R_4NBr series, these salts exert a large effect on the surface tensions of the solvent systems studied and show a marked increase in their solubility in ethanol.

By analyzing the results shown in Table XVIII and perusing Figures 1–12, the validity of applying Equation 2 to the prediction of the salt effects of the R_4NBr salts at various fixed values of x for the ethanol–water system can be assessed. Figures 1–12 indicate the smoothed curves through the experimental points for the salt effects. From these graphs a value for the salt effect parameter, k', could be obtained; however, in some instances it is difficult to justify a linear plot of $\log_{10} \alpha_s/\alpha$ vs. z (Figures 7, 8, 9). The values of k' in Table XVIII were determined by means of a linear least squares calculation and in some cases, as noted in Figures 2, 5–8, 11, the intercept does not pass through the origin. Also in Figures 1–5, 9 the salt additions were made to saturation and the results are denoted by the sharp fall off of the graphical plot, or depicted by the dotted line in some instances. The relative average absolute deviation listed in Table XVIII is a numerical assessment of the feasibility of applying Equation 2 to the systems studied.

As previously mentioned, the values of k' for the ternary systems were calculated with a linear least squares plot of the experimental data. The average absolute deviation of $\log \alpha_s/\alpha$ was then calculated by averaging the absolute deviations of the smoothed curve passing through the experimental points from the linear least squares calculation, at fixed intervals of z. Taking this average absolute deviation of $\log \alpha_s/\alpha$ and finding its relative magnitude by dividing it by the value of the experimental $\log \alpha_s/\alpha$ at the mean z for each system and converting to a percentage will give the R.A.A.D. values listed in Table XVIII. The above procedure was devised in order to eliminate the random errors for k' (which are listed in this table for each system) when assessing the validity of using Equation 2 to predict the salt effects.

For the system $(C_2H_5)_4NBr$–ethanol–water at $x = 0.305$, the actual k' and R.A.A.D. values observed were 1.07 ± 0.03 and 3.5, respectively; however, the shape of the smoothed line passing through the points almost duplicated that for the system with the same components at $x = 0.316$. (Compare Figures 8 and 9.) This provided a means to obtain values of k' and R.A.A.D. for the system at $x = 0.305$ (*see* Table XVIII) that could be used for comparison with the other systems. This procedure was necessary only for the $(C_2H_5)_4NBr$–ethanol–water system since it is impossible to conceive of linear plots of $\log \alpha_s/\alpha$ vs. z for this system (Figures 7, 8, 9). The problem could have been avoided by carrying out the measurements to a higher concentration of salt for the system at $x = 0.305$ (*see* footnote to Table XVIII).

In Table XVIII, there are several trends that can be noted in k' if one proceeds through the R_4NBr series. First of all, k' tends to decrease as the size of the TAA cation increases and, in fact, tetra-n-butylammonium bromide shows a large salting-in effect. This trend is emphatically demonstrated by Figure 13, which shows the smoothed salt effects of the various salts studied in the ethanol–water system at $x = 0.206$. Secondly, it appears that there is a larger salting-out effect as the mole fraction of ethanol increases in the binary solvent mixture.

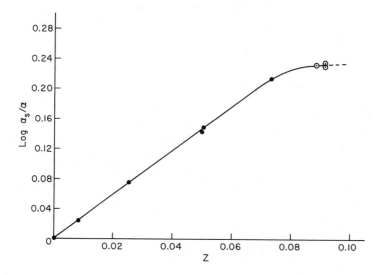

Figure 1. Salt effect of KBr in the ethanol–water system at x
= 0.206

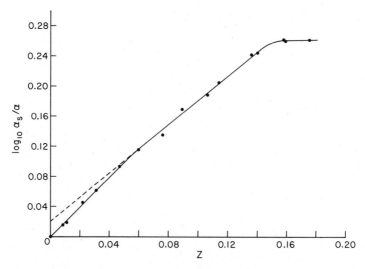

Figure 2. Salt effect of NH₄Br in the ethanol–water system at
x = 0.206

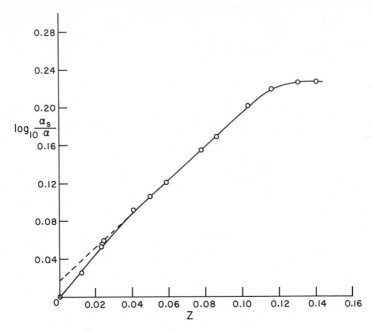

Figure 3. *Salt effect of NH$_4$Br in the ethanol–water system at* x
= *0.309*

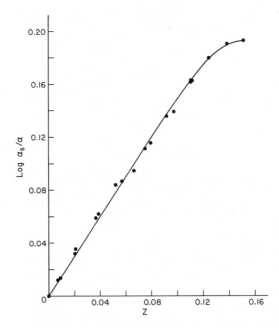

Figure 4. *Salt effect of* $(CH_3)_4NBr$ *in the
ethanol–water system at* x = *0.206*

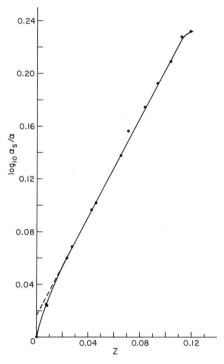

Figure 5. Salt effect of $(CH_3)_4NBr$ *in
the ethanol–water system at* x = 0.305

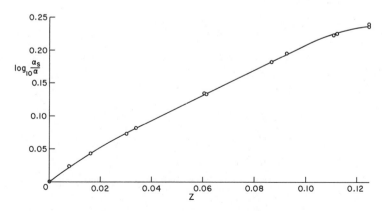

Figure 6. Salt effect of $(CH_3)_4NBr$ *in the ethanol–water system at*
x = 0.311

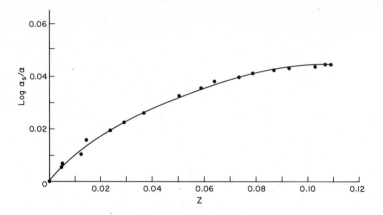

Figure 7. Salt effect of $(C_2H_5)_4NBr$ in the ethanol–water system at x = 0.206

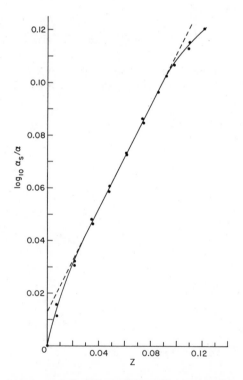

Figure 8. Salt effect of $(C_2H_5)_4NBr$ in the ethanol–water system at x = 0.305

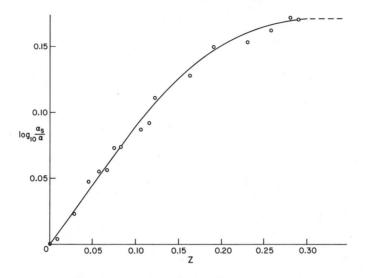

Figure 9. Salt effect of $(C_2H_5)_4NBr$ *in the ethanol–water system at* x = 0.316

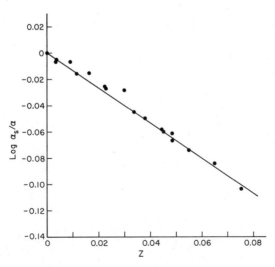

Figure 10. Salt effect of $(n\text{-}C_3H_7)_4NBr$ *in the ethanol–water system at* x = 0.206

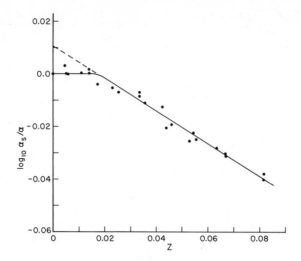

Figure 11. Salt effect of $(n\text{-}C_3H_7)_4NBr$ *in the etha-nol–water system at* x = 0.305

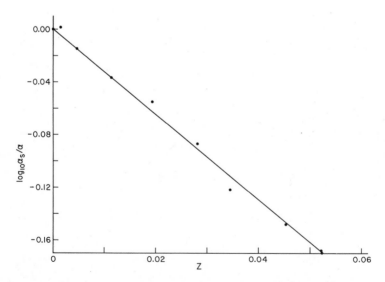

Figure 12. Salt effect of $(n\text{-}C_4H_9)_4NBr$ *in the ethanol–water system at* x = 0.200

Table XIX lists the approximate contributions of the various ions of the salts studied to the salt effect parameter.   Included in this table are the results of similar salt effect studies by other authors (28) as well as the present ones (29) in other work.   Justification for these individual assignments will be discussed later.

Establishment of the error analysis and deviation of the experimentally measured values resulting from random and systematic errors in the present investigation has been made previously by Jaques and Furter (28).   The fluctuations in the barometric pressure are indicated in Tables I–XVI for each system, and

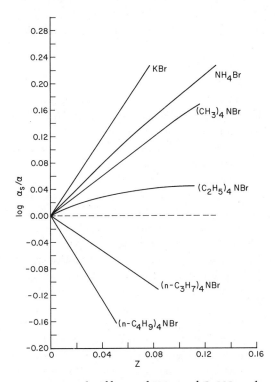

Figure 13. Salt effects of KBr and R_4NBr salts in the ethanol–water system at x = 0.206

the random error in the temperature readings was ±0.1°C.   In all cases, the mole fraction of ethanol in the binary solvent mixture was determined to be within ±0.001.   The values of z, y, and $\log_{10} \alpha_s/\alpha$ are reliable to three significant figures based on deviations in the atmospheric pressure, temperature, and composition change due to evaporation of the more volatile component.   The surface tension results are valid to ±2 × 10^{-4} Pa m, and the solubility errors are indicated for each result in Table XVII.   The error in determining the individual k' values is indicated in Table XVIII and was assessed from the linear least squares plot by considering the standard deviations of the individual points from the best straight line.

Comparisons of some of the results obtained by the present authors (Table XVII) for the surface tensions and solubilities can be made with those in the literature to assess the accuracy of the data. The surface tensions of water and of ethanol at $25°C$ agree within experimental error with those in the literature (39, 40), i.e., 71.97×10^{-3} and 22.33×10^{-3} Pa m, respectively, and 30.45×10^{-3} Pa m for ethanol–water at $x = 0.206$, interpolated from values found in Reference 41. The values for the solubilities of KBr, NH_4Br, and $(C_2H_5)_4NBr$ in water at $25°C$ obtained by other authors (42, pp. 612, 617) are $5.69m$, $7.99m$, and $14.74m$, respectively; and, in ethanol at $25°C$, they are $0.011m$, $0.354m$, respectively. These values compare favorably with those listed in Table XVII.

Discussion

An examination of Figures 1–6 indicates that Equation 1 is valid under conditions of constant x for potassium, ammonium, and tetramethylammonium bromides in ethanol–water mixtures. All three salts show an ability to salt out ethanol from these mixtures (i.e., increase its concentration in the equilibrium vapor) which is verified by their k' values shown in Table XVIII. Also, the results for tetra-n-propylammonium bromide and tetra-n-butylammonium bromide in ethanol–water mixtures reveal that Equation 1 can be used to predict the salt effects of these systems; however, these two salts demonstrate a propensity to salt in ethanol (i.e., decrease its vapor concentration) in ethanol–water mixtures. On the other hand, Figures 7–9 and the data in Table XVIII reveal that Equation 1 cannot be used to correlate the salt effects of tetraethylammonium bromide in ethanol–water. For this system, a linear dependence of log α_s/α vs. z is observed initially; however, a gradual levelling off occurs at higher concentrations.

The above results are not completely unexpected since the tetraalkylammonium cations can be considered as providing a bridge between the typical structure-breaking, electrostrictive alkali and halide ions, and the large organic ions which have been shown to promote structure or increase hydrogen bonding of water (43, 44, 45). Studies (46, 47, 48) on the structural interactions in aqueous solutions have been interpreted as showing that solvent-induced attractive forces exist between two hydrophobic solutes and repulsive forces between a hydrophobic solute and a hydrophilic one. These interpretations would suggest that structural interactions are involved in salting-out of nonelectrolytes, e.g., ethanol, capable of hydrophobic interactions, by hydrophilic salts, e.g., NH_4Br and $(CH_3)_4NBr$; and salting-in by hydrophobic salts, e.g., $(n\text{-}C_3H_7)_4NBr$ and $(n\text{-}C_4H_9)_4NBr$.

The results of the salt effect studies undertaken by Aveyard and Heselden (49) agree with those obtained by the present authors. Aveyard and Heselden measured the changes in free energy, enthalpy, and entropy associated with the transfer of several alkanols at high dilution from water to aqueous solutions of R_4NBr salts. They concluded that butanol was salted in more effectively as the R group increased in size. This observation and that demonstrated in Figure 13 comply with another result found by these authors; that $(n\text{-}C_4H_9)_4NBr$ was more effective in salting in alcohols as the alkyl chain becomes longer. These results tend to confirm that, in fact, two hydrophobic solutes exhibit solvent-induced attractive forces.

The solubility and surface tension results itemized in Table XVII confirm that there is a larger interaction between ethanol and the TAA salts as the size of the cation or organic portion increases. The data show that in spite of the R_4NBr salts becoming more soluble in water as the cation size increases, their solubility increases much more rapidly in ethanol, in fact by a factor of 10 greater in ethanol than in water as the salt series of the present investigation is ascended. As a result, the two highest members of the series, the tetrapropyl and tetrabutyl salts, are actually more soluble in ethanol than in water, while the reverse is true for the lower three. Consequently, on the basis of relative solubilities of the salts studied in both ethanol and water, trends in the salt effect parameters similar to those of this work, based on the vapor-equilibrium studies listed in Table XVIII would be observed.

Bockris, Bowler-Reed, and Kitchener (*50*), in their study of the effect of electrolytes on the solubility of benzoic acid in water at 25°C, found that the salting-in action increases with an increase in the radius of the cation. In this study, the salt effect was measured by the value of k_s in the relationship, $\log S_0/S = k_s c_s$, where S_0 is the solubility of benzoic acid in water and S is its solubility in aqueous electrolyte solution. For a series of R_4NI salts, they observed k_s to be 0.021, -0.256, -0.633, -0.970, and -1.32 for these electrolytes where R is H, CH_3, C_2H_5, $n\text{-}C_3H_7$, and $n\text{-}C_4H_9$, respectively. Gross (*51*) assigned individual salting-out constants for benzoic acid in water, and to I^- he gave a value of -0.02. Considering this value of k_A for I^- and the above results of Bockris et al., it is apparent that the TAA cations show a salting-in behavior which becomes more pronounced as the cation size increases.

As previously stated, Gross (*51*) and Larsson (*52*) suggested that the salt effect is an additive function of two constants characteristic of the cation k_C, and the anion, k_A; i.e., $\log S_0/S = k_s c_s = (k_A + k_C)c_s$. In these studies Larsson assumed $k_{K^+} = k_{Cl^-}$. Individual ion contributions have also been devised in volume studies with the additivity often extending to moderate concentrations (*53*) and enthalpy studies (*54*).

In Table XIX the individual ion contributions to the salt effect parameters obtained by vapor–liquid equilibria are listed. The values are based on the convention that the contributions of the K^+ and Br^- ions to the total salt effect are equal; this assumption is similar to that made by Larsson. This choice appears to be a suitable one, since it seems to yield reasonable relative values for k_C and k_A, but it must be remembered that it is only a convenient convention used mainly for purposes of comparison within the results obtained by the authors. Table XIX shows that to a large degree the data are self-consistent; that is, the k_C for NH_4^+ was calculated to be 0.21, 0.22, and 0.20 for the solutions at $x = 0.20$, 0.25, and 0.31, respectively. Also, at 0.25 the k_C for NH_4^+ was found to be 0.21 and 0.22 using two independent routes to obtain these values. The effectiveness of the cations to salt out is in the order $K^+ \geq Na^+ > NH_4^+ \simeq (CH_3)_4N^+ > (C_2H_5)_4N^+ > (n\text{-}C_3H_7)_4N^+ > (n\text{-}C_4H_9)_4N^+$. For the anions, the order is $Cl^- > Br^- > I^- > {}^-OAc$. These sequences show that the salting out decreases with ion size (compare Table XX) and increasing ability of the ion to alter the degree of structure in water (hydrophobic bonding). The ionic radius for –OAc which appears in Table XX was calculated by means of the method of Conway et al. (57), accounting for the dead air space, and a partial molal volume at infinite dilution of 46.46 cm^3 mole^{-1} (53). The two inconsistent results in the above series for cations are: (a) that $k_{K^+} \geq k_{Na^+}$ which may be due to the ability of the water molecules to form solvent co-spheres about the K^+ ion more easily than the Na^+ ion; and (b) that $k_{NH_4^+} \simeq k_{(CH_3)_4N^+}$ which may be a result of the dissipation of charge on the nitrogen because of electron donation by the hydrogens on the NH_4^+ ion or the ability of water molecules to form clathrate structures around the $(CH_3)_4N^+$ ion; nevertheless, the inconsistency is slight, and possibly the explanation lies in subtle ion-solvent interactions.

Another noteworthy observation to be gleaned from the results in Table XIX is the consistency of k_C for $(C_2H_5)_4N^+$ at the three values of x for the ethanol–water mixtures, whereas those for $(CH_3)_4N^+$ and $(n\text{-}C_3H_7)_4N^+$ increase as x increases. This could be interpreted from the point of view that for $(C_2H_5)_4N^+$ there is a balance between the electrostrictive nitrogen charge center and the hydrophobic alkyl groups, but there is not the same balance existing for the other two cations (54, 58, 59). The remaining ions in the table do not pose the same problem because they are primarily electrostrictive in nature.

Conclusion

In this study it was found that KBr, NH_4Br, and $(CH_3)_4NBr$ were effective in salting out ethanol from an aqueous ethanol solution and hence increasing its concentration in the equilibrium vapor, and that Equation 1 could be applied in order to predict the salt effects of these systems. $(n\text{-}C_3H_7)_4NBr$ and $(n\text{-}C_4H_9)_4NBr$ were effective in salting in ethanol, i.e., decreasing its concentration in the equilibrium vapor in ethanol–water mixtures, and again Equation 1 was found to hold. However, for the systems consisting of $(C_2H_5)_4NBr$–ethanol–water, Equation 1 was unsatisfactory, and the salt was a borderline case with respect to its salt effects in ethanol–water mixtures.

From the results, it can be concluded that large perturbations of phase equilibria may be obtained with relatively small salt concentrations in certain systems, and that the salt effect is specific. Salt effects can become important in a variety of separation processes and biological processes.   Hence, no doubt, further applications will be found.

Nomenclature

k	=	salt effect parameter when natural logarithms are used, determined from vapor–liquid equilibrium studies
k'	=	salt effect parameter when $\log_{10}$ is used, $= k/2.303$
t	=	temperature, °C
x	=	mole fraction of ethanol in the liquid phase, calculated on a salt-free basis
y	=	mole fraction of ethanol in the vapor phase
z	=	mole fraction of salt in the liquid phase
α	=	relative volatility in the absence of salt
α_s	=	relative volatility in the presence of salt, calculated using liquid compositions on a salt-free basis
S_0	=	solubility of a substance in water
S	=	solubility of the same substance in an aqueous electrolyte solution
k_s	=	salt effect of an electrolyte determined from solubility studies
k_A	=	salt effect of the anion
k_C	=	salt effect of the cation
c_s	=	the concentration of the electrolyte in the aqueous solution
R	=	mean radius of the tensiometer ring used in the surface tension studies, cm
r	=	radius of platinum wire in the ring, cm
L	=	mean circumference of the platinum tensiometer ring, cm
γ	=	surface tension, Pa m, $= 10^{+3}$ dyne cm^{-1}
M	=	mass, g
g	=	gravitational force
F	=	correction factor for liquid elevated above the free surface of the liquid by the ring
1 Torr	=	133.3 Nm^{-2} $= 133.3$ Pa
TAA	=	tetraalkylammonium
P_2O_5	=	phosphoric pentoxide

Appendix

Table I.   Isobaric Vapor–Liquid Equilibrium Data for the Potassium Bromide–Ethanol–Water System at $x = 0.206 \pm 0.001$ (760 ± 5 Torr)

z	y	t	$\log_{10} \dfrac{\alpha_s}{\alpha}$
0	0.5346	83.0	0.0000
0	0.5352	83.1	0.0002
0.0085	0.5484	83.4	0.0241
0.0086	0.5492	83.4	0.0286
0.0252	0.5774	83.6	0.0753
0.0500	0.6166	83.4	0.1429
0.0505	0.6180	83.5	0.1486
0.0732	0.6519	83.3	0.2123
0.0885	0.6625	83.2	0.2321
0.0915	0.6614	83.1	0.2305
0.0915	0.6640	83.1	0.2355

Table II.   Isobaric Vapor–Liquid Equilibrium Data for the Potassium Bromide–Ethanol–Water System at $x = 0.309$ (758 ± 3 Torr)

z	y	t	$\log_{10} \dfrac{\alpha_s}{\alpha}$
0	0.5837	82.0	0.0000
0	0.5844	82.0	0.0005
0.0100	0.5983	82.1	0.0263
0.0113	0.6044	82.2	0.0373
0.0218	0.6158	82.2	0.0581
0.0275	0.6233	82.2	0.0719
0.0344	0.6316	82.3	0.0873
0.0438	0.6437	82.6	0.1100
0.0488	0.6508	82.7	0.1235
0.0608	0.6634	82.9	0.1479
0.0687	0.6732	82.9	0.1670
0.0768	0.6839	83.0	0.1885
0.0823	0.6882	83.2	0.1971
0.0881	0.6943	83.5	0.2094
0.0992	0.7013	83.6	0.2238
0.1033	0.7112	83.8	0.2446
0.1187 (sat'd.)	0.7213	84.0	0.2662

Table III.   Isobaric Vapor—Liquid Equilibrium Data for the
Potassium Bromide—Ethanol—Water System at
$x = 0.311$ (758 ± 3 Torr)

z	y	t	$\log_{10} \dfrac{\alpha_S}{\alpha}$
0	0.5839	82.0	0.0000
0.0045	0.5913	82.0	0.0134
0.0128	0.6084	81.9	0.0442
0.0199	0.6204	81.9	0.0662
0.0230	0.6228	82.0	0.0707
0.0339	0.6405	81.8	0.1038
0.0436	0.6571	81.6	0.1354
0.0484	0.6618	81.7	0.1445
0.0577	0.6755	81.5	0.1713
0.0710	0.6761	81.5	0.1724
(sat'd.)			
0.0804	0.6792	81.6	0.1784
(sat'd.)			
0.1375	0.6782	81.4	0.1768
(sat'd.)			

Table IV.   Isobaric Vapor—Liquid Equilibrium Data for the
Ammonium Bromide—Ethanol—Water System at
$x = 0.206$ (755 ± 4 Torr)

z	y	t	$\log_{10} \dfrac{\alpha_S}{\alpha}$
0	0.5351	83.6	0.0000
0	0.5353	83.7	0.0000
0.0083	0.5439	83.8	0.0153
0.0110	0.5458	83.9	0.0188
0.0222	0.5605	84.0	0.0445
0.0312	0.5707	84.0	0.0625
0.0472	0.5877	84.0	0.0929
0.0600	0.6003	84.1	0.1155
0.0765	0.6102	84.3	0.1336
0.0892	0.6293	84.4	0.1688
0.1065	0.6394	84.5	0.1877
0.1144	0.6477	84.6	0.2044
0.1363	0.6674	84.9	0.2414
0.1403	0.6684	85.0	0.2434
0.1581	0.6770	85.2	0.2604
0.1588	0.6766	85.2	0.2595
0.1761	0.6773	85.1	0.2610

Table V.   Isobaric Vapor—Liquid Equilibrium Data for the
Ammonium Bromide—Ethanol—Water System at
$x = 0.305$ (765 ± 4 Torr)

z	y	t	$\log_{10} \dfrac{\alpha_s}{\alpha}$
0	0.5791	83.0	0.0000
0.0129	0.5961	83.1	0.0303
0.0179	0.6107	83.2	0.0568
0.0421	0.6304	83.5	0.0932
0.0464	0.6339	83.9	0.0998
0.0691	0.6537	84.0	0.1372
0.0746	0.6584	84.2	0.1463
0.0947	0.6775	83.9	0.1838
0.0999	0.6832	84.3	0.1950
0.1189	0.6943	84.5	0.2177
0.1243	0.6990	84.9	0.2273
0.1428 (sat'd.)	0.7023	85.2	0.2361
0.1478 (sat'd.)	0.7025	85.5	0.2361

Table VI.   Isobaric Vapor—Liquid Equilibrium Data for the
Ammonium Bromide—Ethanol—Water System at
$x = 0.309$ (758 ± 3 Torr)

z	y	t	$\log_{10} \dfrac{\alpha_s}{\alpha}$
0	0.5819	82.5	0.0000
0.0125	0.5962	82.7	0.0256
0.0231	0.6117	82.7	0.0537
0.0252	0.6147	82.9	0.0594
0.0413	0.6322	82.9	0.0917
0.0503	0.6400	82.9	0.1062
0.0588	0.6477	83.1	0.1210
0.0778	0.6654	83.4	0.1549
0.0865	0.6725	83.6	0.1690
0.1034	0.6883	83.9	0.2005
0.1162	0.6970	84.1	0.2183
0.1305 (sat'd.)	0.7011	84.4	0.2268
0.1407 (sat'd.)	0.7013	84.5	0.2273

Table VII.   Isobaric Vapor–Liquid Equilibrium Data for the
Tetramethylammonium Bromide–Ethanol–Water System at
$x = 0.206$ (757 ± 3 Torr)

z	y	t	$\log_{10} \dfrac{\alpha_s}{\alpha}$
0	0.5346	83.8	0.0000
0	0.5347	83.9	0.0000
0.0069	0.5416	83.8	0.0122
0.0086	0.5426	83.8	0.0139
0.0201	0.5529	84.0	0.0321
0.0203	0.5551	84.0	0.0359
0.0365	0.5683	84.1	0.0592
0.0389	0.5699	84.1	0.0621
0.0517	0.5824	84.1	0.0842
0.0566	0.5838	84.2	0.0867
0.0666	0.5882	84.2	0.0947
0.0742	0.5978	84.3	0.1119
0.0797	0.6000	84.3	0.1159
0.0912	0.6112	84.5	0.1363
0.0973	0.6128	84.5	0.1391
0.1108	0.6250	84.7	0.1616
0.1109	0.6256	84.8	0.1628
0.1109	0.6251	84.8	0.1619
0.1249	0.6346	85.2	0.1795
0.1384	0.6422	85.2	0.1938
0.1515	0.6416	85.0	0.1926
(sat'd.)			

Table VIII.   Isobaric Vapor–Liquid Equilibrium Data for the
Tetramethylammonium Bromide–Ethanol–Water System at
$x = 0.305$ (758 ± 3 Torr)

z	y	t	$\log_{10} \dfrac{\alpha_s}{\alpha}$
0	0.5791	83.0	0.0000
0.0078	0.5928	83.5	0.0244
0.0079	0.5926	83.6	0.0240
0.0232	0.6123	83.5	0.0599
0.0277	0.6169	83.7	0.0683
0.0428	0.6321	83.8	0.0964
0.0461	0.6348	83.9	0.1014
0.0652	0.6538	83.7	0.1375
0.0710	0.6635	83.9	0.1562
0.0836	0.6727	83.7	0.1742
0.0934	0.6819	83.8	0.1926
0.1037	0.6900	83.9	0.2088
0.1117	0.6995	84.0	0.2283
0.1190	0.7012	84.2	0.2318
(sat'd.)			

Table IX. Isobaric Vapor–Liquid Equilibrium Data for the Tetramethylammonium Bromide–Ethanol–Water System at $x = 0.309$ (760 ± 4 Torr)

z	y	t	$\log_{10} \dfrac{\alpha_s}{\alpha}$
0	0.5819	82.0	0.0000
0	0.5820	82.0	0.0004
0.0079	0.5947	82.2	0.0229
0.0160	0.6055	82.3	0.0426
0.0300	0.6217	82.3	0.0723
0.0341	0.6268	82.3	0.0817
0.0607	0.6547	82.6	0.1342
0.0614	0.6542	82.7	0.1334
0.0868	0.6790	82.8	0.1817
0.0924	0.6858	83.3	0.1955
0.1112	0.6994	83.5	0.2231
0.1123	0.6999	83.5	0.2242
0.1253 (sat'd.)	0.7089	83.6	0.2430
0.1257 (sat'd.)	0.7056	83.8	0.2361

Table X. Isobaric Vapor–Liquid Equilibrium Data for the Tetraethylammonium Bromide–Ethanol–Water System at $x = 0.206$ (758 ± 3 Torr)

z	y	t	$\log_{10} \dfrac{\alpha_s}{\alpha}$
0	0.5346	84.0	0.0000
0	0.5345	84.0	0.0000
0.0049	0.5378	84.1	0.0055
0.0050	0.5387	84.2	0.0071
0.0126	0.5406	84.5	0.0105
0.0147	0.5436	84.6	0.0158
0.0240	0.5457	84.9	0.0194
0.0293	0.5475	85.1	0.0225
0.0377	0.5495	85.5	0.0261
0.0441	0.5497	85.8	0.0264
0.0511	0.5532	86.2	0.0325
0.0590	0.5549	86.2	0.0355
0.0642	0.5563	86.3	0.0380
0.0735	0.5573	86.8	0.0397
0.0788	0.5580	87.2	0.0410
0.0869	0.5586	88.0	0.0421
0.0933	0.5592	88.2	0.0431
0.1029	0.5596	89.0	0.0438
0.1071	0.5601	89.2	0.0447
0.1089	0.5599	89.3	0.0444

Table XI.   Isobaric Vapor–Liquid Equilibrium Data for the
Tetraethylammonium Bromide–Ethanol–Water System at
$x = 0.245$ (758 ± 3 Torr)

z	y	t	$\log_{10} \dfrac{\alpha_s}{\alpha}$
0	0.5797	83.0	0.0000
0.0066	0.5810	83.0	0.0023
0.0159	0.5873	83.3	0.0136
0.0270	0.5923	83.6	0.0225
0.0296	0.5934	83.8	0.0246
0.0513	0.6013	84.0	0.0388
0.0523	0.6045	84.3	0.0446
0.0764	0.6140	85.6	0.0620
0.0986	0.6211	86.8	0.0750
0.1409	0.6293	89.8	0.0903
0.1796	0.6389	93.5	0.1082
0.2148	0.6438	96.3	0.1173
0.2273	0.6480	97.7	0.1254

Table XII.   Isobaric Vapor–Liquid Equilibrium Data for the
Tetraethylammonium Bromide–Ethanol–Water System at
$x = 0.305$ (755 ± 4 Torr)

z	y	t	$\log_{10} \dfrac{\alpha_s}{\alpha}$
0	0.5791	82.5	0.0000
0.0072	0.5878	82.6	0.0156
0.0074	0.5855	82.7	0.0114
0.0212	0.5972	83.0	0.0323
0.0217	0.5962	83.0	0.0305
0.0348	0.6059	83.2	0.0481
0.0354	0.6048	83.5	0.0462
0.0481	0.6116	83.9	0.0586
0.0485	0.6128	84.0	0.0607
0.0613	0.6195	84.1	0.0731
0.0618	0.6194	84.3	0.0728
0.0738	0.6267	84.7	0.0864
0.0745	0.6256	84.8	0.0844
0.0859	0.6322	85.2	0.0966
0.0922	0.6355	85.7	0.1067
0.0979	0.6376	86.1	0.1067
0.1090	0.6409	86.7	0.1130
0.1097	0.6424	86.8	0.1157
0.1213	0.6455	87.0	0.1216

Table XIII.   Isobaric Vapor–Liquid Equilibrium Data for the Tetraethylammonium Bromide–Ethanol–Water System at $x = 0.316$ (758 ± 3 Torr)

z	y	t	$\log_{10} \dfrac{\alpha_s}{\alpha}$
0	0.5987	82.0	0.0000
0.0094	0.6007	82.3	0.0036
0.0294	0.6112	83.0	0.0227
0.0460	0.6246	83.7	0.0474
0.0581	0.6287	83.9	0.0550
0.0676	0.6293	84.0	0.0562
0.0762	0.6383	84.2	0.0729
0.0834	0.6383	84.5	0.0729
0.1075	0.6473	85.8	0.0900
0.1073	0.6509	86.1	0.0968
0.1159	0.6532	87.5	0.1013
0.1636	0.6671	90.0	0.1281
0.2062	0.6777	94.8	0.1491
0.2313	0.6797	98.5	0.1530
0.2588	0.6842	101.0	0.1620
0.2804	0.6892	103.4	0.1722

Table XIV.   Isobaric Vapor–Liquid Equilibrium Data for the Tetra-n-propylammonium Bromide–Ethanol–Water System at $x = 0.206$ (758 ± 3 Torr)

z	y	t	$\log_{10} \dfrac{\alpha_s}{\alpha}$
0	0.5346	83.7	0
0	0.5347	83.8	0
0.0034	0.5310	84.0	−0.0062
0.0040	0.5317	84.0	−0.0050
0.0092	0.5293	84.7	−0.0063
0.0115	0.5251	84.9	−0.0166
0.0164	0.5252	85.3	−0.0153
0.0227	0.5187	85.9	−0.0254
0.0229	0.5182	86.0	−0.0285
0.0301	0.5173	86.6	−0.0277
0.0338	0.5088	86.8	−0.0449
0.0381	0.5060	87.4	−0.0498
0.0443	0.5012	88.0	−0.0582
0.0450	0.5001	88.1	−0.0600
0.0485	0.4995	88.5	−0.0610
0.0485	0.4961	88.6	−0.0670
0.0550	0.4922	89.1	−0.0738
0.0658	0.4865	89.9	−0.0837
0.0754	0.4752	90.9	−0.1034

Table XV.   Isobaric Vapor–Liquid Equilibrium Data for the
Tetra-*n*-propylammonium Bromide–Ethanol–Water System at
$x = 0.305$ (758 ± 3 Torr)

z	y	t	$\log_{10} \dfrac{\alpha_S}{\alpha}$
0	0.5791	82.8	0.0000
0.0046	0.5809	83.0	0.0032
0.0050	0.5792	83.1	0.0001
0.0057	0.5791	83.3	−0.0001
0.0112	0.5793	83.5	0.0003
0.0139	0.5800	83.6	0.0017
0.0143	0.5793	83.6	0.0003
0.0222	0.5768	83.9	−0.0042
0.0228	0.5760	83.9	−0.0055
0.0252	0.5750	84.0	−0.0072
0.0335	0.5742	84.2	−0.0086
0.0335	0.5753	84.2	−0.0069
0.0357	0.5728	84.3	−0.0111
0.0427	0.5721	84.8	−0.0125
0.0441	0.5677	84.9	−0.0203
0.0459	0.5682	85.1	−0.0193
0.0532	0.5648	85.4	−0.0255
0.0545	0.5665	85.4	−0.0224
0.0556	0.5651	85.7	−0.0248
0.0636	0.5632	85.9	−0.0282
0.0666	0.5619	86.0	−0.0304
0.0669	0.5616	86.3	−0.0310
0.0815	0.5567	87.0	−0.0397
0.0815	0.5578	87.0	−0.0377

Table XVI.   Isobaric Vapor–Liquid Equilibrium Data for the
Tetra-*n*-butylammonium Bromide–Ethanol–Water System at
$x = 0.200$ (762 ± 2 Torr)

z	y	t	$\log_{10} \dfrac{\alpha_S}{\alpha}$
0	0.5253	83.9	0.0000
0.0017	0.5263	84.0	0.0018
0.0048	0.5170	84.2	−0.0145
0.0116	0.5040	84.8	−0.0370
0.0197	0.4937	85.3	−0.0549
0.0283	0.4752	86.5	−0.0872
0.0344	0.4547	87.7	−0.1229
0.0453	0.4403	88.3	−0.1482
0.0522	0.4281	89.1	−0.1697
0.0522	0.4291	89.0	−0.1679

Table XVII.   Physical Properties of Saturated Solutions of
Potassium Bromide and Various Tetraalkylammonium Bromides
in Water, Ethanol, and 0.206 Mole Fraction Ethanol–Water at 25°C.

| | | Surface Tension | Solubility | |
| | | $Pa\ m \times 10^3$ | m | |
Solvent	Salt	(±0.2)	(mol kg⁻¹)	z
H_2O		72.0		
	KBr	73.4	5.6 ± 0.1	0.092
	NH_4Br	60.7	8.0 ± 0.2	0.126
	$(CH_3)_4NBr$	56.0	6.3 ± 0.2	0.102
	$(C_2H_5)_4NBr$	54.6	15.0 ± 0.2	0.213
	$(n\text{-}C_3H_7)_4NBr$	51.2	10.4 ± 0.2	0.158
	$(n\text{-}C_4H_9)_4NBr$	42.2	>15	0.213
C_2H_5OH		22.3		
	KBr	22.3	0.01	0.0004
	NH_4Br	22.4	0.33 ± 0.02	0.015
	$(CH_3)_4NBr$	22.3	0.02 ± 0.01	0.0009
	$(C_2H_5)_4NBr$	25.6	2.8 ± 0.1	0.114
	$(n\text{-}C_3H_7)_4NBr$	27.2	5.0 ± 0.1	0.187
	$(n\text{-}C_4H_9)_4NBr$	29.2	10.5 ± 0.3	0.326
0.206 m.f.[a]		30.5		
	KBr	29.5	1.9 ± 0.1	0.043
	NH_4Br	31.3	3.5 ± 0.2	0.076
	$(CH_3)_4NBr$	30.3	2.3 ± 0.2	0.052
	$(C_2H_5)_4NBr$	36.7	10.3 ± 0.2	0.196
	$(n\text{-}C_3H_7)_4NBr$	36.6	8.3 ± 0.2	0.164
	$(n\text{-}C_4H_9)_4NBr$	41.0	13.5 ± 0.3	0.242

[a] 0.206 mole fraction C_2H_5OH–H_2O is 44.8% C_2H_5OH by volume.

Table XVIII.   Salt Effect Parameters and Reliability of Equation 1
to Predict the Salt Effect for Potassium Bromide and
Tetraalkylammonium Bromides in Ethanol–Water Mixtures
at Various Values of x

Salt	x	k'	R.A.A.D.[a] (%)
KBr	0.206	2.89 ± 0.04	1.3
	0.311	3.02 ± 0.09	2.5
NH_4Br	0.206	1.61 ± 0.07	3.9
	0.246	1.67 ± 0.06	3.0
	0.305	1.83 ± 0.10	5.8
	0.309	1.94 ± 0.11	4.5
$(CH_3)_4NBr$	0.206	1.44 ± 0.03	1.3
	0.305	1.97 ± 0.06	2.4
	0.309	1.98 ± 0.10	5.9

Table XVIII. (*Continued*)

Salt	x	k'	R.A.A.D.[a] (%)
$(C_2H_5)_4NBr$	0.206	0.41 ± 0.07	10.6
	0.245	0.58 ± 0.07	11.1
	0.305	0.75 ± 0.10[b]	11.5[b]
	0.316	0.65 ± 0.08	11.5
$(n\text{-}C_3H_7)_4NBr$	0.206	-1.34 ± 0.03	1.3
	0.305	-0.53 ± 0.03	9.3
$(n\text{-}C_4H_9)_4NBr$	0.200	-3.33 ± 0.10	1.6

[a] R.A.A.D. is the relative average absolute deviation (*see* Results).
[b] These values are approximated for purposes of comparison since this system was investigated only to $z = 0.12$ as compared with 0.27 for the other $(C_2H_5)_4NBr$–Ethanol–Water systems (*see* Results).

Table XIX. Contributions of Various Ions to the Salt Effect Parameter

x	Salt	k'	Ion Contributions[a]	
			k_C	k_A
0.20	KBr	2.89	1.50	1.39
	NH_4Cl	2.30[b]	0.21	2.17
	NH_4Br	1.61	0.21	1.39
	$(CH_3)_4NBr$	1.44	0.05	1.39
	$(C_2H_5)_4NBr$	0.41	−0.98	1.39
	$(n\text{-}C_3H_7)_4NBr$	−1.34	−2.73	1.39
	$(n\text{-}C_4H_9)_4NBr$	−3.33	−4.72	1.39
0.25	NaCl	3.54[c]	1.45	2.09
	NaBr	2.90[c]	1.45	1.45
	KBr	2.95[d]	1.50	1.45
	KOAc	2.11	1.50	0.61
	NH_4Cl	2.30[b]	0.21	2.09
	NH_4Br	1.67	0.22	1.45
	$(CH_3)_4NBr$	1.70	0.25	1.45
	$(C_2H_5)_4NBr$	0.58	−0.97	1.45
0.31	KBr	3.02	1.50	1.52
	KI	2.33	1.50	0.83
	NaOAc	2.05	1.30	0.75
	KOAc	2.25	1.50	0.75
	NH_4Cl	2.20	0.20	2.00
	NH_4Br	1.90	0.20	1.52
	$(CH_3)_4NBr$	1.98	0.46	1.52
	$(C_2H_5)_4NBr$	0.65	−0.97	1.52
	$(n\text{-}C_3H_7)_4NBr$	−0.53	−2.05	1.52

[a] Based on the convention that $k_C = k_A$ for KBr; that is $k_{K^+} = 1.50$.
[b] Interpolated from work presented in Reference *28*.
[c] Averaged from work presented in Reference *28*.
[d] Interpolated from present work.

Table XX.   Formal Radii of Various Ions at 25°C in Å

Ion	Radii	Reference
Na^+	0.96	55,56
K^+	1.33	55,56
NH_4^+	1.45	55,56
$(CH_3)_4N^+$	2.85	57
$(C_2H_5)_4N^+$	3.48	57
$(n\text{-}C_3H_7)_4N^+$	3.98	57
$(n\text{-}C_4H_9)_4N^+$	4.37	57
Cl^-	1.79	55,57
Br^-	1.96	55,57
I^-	2.20	55,57
OAc^-	2.25	(see Discussion)

Literature Cited

1. Prausnitz, J. M., Targovnik, J. H., *Ind. Eng. Chem. Data Ser.* (1958) **3**, 234.
2. Battino, R., Clever, H. L., *Chem. Rev.* (1966) **66**, 395.
3. Sergeeva, V. F., *Russ. Chem. Rev.* (1965) **34**, 309.
4. Conway, B. E., Desnoyers, J. E., Smith, A. C., *Philos. Trans. R. Soc. London* (1964) **A256**, 389.
5. Gubbins, K. E., Tiepel, E. W., *Ind. Eng. Chem. Fundam.* (1973) **12**, 18.
6. Tiepel, E. W., Ph.D. Thesis, University of Florida, 1971.
7. Long, F. A., McDevit, W. F., *Chem. Rev.* (1952) **51**, 119.
8. Furter, W. F., Cook, R. A., *Int. J. Heat Mass Transfer* (1967) **10**, 23.
9. Costa Novella, E., Moragues Tarraso, J., *An. R. Soc. Esp. Fis. Quim Madrid* (1952) **48B**, 441.
10. Ramalho, R. S., Edgett, N. S., *J. Chem. Eng. Data* (1964) **9**, 324.
11. Yoshida, F., Yasunishi, A., Hamada, Y., *Chem. Eng. Tokyo* (1964) **28**, 133.
12. Meranda, D., Furter, W. F., *Can. J. Chem. Eng.* (1966) **44**, 298.
13. Meranda, D., Furter, W. F., *AIChE. J.* (1971) **17**, 38.
14. Meranda, D., Furter, W. F., *AIChE. J.* (1972) **18**, 111.
15. Meranda, D., Furter, W. F., *AIChE. J.* (1974) **20**, 103.
16. Dobroserdov, L. L., Il'yina, V. P., *Tr. Leningr. Tekhnol. Inst. Pishch. Prom.* (1958) **14**, 139.
17. Klar, R., Sliwka, A., *A. Phys. Chem. Frankfurt* (1958) **28**, 133.
18. Wen, W-Y, Hung, J. H., *J. Phys. Chem.* (1970) **74**, 170.
19. Desnoyers, J. E., Pelletier, G. E., Jolicoeur, C., *Can. J. Chem.* (1965) **43**, 3232.
20. Saito, S., Lee, M., Wen, W. Y., *J. Amer. Chem. Soc.* (1966) **88**, 5107.
21. Kozak, J. J., Knight, W. S., Kauzmann, W., *J. Chem. Phys.* (1968) **48**, 675.
22. Furter, W. F., Ph.D., Thesis, University of Toronto, Toronto, Ontario, 1958.
23. Johnson, A. I., Furter, W. F., *Can. J. Chem. Eng.* (1960) **38**, 78.
24. Johnson, A. I., Furter, W. F., *Can. J. Chem. Eng.* (1965) **43**, 356.
25. Ciparis, J. N., "Data of Salt Effect in Vapor-Liquid Equilibrium," Ed. Lithuanian Agr. Acad. Kaunas, USSR, 1966.
26. Jaques, D., Furter, W. F., *Can. J. Chem. Eng.* (1972) **50**, 502.
27. Sada, E., Kito, S., Morisue, T., *Kagaku Kogaku* (1973) **37**, 983.
28. Jaques, D., Furter, W. F., *Ind. Eng. Chem. Fund.* (1974) **13**, 238.
29. Burns, J. A., Furter, W. F., data not yet published.
30. Franks, F., Ives, D. J. G., *Quart. Rev. Chem. Soc. London* (1966) **20**, 1.
31. Franks, F., "Physico-Chemical Processes in Mixed Aqueous Solvents," Heinemann Educational Books Ltd., London, 1967.

32. Conway, B. E., Verrall, R. E., Desnoyers, J. E., Z. *Physik Chem. Liepzig,* Falkenhagen Anniversary Papers (1965) **230,** 157.
33. Lindenbaum, S., Boyd, G. E., *J. Phys. Chem.* (1964) **68,** 911.
34. Othmer, D. F., *Anal. Chem.* (1948) **20,** 763.
35. Johnson, A. I., Furter, W. F., *Can. J. Technol.* (1957) **34,** 413.
36. "International Critical Tables," Vol. III, National Research Council, McGraw-Hill, New York, 1928, p 116.
37. Bulletin 101, Central Scientific Company, Chicago, Ill., p 6.
38. Cook, R. A., Furter, W. F., *Can. J. Chem. Eng.* (1968) **46,** 119.
39. Richards, F. W., Combs, L. B., *J. Am. Chem. Soc.* (1915) **37,** 1656.
40. Harkins, W. D., Brown, F. E., *J. Am. Chem. Soc.* (1919) **41,** 499.
41. "Handbook of Chemistry and Physics," 51st Ed., R. C. Weast, 1970–71, p. F-30.
42. Linke, W. F., "Seidell's Solubilities: Inorganic and Metal Organic Compounds," Vol. II, 4th Ed., Am. Chem. Soc., Wash., D. C., 1965.
43. Frank, H. S., *Z. Phys. Chem. Liepzig* (1965) **228,** 364.
44. Lindenbaum, S., *J. Phys. Chem.* (1966) **70,** 814.
45. Narten, A. H., Lindenbaum, S., *J. Chem. Phys.* (1969) **51,** 1108.
46. Desnoyers, J. E., Arel, M., Perron, G., Jolicoeur, C., *J. Phys. Chem.* (1969) **73,** 3346.
47. Desnoyers, J. E., Ichhaporia, F. M., *Can. J. Chem.* (1969) **47,** 4639.
48. Desnoyers, J. E., Arel, M., *Can. J. Chem.* (1967) **45,** 359.
49. Aveyard, R., Heselden, R., *J. Chem. Soc., Faraday Trans. 1* (1974) **70,** 1953.
50. Bockris, J. O'M., Bowler-Reed, J., Kitchener, J. A., *Trans. Far. Soc.* (1951) **47,** 184.
51. Gross, P., *Chem. Rev.* (1933) **13,** 91.
52. Larsson, E., *Z. Phys. Chem.* (1930) **148,** 148; (1927) **124,** 233; (1931) **153,** 299; (1930) **148,** 304.
53. Millero, F. J., *Chem. Rev.* (1971) **71,** 147.
54. Krishnan, C. V., Friedman, H. L., *J. Phys. Chem.* (1970) **74,** 2356; (1969) **73,** 3934.
55. "Handbook of Chemistry and Physics," 51st Ed., R. C. Weast, 1970–71, p. F-152.
56. Glasstone, S., "Elements of Physical Chemistry," D. Van Nostrand, New York, 1946.
57. Conway, B. E., Verrall, R. E., Desnoyers, J. E., *Trans. Far. Soc.* (1966) **62,** 2738.
58. Kay, R. L., Vituccio, T., Zawoyski, C., Evans, D. F., *J. Phys. Chem.* (1966) **70,** 2336.
59. Kay, R. L., Evans, D. F., *J. Phys. Chem.* (1966) **70,** 2325.

RECEIVED July 14, 1975. Research financially supported through Grant No. 9530-142 from the Defence Research Board of Canada.

9

The Extractive Distillation Process for Nitric Acid Concentration Using Magnesium Nitrate

J. G. SLOAN

Imperial Chemical Industries Ltd., Organics Division, Stevenston, Ayrshire, Scotland, KA20, 3 LN

Enhancement of relative volatility in the nitric acid–water system by the presence of magnesium nitrate as dissolved salt component makes possible an economic and reliable process for making high strength nitric acid. The process has a continuous extractive distillation stage producing 90% HNO_3 vapor which is further rectified to nearly 100% concentration. Diluted magnesium nitrate solution from the still base is readily reconcentrated to the preferred feed strength of 72% $Mg(NO_3)_2$ by vacuum evaporation. Steam provides process heat supply for the distillation and evaporation sections, some 2.5 parts being needed to concentrate one part HNO_3 from 60% to 99.5% concentration. Extended commercial operation has confirmed that this is a robust and satisfactory process.

Though there are commercial processes for the production of high strength nitric acid (98–100 wt% HNO_3) directly from ammonia using oxygen, the bulk of such acid is made from the weaker aqueous solutions, ca. 60 wt% HNO_3, which are produced by conventional air oxidation plants. Since the nitric acid–water system has an azeotrope of 68.2 wt% HNO_3 at 760 mm Hg, concentration of weaker acids by direct distillation is not possible. For many years, when ammonia oxidation plants operated at atmospheric pressure and produced acid containing 45–55 wt% HNO_3, the high strength product required by the explosives and dyestuffs industries was made by extractive distillation with sulfuric acid as the third component. The ammonia oxidation plant was thus associated with a sulfuric acid concentration plant in which essentially the water present in the original aqueous nitric acid was removed, using such commercial acid concentrating units as the Pauling Pot, Gaillard tower, drum concentrator, or Mantius concentrator. Such processes still find commercial acceptance.

As ammonia oxidation technology developed, it became possible to produce aqueous weak nitric acid of super-azeotropic composition, typically 69–73 wt% HNO_3. Such acid is theoretically distillable to higher concentration in one step. At the same time, acid of azeotropic composition is returned to the plant absorbers; the higher the weak acid concentration, the lower is the recycle of azeotrope.

Extractive distillation processes are still widely used for nitric acid concentration. Because the operational and maintenance problems associated with sulfuric acid concentration plants are considerable, and their capital cost substantial, attention has been directed periodically to the use of extractive agents other than sulfuric acid. Phosphoric acid (1) acts like sulfuric acid but poses similar problems of reconcentration. Solutions of certain metallic salts, in particular metallic nitrates, permit similar enhancement of relative volatility and are readily reconcentrated in straightforward evaporation equipment, offering the possibility of a compact integrated concentration process.

Extractive Distillation of Nitric Acid in the Presence of Metal Salts. The nitrates of magnesium (2, 3, 4), lithium (4, 5), potassium, calcium (3, 4), sodium, aluminum, iron (3), barium (4), and zinc (6) have been studied for this purpose, and in general terms it has been found that:

1. Potassium, sodium and barium nitrate produce very little alteration in the relative volatility of nitric acid–water.

2. Magnesium, zinc and lithium nitrates have a much greater effect on the equilibrium.

3. Calcium nitrate has less effect, at a given concentration, than magnesium, zinc, or lithium but, being more soluble, it can have greater overall effect.

Table I summarizes the effect of such nitrates on the azeotropic composition in the nitric acid–water system. Addition of potassium nitrate increases the azeotropic composition, but other nitrates decrease it, the amount by which the azeotrope is displaced being proportional to the amount of nitrate added. The azeotrope is eliminated completely at salt concentrations of 45, 48, 54, or 64% by weight for addition of magnesium, zinc, lithium, or calcium nitrates respectively.

For a continuous extractive distillation process to be possible there must be adequate enhancement of the nitric acid–water relative volatility, and a system equilibrium which permits virtually complete separation of nitric acid from magnesium nitrate, the latter taking up the water content of the weak acid feedstock. This requires addition to the weak nitric acid of solutions of magnesium nitrate usually containing 60 wt% or more of $Mg(NO_3)_2$. Under these conditions a nitric acid–water relative volatility of greater than 2.0 is obtained at the low end of the liquid phase concentration at a nitric acid mole fraction below 0.05 (4, 7).

Table I.   Effect of Nitrates on Azeotropic

Weight of metal nitrate in liquid phase	Wt % of HNO₃ in the azeotrope			
	KNO_3	$NaNO_3$	$Fe(NO_3)_3$	$Al(NO_3)_3$
0	68	68	68	68
10	75	67	60	61
20	78	63	—	52
30	81	57		Other azeo-
40	85	45		tropes appear
50				
60				
70			48	

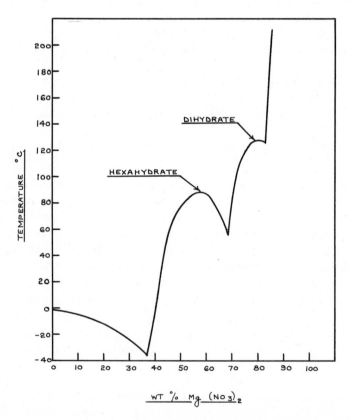

Figure 1.   Magnesium nitrate–water solubility diagram

Composition in HNO_3-H_2O
in the presence of:

$LiNO_3$	$Ca(NO_3)_2$	$ZnNO_3$	$Mg(NO_3)_2$
68	68	68	68
58	60	52	58
47	52	45	44
37	42	30	32
25	32	10	14
8	20	0	0
0	8		
	0		

Magnesium Nitrate Solutions. A number of hydrates exist in the magnesium nitrate–water system. Figure 1 is a solubility diagram showing, for instance, a hexahydrate melting at 89.9°C and a dihydrate melting at 130.9°C with intervening eutectic mixtures (8). Progressive thermal dehydration by melting and evaporation of water forms lower hydrates but at temperatures far above 120°C, and if the time of heating is prolonged, hydrolysis takes place with loss of nitrogen oxides and the formation of basic compounds.

Aqueous solutions of magnesium nitrate are appreciably denser and more viscous than water. Table II illustrates data (9) on the densities (in g/ml) of concentrated solutions at high temperatures. Figure 2 illustrates the viscosity variations in concentrated solutions (9).

Freezing points of aqueous solutions may be obtained from the solubility diagram, Figure 1. The boiling point at 760 mm Hg is shown in Figure 3. It will be seen from these graphs that at solution concentrations above 70 wt% $Mg(NO_3)_2$, the freezing point rises rapidly (more rapidly than the boiling point) and the viscosity rises rapidly also. For ease of handling therefore, solution

Table II.   Viscosity Variations in Concentrated Solutions (9)

Temp (°C)	% by weight $Mg(NO_3)_2$						
	60	62	64	66	68	70	72
100	1.564	1.588	1.612	1.636	1.660	1.684	1.708
120	1.553	1.575	1.601	1.624	1.648	1.672	1.696
140		1.564	1.588	1.612	1.637	1.662	1.686
150					1.631	1.657	1.680

Zhurnal Priktal Khimie

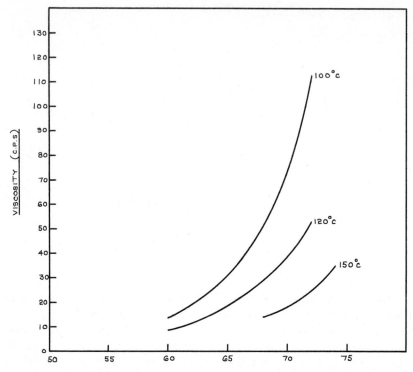

Figure 2.   Viscosity of magnesium nitrate solutions

concentrations used in the extractive distillation process do not normally exceed 72–74 wt% $Mg(NO_3)_2$.

The specific heat of solid anhydrous $Mg(NO_3)_2$ may be calculated from the equation:

$$C_p \text{ (cal/g mole)} = 10.68 + 71.2 \times 10^{-3}T + 1.79 \times 10^5 T^{-2}$$

$$(T \text{ in } °K) \ (10)$$

The equivalent values in cal/g are found in Table III.

Table III.   Specific Heat of Anhydrous $Mg(NO_3)_2$

Temperature (° C)	C_p (Cal/g)
25	0.229
100	0.260
120	0.269
140	0.278
160	0.287
180	0.296
200	0.304

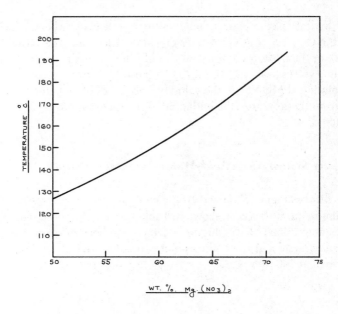

Figure 3. Boiling point of magnesium nitrate so-lutions at 760 mm

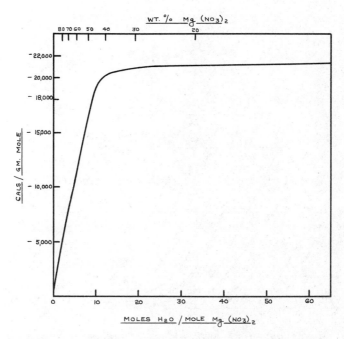

Figure 4. Integral heat of solution of magnesium nitrate

The hydration of anhydrous magnesium nitrate evolves heat, 25,730 cal/g mole $Mg(NO_3)_2 \rightarrow Mg(NO_3)_2 \cdot 6H_2O$ (11).   Likewise, the dissolution of $Mg(NO_3)_2$ or the hydrates in water or the addition of further water to these solutions also evolves heat (12, 13, 14, 15).   Figure 4 illustrates the molar integral heat of solution of $Mg(NO_3)_2$, the value for infinite dilution being 21,575 cal/g mole.   From these figures, the enthalpies of magnesium nitrate solutions may be computed.

The Ternary System Nitric Acid–Water–Magnesium Nitrate

The displacement of the azeotropic composition by progressive addition of magnesium nitrate has been shown in Table I above.   Vapor–liquid equilibria have been determined (3, 4).   Figure 5 depicts equilibrium vapor compositions in the ternary system at the boiling point, while Figure 6 shows boiling points in the system at 760 mm Hg.

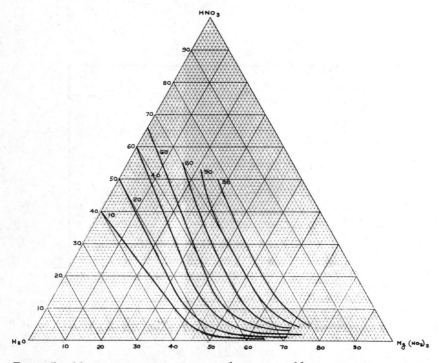

Figure 5.   Magnesium nitrate–nitric acid–water equilibrium vapor composition wt % HNO_3

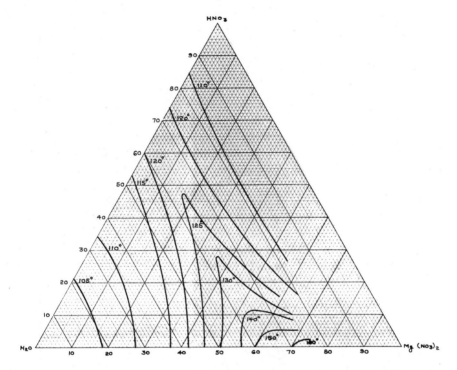

Figure 6.   Magnesium nitrate–nitric acid–water boiling points (760 mm)

Since in an extractive distillation process based on this ternary system the extractive agent is nonvolatile and remains in the liquid phase, and since because of the similarity of the molar latent heats of nitric acid and water there is substantially constant molar liquid overflow, the mole fraction of magnesium nitrate remains almost constant throughout the process.   It is appropriate to represent the equilibrium situation as a pseudo-binary system for each magnesium nitrate concentration, and Figure 7 shows vapor–liquid equilibria on a nitric acid–water basis at a series of magnesium nitrate concentrations from zero to 0.25 mole fraction in the liquid phase.

Figure 5 shows that when nitric acid solutions containing 50–60 wt% HNO_3 are mixed with magnesium nitrate solutions containing 60–70 wt% $Mg(NO_3)_2$, the equilibrium vapor composition at the boiling point does not exceed 85–90 wt% HNO_3.   Thus, to achieve concentrations higher than this the process must provide for rectification of the top vapor.

Thermal data for the ternary system have not been widely reported, but may be evaluated as required for process calculations from available data for the nitric acid–water and magnesium nitrate–water binary systems.

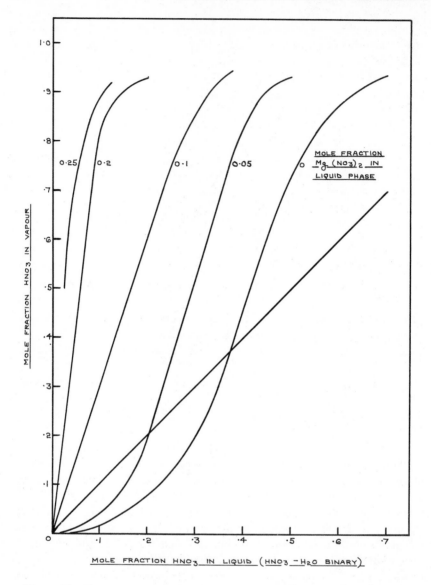

Figure 7. System magnesium nitrate–nitric acid–water, liquid–vapor equilibrium, pseudo binary basis

Process Conditions for Extractive Distillation

Consider the concentration of 60 wt% nitric acid by extractive distillation with 70–75 wt% solutions of magnesium nitrate. It may be seen from Figure 5 that the vapor composition above boiling solutions is at the level 85–90 wt% HNO_3 over a wide range of mixtures, from 3.5 parts of magnesium nitrate solution per part of weak nitric acid up to about 8 parts.   In the particular case of 5 parts of 72 wt% $Mg(NO_3)_2$ solution and 1 part of 60 wt% nitric acid, the mixture has the composition 60 wt% $Mg(NO_3)_2$, 10 wt% HNO_3, and 30 wt% H_2O with an equilibrium vapor composition of 88 wt% HNO_3 at the boiling point.   A rectifying section of the column gives a top product of nearly 100 wt% HNO_3, while the nitric acid content at the base of the distillation column may be taken as zero. A suitable reflux ratio may be determined by trial and error.   For the case of a 3:1 ratio with the heat requirement supplied by an external reboiler, the reboiler outlet liquor composition is 66.6 wt% $Mg(NO_3)_2$.   Assuming constant molar overflow, the vapor rate is 1.15 parts per part of HNO_3 distilled, making the column base composition 59 wt% $Mg(NO_3)_2$.   This molar composition (0.15 mole fraction) applies throughout the stripping section of the column, and the appropriate equilibrium curve is selected from Figure 7.   Graphical methods (*16, 17*) may be used to calculate the number of theoretical plates required for the separation.

A variety of feed compositions and reflux ratios may be thus examined, preferably by carrying out detailed plate-to-plate equilibrium calculations with check heat balances, as the thermal effects are substantial, aimed at optimizing the reflux ratio (representing operating cost) against the number of theoretical plates (representing capital cost).   In particular terms, ICI has found that feed ratios between 4:1 and 7:1 (parts of magnesium nitrate solution per part of weak nitric acid feed) are possible, with reflux ratios in the range 2:1 to 4:1.   The theoretical plate requirement for the complete column is between 15 and 20. Within this range the process will concentrate 60 wt% HNO_3 to 99.5 wt% using 72 wt% $Mg(NO_3)_2$ as extractive agent and denitrating it to less than 0.1 wt% HNO_3.

Heat Requirement of the Process.   Heat is required for vaporization in the extractive distillation column, and for the reconcentration of magnesium nitrate solution.   Overall thermal effects caused by the magnesium nitrate cancel out, and the heat demand for the complete process depends on the amount of water being removed, the reflux ratio employed, and the terminal (condenser) conditions in distillation and evaporation.   The composition and temperature of the mixed feed to the still influence the relative heat demands of the evaporation and distillation sections.   For the concentration of 60 wt% HNO_3 to 99.5 wt% HNO_3 using a still reflux ratio of 3:1, a still pressure of 760 mm Hg, and an evaporator pressure of 100 mm Hg, the theoretical overall heat requirement is 1,034 kcal/kg HNO_3.

ICI's Commercial Process.   In 1960 ICI constructed a concentration plant using this extractive distillation process (*18*) with a capacity of 16,000 tonnes/ annum of product acid (99.5 wt% HNO_3) which has subsequently been extended. A flowsheet is given in Figure 8, and the process description is as follows.

Weak nitric acid (normally 60 wt% HNO_3) and concentrated magnesium nitrate solution (72 wt% $Mg(NO_3)_2$) enter at the feed point of an extractive distillation column.   The rectifying section above the feed point has a water-cooled

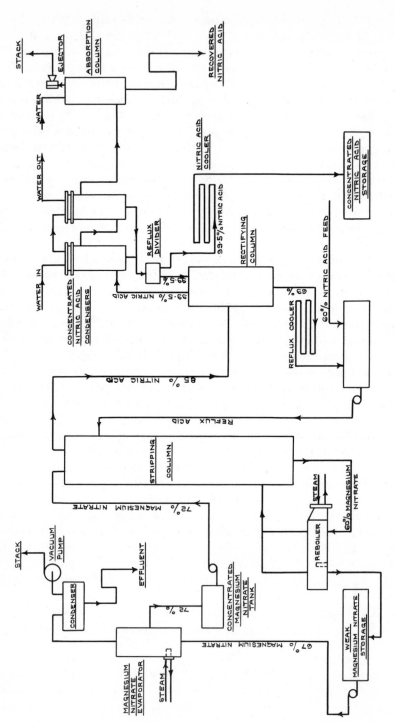

Figure 8.   Flowsheet of the concentration of nitric acid by the magnesium nitrate process

condenser and reflux supply and gives concentrated top product, normally 99.5 wt% HNO_3. The stripping section below the feed point has an external steam-heated reboiler, and its bottom product is diluted magnesium nitrate solution, normally 65–67 wt% $Mg(NO_3)_2$, containing less than 0.1 wt% HNO_3. This weaker magnesium nitrate solution is reconcentrated to the feed strength in a steam-heated vacuum evaporator.

The original plant had the facility to fractionate the vapors from this evaporator to recover nitric acid present in the magnesium nitrate stream, but the high degree of denitration obtainable in the extractive distillation column made this recovery operation unnecessary and it is no longer used.

Water vapor from the evaporator passes to a water-cooled condenser, and the weakly acidic condensate is discharged as liquid effluent. The extractive distillation section of the plant is maintained under slight negative pressure, and the condenser off-gases are scrubbed with water and discharged as gaseous effluent.

In the distillation system used in the ICI plants, the stripping section is designed as a packed column, constructed as a tile-lined shell with silicon iron internal fittings, end cover, and ring packing. The feed point temperature is 120°C and the bottom, 160°C; operating pressure is atmospheric or just below. The separate rectifying section is a silicon iron bubblecap column. The columns were sized by conventional fractionator design methods, and both types of construction have given good service in practice.

Concentrated nitric acid vapor from the top of the rectifying section passes to vertical silicon iron condensers with water-cooled U-tube bundles in high purity aluminum. Condensed acid is refluxed through a divider system, the product fraction being cooled in titanium coolers. Off-gases are scrubbed with water in a tail gas absorber with final discharge to atmosphere through a steam ejector. The magnesium nitrate concentrator is a vertical long-tube natural circulation vacuum evaporator of standard design having a water-cooled shell and tube condenser, vacuum being maintained by a water ring pump.

Materials of Construction

The handling of boiling saturated salt solutions and concentrated nitric acid liquid and vapor, in the presence of nitrous gases, restricts the choice of constructional materials and places constraints on the chemical engineering of the process. High-silicon iron is an exceptionally corrosion-resistant material used successfully for column stills, packing, lines, and ducting, but it is relatively brittle and has a high coefficient of thermal expansion so that adequate thought must be given to the supporting and jointing of equipment, with expansion bellows where appropriate.

When the process is used to make nitric acid of over 98.5 wt% concentration, particular care is necessary in the selection of materials for handling the liquid product. Unstabilized stainless steels with extra low carbon content (<0.03%

C) gave reasonable service at normal temperature, but for severe jobs it may be appropriate to use the more exotic metals such as tantalum. Zirconium or titanium may also be used, though in high strength acids with an appreciable content of dissolved nitrogen oxides these metals can develop unstable, potentially hazardous oxide films. The newer varieties of high silicon austenitic stainless steels may be of service.

Process Operation

Successful operation of this concentration process demands adequate and consistent denitration in the stripping section of the extractive distillation column. If this is not achieved, instability develops, as evidenced by fluctuating flows and pressure drops in the column. The design value for HNO_3 content in the bottoms (magnesium nitrate stream) is 0.1 wt% and this is readily achievable. Higher values, besides causing instability, promote corrosion in the reboiler and the magnesium nitrate system.

Stable operation of the process is assisted greatly by attention to the magnesium nitrate solution. Its concentration is important and should not fall significantly below 70 wt% $Mg(NO_3)_2$ in the evaporator for good denitration in the stripping column. Over-concentration can lead to handling difficulties associated with high freezing points, with reduced heat transfer in the evaporator because of increased viscosity, or with foaming or liquid carryover from the stripping column. Extraneous impurities such as iron salts, siliceous matter, or chlorides can adversely affect the equilibrium pattern and give rise to handling and measurement problems. These can be avoided by selecting high quality magnesium salts.

Process performance is straightforward and predictable. While commercial plants are comprehensively instrumented, operating features are readily recognizable from simple measurements, particularly of temperatures. Conditions, for instance, around the base of the column and the reboiler allow interpretation of process parameters, reasoning as follows:

For the concentration of 60–100 wt% HNO_3:

let C = wt fraction of $Mg(NO_3)_2$ in feed to column,
let R_1 = ratio of feeds (parts magnesium nitrate solution per part of weak acid), and
let R_2 = column reflux ratio,

then the weight fraction of $Mg(NO_3)_2$ is the magnesium nitrate solution leaving the reboiler $= CR_1/R_1 + 0.4$

For constant molar vapor flow in the column system, the vapor flow rate leaving the reboiler, per part of weak acid feed, is $0.6 (1 + R_2) 18/63$ and the weight fraction of $Mg(NO_3)_2$ in the magnesium nitrate solution leaving the column base

$$= \frac{CR_1}{R_1 + 0.4 + 0.6 (1 + R_2) 18/63} = \frac{CR_1}{R_1 + 0.17R_2 + 0.57}$$

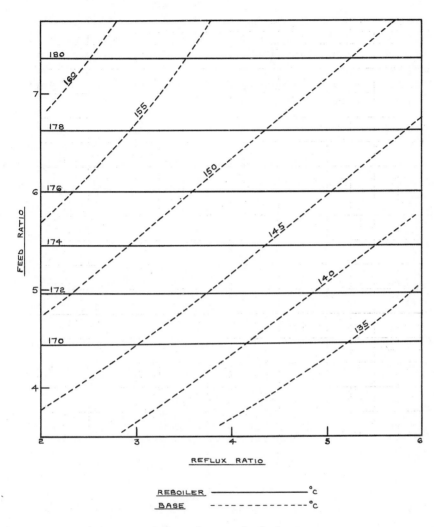

Figure 9.   Column base and reboiler temperatures

Figure 9 presents the equivalent boiling solution temperatures at the reboiler and the column base calculated for a series of feed ratios and reflux ratios, from which the appropriate process conditions and temperatures may be related.

Conclusions

This concentration process has been operated on a commercial scale for some fifteen years and has proved robust and reliable.   Weak nitric acid feedstocks with concentrations varying from 55 to 70 wt% HNO_3 have been handled satisfactorily with product strengths of 98–99.6 wt% HNO_3 according to demand.

Losses of nitric acid and magnesium salts are trivial, and the process can be run in a straightforward manner with minimum manning. The major operational cost is for the steam required for the process heat supply; this depends on the concentration range required but will be ca. 2.0—2.5 parts per part of product acid. The capital cost of a commercial plant compares favorably with that for alternative concentration processes.

Acknowledgment

The author thanks the Directors of the Organics Division of Imperial Chemical Industries for permission to publish this paper.

Literature Cited

1. Flatt, R., Bonnet, J., *Chimia* (1958) **11**, 343–356.
2. Bechtel, R. J., U.S. Patent **2716631**, 1955.
3. Wolf, F., Schier, K. H., *Chem. Technol.* (1967), **19**, 339–343.
4. Cigna, R., diCave, S., Giona, A. R., Mariani, E., *Chim. Ind.* (1964) **46**, 36–45.
5. Allied Chemical Corp.,·UK Patent **1,146,338** (1968).
6. Baranov, A. V., Karen, V. G., *Zh. Prikl. Khim.* (1966) **39**, 1642–1644.
7. Baranov, A. V., Liberzon, E. A., Klyuchnik, L. A., *Tr. Sib. Tekhnol. Inst.* (1963) **36** 50–52.
8. Gmelin's "Handbook de Anorganishen Chemie," 8th ed., **27B**, Verlag Chemie GmbH, Berlin, 1939, 78–96.
9. Shneerson A., Filippova, K. M., Miniovich, M. A., *Zh. Prikl. Khim.* (1965) **38a**, 2110–2112.
10. Shomate, *J. Amer. Chem. Soc.* (1944) **66**, 928.
11. Kelley, K. K., U.S. *Bureau of Mines Report* **3776** (1944).
12. Dunnington, F. P., Hoggard, T., *Amer. Chem. J.* (1899) **22**, 207–211.
13. Schlunder, E. U., *Chem. Ing. Tech.* (1963) **35**, 482–487.
14. Ewing, W. W., Klinger, E., Brandner, J. D., *J. Amer. Chem. Soc.* (1934) **56**, 1053–1057.
15. Hammerschmid, H., Lange, E., *Z. Phys. Chem.* (1932) **160**, 445–465.
16. Robinson, C. S., Gilliland, E. R. "Elements of Fractional Distillation," 4th ed., McGraw-Hill, New York, 1950, 296–312.
17. Doig, I. D., *British Chem. Eng.* (1963) **8**, 688–690.
18. Sloan, J. G., Jamieson, J. M., *Ind. Chem.* (1960) 165–169.

RECEIVED June 13, 1975.

Use of Magnesium Nitrate in the Extractive Distillation of Nitric Acid

JOSEPH A. VAILLANCOURT

Hercules Inc., 910 Market St., Wilmington, Del. 19899

*The maximum boiling azeotrope of nitric acid and water (68%
nitric acid) requires extractive distillation to produce concentrated nitric acid when starting with acid that is weaker than
the azeotrope. Sulfuric acid has been used for this extractive
distillation, but its use requires high investment and maintenance costs. Magnesium nitrate is being used in several commercial plants in the extractive distillation of nitric acid.
Magnesium nitrate was selected rather than other nitrate salts
because it has the most favorable combination of physical
properties. Although magnesium nitrate requires slightly
more steam than sulfuric acid in the extractive distillation of
nitric acid, this disadvantage is offset by the lower capital and
maintenance costs associated with the magnesium nitrate process.*

The demand for concentrated nitric acid 98+% strength, primarily
for use in nitration, led to the use of sulfuric acid in the extractive distillation
of nitric acid. The maximum boiling azeotrope of 68 wt % nitric acid prevents
distilling the 55–60 wt % nitric acid produced in an ammonia oxidation plant
(AOP) to a strength greater than 68 wt % unless extractive distillation is used.

The Sulfuric Acid Process

In the process using sulfuric acid (*see* Figure 1) this acid was, and in many
instances still is, added to the weak nitric acid produced by an AOP before the
mixed acid was fed to the top of a distillation column. The feed has been pre-
heated in some processes to minimize the vapor load in the distillation column.
Enough sulfuric acid was added to the feed so that the vapor leaving the top of
the column was at least 98% nitric acid. Live steam was added to the base of the
column to provide the heat for the column and the stripping vapor required to

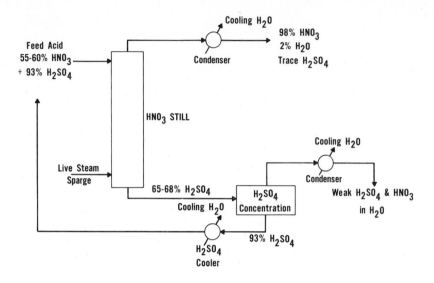

Figure 1.   HNO₃ concentration using sulfuric acid

minimize the nitric acid content of the sulfuric acid leaving the base of the distillation column.   Any nitric acid leaving the base of the column represented a loss of desirable product.   Most of the sulfuric acid and water, including the live steam, that was fed to the distillation column left the base of the column as 65–68 wt % sulfuric acid.   This sulfuric acid was then concentrated via vacuum or submerged combustion evaporation to a strength of about 93% sulfuric acid.   The sulfuric acid was then cooled and recycled to make up the feed for the distillation column.   Any nitric acid present in the bottoms from the distillation column was lost with the water removed in the sulfuric acid concentrator.   A small amount of the sulfuric acid fed to the distillation column was entrained with the product 98 wt % nitric acid.   The entrained sulfuric acid was not a problem for most of the captive uses of the concentrated nitric acid in nitration.   However, the entrained sulfuric acid was a problem in meeting some specifications of concentrated nitric acid for sale on the open market.

The Magnesium Nitrate Process

In 1957 Hercules Inc. started the first unit that produced concentrated nitric acid for commercial sales using magnesium nitrate as the extractive agent.   In this process (*see* Figure 2) the weak nitric acid product from an AOP is fed to the appropriate tray of a distillation column.   A concentrated solution of magnesium nitrate and water is fed to the proper tray in sufficient quantity to enrich the vapors to a concentration greater than 68 wt % nitric acid.   The overhead product from the column is concentrated (98–99.5 wt %) nitric acid.   A portion of the concentrated nitric acid is returned as reflux to aid in rectification.   The bulk of the water entering with the AOP feed acid is removed from the base of the column in the 65–68 wt % magnesium nitrate solution. This quantity of water is then removed in the magnesium nitrate evaporator which operates under vacuum. The concentrated (72 wt %) magnesium nitrate solution produced in the magnesium nitrate evaporator is returned to the distillation column. No stripping steam is used in the column since water evaporated in the column reboiler serves as stripping vapor. The overhead product from the column is a high purity distillate free of sulfates and very low in metal content, chlorides, and other impurities.

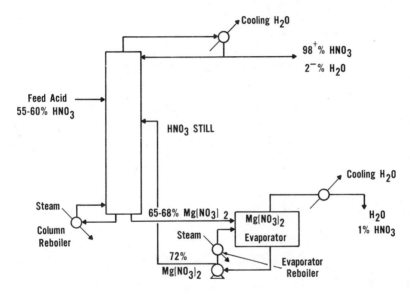

Figure 2. HNO₃ concentration using magnesium nitrate

Vapor–Liquid Equilibrium

The magnesium nitrate, as well as other nitrate salts, enhances the volatility of the nitric acid while suppressing the volatility of water. This effect is similar to that achieved with sulfuric acid. Contrary to most commercial extractive distillations in which the extractive agent enhances the volatility of the more volatile component in the feed, extractive agents for nitric acid distillation must enhance the volatility of the less volatile component in the feed (nitric acid). Magnesium nitrate was selected from among several nitrate salts since it had a large, favorable effect on the relative volatility of nitric acid as well as for its thermal stability and good physical properties in water solution.

The effect of the magnesium nitrate on the vapor–liquid equilibrium of nitric acid and water can be seen in Figure 3. As the concentration of magnesium nitrate in the liquid increases, the volatility of nitric acid also increases.

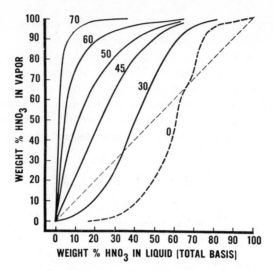

*Figure 3. Vapor–liquid equilibria. Values
shown are at constant wt % $Mg(NO_3)_2$ in the
liquid—on an acid-free basis.*

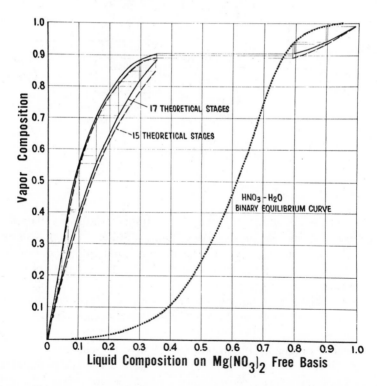

Figure 4.   Weight fraction of HNO_3 in vapor vs. liquid

Therefore, the vapors emitted from the liquid of a given composition are richer in nitric acid as the concentration of the magnesium nitrate increases. Though this might suggest that the highest possible concentration of magnesium nitrate should be used, consideration of the distillation column conditions shows that this is not the case. In a distillation column, as the magnesium nitrate concentration is increased, the liquid rate increases more rapidly than the vapor rate; therefore, the slope of the operating line (liquid rate/vapor rate) increases. Since the number of theoretical stages in a distillation column is a function of the degree of separation between the vapor–liquid equilibrium curve and the operating line, as well as the steepness of the vapor–liquid equilibrium curve, there is an optimum range for the magnesium nitrate concentration in the liquid. If the magnesium nitrate concentration is too low, the vapor–liquid equilibrium curve is not steep enough, and too many stages are required to effect the distillation. If the magnesium nitrate concentration is too high, the separation between the vapor–liquid equilibrium curve and the operating line is too small; therefore, too many stages are required. The effect of a 1% increase in the magnesium nitrate content of the liquid in the distillation column is shown in Figure 4. Since the magnesium nitrate is not a volatile component, the weight fraction HNO_3 is plotted on a magnesium nitrate-free basis. The concentration of magnesium nitrate in the liquid does not vary appreciably between the feed tray and the base of the stripping section.

The design of a distillation column for the concentration of nitric acid using magnesium nitrate as the extractive agent is rather specific to the strength of the feed acid produced in the AOP. As the feed acid concentration changes, a different number of theoretical stages is required to produce the desired overhead and bottoms purity. In the initial design the reflux ratio and the concentration of magnesium nitrate are adjusted to minimize the number of theoretical stages in the column. Since the cost of the equipment and the cost of energy both increase as the vapor rate in the column is increased, the optimum design is a compromise that must consider economics as well as vapor–liquid equilibria. Also some allowance must be made for changes in the feed acid concentration, as these changes do occur to some extent in the operation of the AOP. The final design is one that tolerates changes in feed composition, reflux rate, and magnesium nitrate concentration in a column containing a fixed number of theoretical stages.

Economics

Magnesium nitrate has been used instead of sulfuric acid in new commercial plants since it requires lower capital investment and has lower overall operating costs. The lower capital investment results from the use of metal equipment for all the components as well as from a more compact plant layout. The upper portion of the distillation column and the condenser and piping for handling the concentrated nitric acid are fabricated from the same materials whether sulfuric acid or magnesium nitrate is used as the extractive agent. The savings in equipment costs are realized in the base of the column, the column reboiler, the extractive agent concentrator, and the piping that handles the hot extractive agent. When magnesium nitrate is the extractive agent, all of this equipment can be fabricated from conventional stainless steel. When sulfuric acid is the extractive agent, all of this equipment must be fabricated from lined steel equipment that utilizes glass, teflon, or brick as the corrosion resistant barrier. The difference in the cost of this portion of the plant results in a sizeable capital saving when magnesium nitrate is used as the extractive agent. Some savings in structural steel and foundation costs are also realized from reductions in the equipment weights. A single stage vacuum evaporator can be used to remove the water from the magnesium nitrate compared with the two or more vacuum or submerged combustion stages required for sulfuric acid. The more compact layout results in reduction of costs for building steel, foundations, and shorter runs of process and utility piping.

The lower operating costs of the magnesium nitrate–nitric acid–water extractive distillation process are caused primarily by lower maintenance costs compared with the sulfuric acid–nitric acid–water extractive distillation. The hot magnesium nitrate solutions are less corrosive to the equipment than the hot sulfuric acid solutions despite the more exotic materials of construction used to handle the sulfuric acid. In addition, when repairs or changes are required in the equipment and piping, they are easier and cheaper to make in the magnesium nitrate process than in the sulfuric acid process. When magnesium nitrate is used, the stainless steel equipment requires no special skills and less advanced planning compared with the lined equipment required when sulfuric acid is used. Fewer and less complicated equipment and piping spare parts are required for magnesium nitrate compared to sulfuric acid.

An additional operating saving can be realized by reusing a portion of or all of the water evaporated in the magnesium nitrate concentrator. This water can generally be reused as absorption water in the AOP or as wash water in a nitration process. The sulfuric acid content of the water evaporated from the sulfuric acid concentrators or the corrosivity of this stream generally negates its reuse in other portions of the plant. This ability to reuse water when magnesium nitrate is the extractive agent generally results in savings in the waste treatment costs compared with the use of sulfuric acid as the extractive agent.

The use of magnesium nitrate as the extractive agent does, however, have some disadvantages in comparison with sulfuric acid. Since magnesium nitrate is a less efficient extractive agent than sulfuric acid, the distillation column generally requires more theoretical stages and a slightly higher reflux ratio when magnesium nitrate is used as the extractive agent. The higher reflux ratio for magnesium nitrate use means a higher steam consumption in the column reboiler. However, this is offset partially by a lower steam consumption in the equipment used to evaporate water from the extractive agent. The slightly higher steam consumption when magnesium nitrate is used requires a correspondingly higher electrical consumption in operating the cooling tower if used, or higher consumption of electricity to pump cooling water if it is used on a once-through basis.

Future Outlook

Another disadvantage of using magnesium nitrate instead of sulfuric acid would be the disposal of the evaporated water if all of it cannot be reused in other processes. When the evaporated water is neutralized prior to disposal, the magnesium nitrate process results in nitrate salts, whereas the sulfuric acid process results in sulfate salts being discharged. In the past, sulfate salts have been more acceptable in plant effluents than nitrate salts. Future environmental considerations may prohibit the disposal of either of these salts, thereby negating this advantage of the use of sulfuric acid.

In the recent past new commercial facilities have been built using magnesium nitrate in extractive distillation to produce concentrated nitric acid from AOP acid. To the author's knowledge no new commercial facilities have been built using sulfuric acid as the extractive agent for concentrating AOP acid. New facilities generally use sulfuric acid to concentrate spent nitration acids only if they already contain sulfuric acid.

Conclusions

The preference for magnesium nitrate in commercial operations indicates that the advantages of using magnesium nitrate outweigh the disadvantages. That is, the lower investment and maintenance costs inherent in the use of magnesium nitrate more than offset the increased steam consumption, even in areas of the world where energy costs have been high historically.

RECEIVED June 23, 1975.

Behavior of
Other Properties

11

Thermodynamics of Preferential Solvation of Electrolytes in Binary Solvent Mixtures

A. K. COVINGTON

Department of Physical Chemistry, University of Newcastle, Newcastle-upon-Tyne, NE1 7RU, U.K.

K. E. NEWMAN

Department of Chemistry, University of Keele, Keele, Staffs, ST5 5BG, U.K.

The solvation changes of an ion, as the composition of a binary solvent is varied, can be treated on the basis of n *successive equilibria where* n *is the solvation number. From a detailed consideration of the thermodynamics of preferential solvation, it is possible to define a free energy of preferential solvation* $\Delta G_{ps}^{\ominus}$ *and relate it to the more familiar free energy of transfer,* $\Delta G_t^{\ominus}$, *determinable for a neutral combination of ions from emf measurements. The treatment can be modified to include a case of change of solvation number and to non-statistical distribution of the solvated species. When separate solvent NMR signals from solvation shells can be detected or mixed solvates isolated, a further development enables a satisfactory explanation to be given for the observed distribution of solvated species.*

The problem of understanding the processes which occur when a gaseous ion is put into a solvent such as water has commanded the attention of chemists for more than 80 years. It is a key to the understanding of the properties of solutions and is of far reaching technological and theoretical importance. If the solvent is made up of two or more components, preference may be shown for one of these components by the cations, anions, or both. This is the problem of preferential solvation or of solvent-sorting in the immediate vicinity, the co-spheres (1), of the ions.

Thermodynamic methods, per se, are of limited use, although they have been used extensively to unravel problems as complex as these. Nevertheless, studies must be made on a sound thermodynamic, or statistical thermodynamic, framework. It is the purpose of this review to provide first the thermodynamic basis for such studies and then to indicate how progress can be made in understanding ion–solvent interactions (2, 3, 4, 5, 6) using spectroscopic methods, particularly NMR in combination with classical thermodynamic studies.

Basic Thermodynamic Treatment

Grunwald, Baughman and Kohnstam (GBK), in an appendix to a classic paper on vapor pressure studies of solvation in dioxane and water mixtures (7), presented an outline of a thermodynamic treatment of the solvation of ions in a mixed solvent. It is convenient to start from their general treatment but to adopt a different nomenclature, used previously (3).

Consider a homogeneous solution of w° moles of water, p° moles of a co-solvent P and a moles of a solute X. The Gibbs free energy of the system at constant temperature and pressure is given formally by

$$G = a\mu_X + w^\circ\mu_W + p^\circ\mu_P \tag{1}$$

and is independent of whether the solute is considered solvated or not. However, this is not true of the solute chemical potential. If the solvation number of X is n_W for pure water and n_P for pure P, the general solvated species can be written as XW_jP_i where $0 \leqslant j \leqslant n_W$ and $0 \leqslant i \leqslant n_P$. The fraction of X existing as this general ij species is denoted by ϕ_{ij}, so that average solvation numbers for water and P can be defined

$$h_W = \sum_{j=0}^{n_W} \sum_{i=0}^{n_P} j\phi_{ij} \tag{2}$$

$$h_P = \sum_{j=0}^{n_W} \sum_{i=0}^{n_P} i\phi_{ij} \tag{3}$$

and $\Sigma_j \Sigma_i \phi_{ij} = 1$.

A general treatment based on different solvation numbers for water and P, although physically very plausible, becomes impossibly difficult to handle later and some simplification is necessary. It will be assumed that the solvation number is the same in both pure solvents, i.e., $n_W = n_P = n$. The general species now referred to as the ith species, can be written as $XW_{n-i}P_i$ and $0 \leqslant i \leqslant n$.

Also the average solvation number $= h = \sum_i i\phi_i \tag{4}$

$$\text{and} \quad \sum_{i=0}^{n} \phi_i = 1 \tag{5}$$

Considering now the variously solvated species i with chemical potentials μ_i, the free energy of the system G, can now be expressed as

$$G = a\Sigma\phi_i\mu_i + p\mu_P + w\mu_W \qquad (6)$$

where μ_W, μ_P are the chemical potentials of W and P respectively.

But

$$w = w° - a\Sigma(n - i)\phi_i \qquad (7)$$

$$p = p° - a\Sigma i\phi_i \qquad (8)$$

since the solvent has lost molecules which are considered part of the solute species.

Substituting Equations 7 and 8 into 6 gives

$$G = a\Sigma\phi_i\mu_i + (p° - a\Sigma i\phi_i)\mu_P + (w° - a\Sigma(n - i)\phi_i)\mu_W \qquad (9)$$

which is equivalent to GBK's Equation 49.

Adopting the mole fraction scale and introducing standard chemical potentials and activity coefficients as follows

$$\mu_i = \mu_i{}^\ominus + RT \ln \phi_i f_i$$

$$\mu_W = \mu_W{}^\ominus + RT \ln x_W f_W$$

$$\mu_P = \mu_P{}^\ominus + RT \ln x_P f_P \qquad (10)$$

Equations 9 and 1 for the Gibbs free energy of the system can be equated (7) to give

$$\mu_X = \mu_X{}^\ominus + RT \ln x_X f_X$$
$$= \Sigma\phi_i\mu_i{}^\ominus + \Sigma\phi_i RT \ln \phi_i f_i - \Sigma i\phi_i[\mu_P{}^\ominus + RT \ln x_P f_P] - \Sigma(n - i)\phi_i$$
$$\times [\mu_W{}^\ominus + RT \ln x_W f_W] \qquad (11)$$

Now μ_X is the conventionally defined chemical potential of X (disregarding solvation) and the standard (conventional) chemical potential is given by

$$\mu_X{}^\ominus = \lim_{a \to 0} (\mu_X - RT \ln x_X) \qquad (12)$$

Noting that as $a \to 0$, $f_i \to 1$ and $f_W \to f_W°$, $f_p \to f_P°$ (activity coefficients for the binary mixture in the absence of solute) also $\phi_i \to \phi_i°$ (the solvated species mole fraction when the solute concentration is vanishingly small), and so the solvent mole fractions can be equated to the stoichiometric mole fractions, i.e. $x_W = w°/(w° + p°)$ and $x_P = p°/(w° + p°)$, then

$$\mu_X{}^\ominus = \Sigma\phi_i°\mu_i{}^\ominus + RT\Sigma\phi_i° \ln \phi_i° - \Sigma(n - i)\phi_i°[\mu_W{}^\ominus + RT \ln w°f_W°]$$
$$- \Sigma i\phi_i°[\mu_P{}^\ominus + RT \ln p°f_P°] + nRT \ln (w° + p°) \qquad (13)$$

Equation 13 is Equation 4 of the previously given treatment (3) and derived there by an alternative but equivalent method (noting that $w° + p° = 55.5$ if an aquamolality scale is adopted as henceforth). It is also similar to GBK's Equation 53.

In pure solvent P, since all $\phi_i° = 0$ except $\phi_n°$ which is unity,

$$\mu_{X(P)}^{\ominus} = \mu_{n(P)}^{\ominus} - n\mu_P^{\ominus} \tag{14}$$

and in pure water, since all $\phi_i° = 0$ except $\phi_0°$ which is unity,

$$\mu_{X(W)}^{\ominus} = \mu_{0(W)}^{\ominus} - n\mu_W^{\ominus} \tag{15}$$

From Equations 14 and 15 the free energy of transfer of X from pure water to pure solvent P is given by

$$\Delta G_{t(X)}^{\ominus} \equiv \mu_{X(P)}^{\ominus} - \mu_{X(W)}^{\ominus} = \mu_{n(P)}^{\ominus} - \mu_{0(W)}^{\ominus} - n(\mu_P^{\ominus} - \mu_W^{\ominus}) \tag{16}$$

However, for the free energy of transfer (aquamolality standard state) of X from water to a mixed solvent of P mole fraction x_P, from Equations 13 and 15

$$\Delta G_{t(X)}^{\ominus\ \text{mix}} = \mu_X^{\ominus} - \mu_{X(W)}^{\ominus} = \Sigma\phi_i°\mu_i^{\ominus} + RT\Sigma\phi_i° \ln \phi_i° - \mu_{0(W)}^{\ominus}$$
$$- \Sigma i\phi_i°(\mu_P^{\ominus} - \mu_W^{\ominus}) - RT\Sigma i\phi_i° \ln x_P f_P$$
$$- RT\Sigma(n - i)\phi_i° \ln(1 - x_P)f_W \tag{17}$$

Both $\Delta G_{t(X)}^{\ominus}$ and $\Delta G_{t(X)}^{\ominus\ \text{mix}}$ refer to a neutral solute species or a neutral combination of ions X. These quantities may be determined thermodynamically from emf measurements on suitable cells or from solubility studies in the mixed or pure solvents. The splitting of these quantities into separate contributions from the ions has been the subject of much speculation (8). We may regard Equations 16 and 17 as valid if the species X bears a charge, provided the ionic contributions are added to obtain the values for a neutral combination of ions.

The free energy of transfer $\Delta G_{t(X)}^{\ominus}$ refers to the removal of X from pure water with the breaking of all ion–water interactions, and its transfer to the pure solvent with the formation of a new set of ion–solvent interactions. However, a simpler process can be envisaged in which the solvation shell of X containing n molecules of water of solvation is progressively, step by step, changed until it contains n molecules of P of solvation. The general step in such a process occurring in a given solvent mixture can be written as (2, 3)

$$XW_{n+1-i}P_{i-1} + P \overset{K_i}{\rightleftharpoons} XW_{n-i}P_i + W \tag{18}$$

where K_i is an equilibrium constant for the formation of the ith species $XW_{n-i}P_i$ in solvent of mole fraction x_P. A similar approach has been used by other workers (9, 10).

The free energy change for this general step is

$$\Delta G_i = \mu_i + \mu_W - \mu_{i-1} - \mu_P = \mu_i^{\ominus} + RT \ln m_i f_i - \mu_{i-1}^{\ominus} - RT \ln m_{i-1}f_{i-1}$$
$$+ \mu_W^{\ominus} + RT \ln(1 - x_P)f_W - \mu_P^{\ominus} - RT \ln x_P f_P \tag{19}$$

At equilibrium $\Delta G_i = 0$, hence

$$\Delta G_i^{\ominus} = \mu_i^{\ominus} - \mu_{i-1}^{\ominus} + \mu_W^{\ominus} - \mu_P^{\ominus} = -RT \ln(m_i/m_{i-1})(f_i/f_{i-1})$$
$$- RT \ln[(1 - x_P)/x_P](f_W/f_P) = -RT \ln K_i \tag{20}$$

where K_i takes values between K_1 and K_n.   The equilibrium constant for the overall process, K, in a solvent mixture of mole fraction x_P,

$$XW_n + nP \overset{K}{\rightleftharpoons} XP_n + nW \tag{21}$$

is related to the K_i values by $K = \prod_{i=1}^{n} K_i$ and the free energy change for the process

represented by Equation 21, which will be defined as the free energy of preferential solvation, $\Delta G_{ps}^{\ominus}$ is given by

$$\Delta G_{ps}^{\ominus} = \sum_{i=1}^{n} \Delta G_i^{\ominus} \tag{22}$$

To obtain ϕ_i° values, Equation 20 is used, in the form

$$m_i/m_{i-1} = K_i Y \text{ where } Y = x_P f_P / x_W f_W,$$

and assuming that $f_i = f_{i-1}$ which is true at infinite dilution.   Thus $m_i/m_0 = K_i Y$,

$m_2/m_1 = K_2 Y$, etc., so that $m_i/m_0 = Y^i \prod_{1}^{i} K_i$ for $1 \leqslant i \leqslant n$

Now $\phi_i^{\circ} = m_i / \sum_{0}^{n} m_i$ for $0 \leqslant i \leqslant n$

$$= \frac{m_0 Y^i \Pi K_i}{m_0 + m_0 \underset{n \text{ terms}}{\sum} Y^i \Pi K_i} = \frac{Y^i \Pi K_i}{1 + \underset{n \text{ terms}}{\sum} Y^i \Pi K_i} \qquad \text{for } 1 \leqslant i \leqslant n \tag{23}$$

and $\quad \phi_0^{\circ} = \dfrac{1}{1 + \underset{n \text{ terms}}{\sum} Y^i \Pi K_i} \tag{24}$

To proceed further it is necessary to make some assumptions about the interrelations of the $\mu_i^{\ominus}$ terms.   Such assumptions will be extrathermodynamic in nature.   It will be assumed that each μ_i term can be split into an intrinsic contribution from the bare ion and a term from the solvation process.   Thus

$$\mu_i^{\ominus} = \mu_{int}^{\ominus} + \mu_i^{\ominus \, solv} \tag{25}$$

It is immediately apparent that because only differences in chemical potential between states can be measured, the intrinsic term will always disappear.   It will be further assumed that long-range and short-range contributions to the solvation process may be separated out.   Thus

$$\mu_i^{\ominus \, solv} = \mu_i^{\ominus \, elec} + \mu_i^{\ominus \, chem} \tag{26}$$

where $\mu_i^{\ominus \, elec}$ represents the long range interactions (which are assumed to be electrostatic in origin and are normally treated by Born theory involving the radius of the solvated ion and the dielectric constant of the solvent) and $\mu_i^{\ominus \, chem}$ represents the short range interactions which will be referred to as chemical, although no particular inference as to the type of short-range interaction is implied.

The term $\mu_i{}^{\ominus\,\text{chem}}$ defines the effect on the chemical potential of the solvated ion of i molecules of P and $(n-i)$ molecules of W in the solvation shell of the ion X. It may then be treated as a free energy of mixing, with the important difference that the statistical contribution, the number of ways of mixing i molecules of P and $(n-i)$ molecules of W, is given by $RT \ln n!/(n-i)!i!$ and not by the more familiar $RTx \ln x$ form which is valid only when n and i are very large.

Thus

$$\mu_i{}^{\ominus\,\text{chem}} = (n-i)\mu_W{}^{\ominus\,\text{chem}} + i\mu_P{}^{\ominus\,\text{chem}} + (n-i)RT \ln f_W' + iRT \ln f_P'$$
$$- RT \ln [n!/(n-i)!i!] \quad (27)$$

where $\mu_P{}^{\ominus\,\text{chem}}$, $\mu_W{}^{\ominus\,\text{chem}}$ are chemical potentials of P and W respectively in the solvation shell in pure P or pure W and f_P' and f_W' are activity coefficients pertaining to the solvation shell, which go to unity in their respective pure solvents and are introduced in an attempt to take into account possible non-ideality in the solvation shell. Substituting Equations 25, 26, and 27 into Equation 17 gives

$$\Delta G_{t(X)}{}^{\ominus\,\text{mix}} = \Sigma \phi_i{}^\circ \mu_i{}^{\ominus\,\text{elec}} - \mu_{0(W)}{}^{\ominus\,\text{elec}} + \Sigma i \phi_i{}^\circ (\mu_P{}^{\ominus\,\text{chem}} - \mu_W{}^{\ominus\,\text{chem}})$$
$$- RT \Sigma i \phi_i{}^\circ [\ln x_P + \ln f_P{}^\circ/f_P'] - RT \Sigma (n-i)\phi_i{}^\circ [\ln (1-x_P) + \ln f_W{}^\circ/f_W']$$
$$- \Sigma i \phi_i{}^\circ (\mu_P{}^\ominus - \mu_W{}^\ominus) - RT \Sigma \phi_i{}^\circ \ln n!/(n-i)!i! + RT \Sigma \phi_i{}^\circ \ln \phi_i{}^\circ \quad (28)$$

Likewise from Equation 16, referring to transfer between pure solvents,

$$\Delta G_{t(X)}{}^\ominus = \mu_{n(P)}{}^{\ominus\,\text{elec}} - \mu_{0(W)}{}^{\ominus\,\text{elec}} + n(\mu_P{}^{\ominus\,\text{chem}} - \mu_W{}^{\ominus\,\text{chem}})$$
$$- n(\mu_P{}^\ominus - \mu_W{}^\ominus) \quad (29)$$

Assuming that the electrostatic contributions are given by Born theory and that the solvated ions, irrespective of the composition of the solvation shell, have the same radii, then Equation 19, utilizing the assumptions embodied in Equations 25, 26, and 27, simplifies to

$$\Delta G_i{}^\ominus = \mu_P{}^{\ominus\,\text{chem}} - \mu_W{}^{\ominus\,\text{chem}} - (\mu_P{}^\ominus - \mu_W{}^\ominus) - RT \ln [(n+1-i)/i]$$
$$- RT \ln f_W' f_P' \quad (30)$$

and Equation 22 with Equation 30 yields

$$\Delta G_{ps}{}^\ominus = n\mu_P{}^{\ominus\,\text{chem}} - n\mu_W{}^{\ominus\,\text{chem}} - n(\mu_P{}^\ominus - \mu_W{}^\ominus) - RT \ln \prod_{i=1}^{n} (n+1-i)/i$$
$$- nRT \ln f_W'/f_P' = n(\mu_P{}^{\ominus\,\text{chem}} - \mu_W{}^{\ominus\,\text{chem}}) - n(\mu_P{}^\ominus - \mu_W{}^\ominus)$$
$$- nRT \ln f_W'/f_P' \quad (31)$$

Therefore, comparing Equations 31 and 30,

$$\Delta G_i{}^\ominus = \Delta G_{ps}{}^\ominus/n - RT \ln [(n+1-i)/i] \quad (32)$$

or in the form of equilibrium constants

$$K_i = K^{1/n}(n+1-i)/i \quad (33)$$

Equation 33 is the familiar statistical relation between the equilibrium constants in a series of stepwise equilibria as derived by N. Bjerrum (*11*) for polyprotic acids and applied by J. Bjerrum (*12*) to complex ion equilibria. Substituting Equation 33 for K_i into Equations 23 and 24 for ϕ_i° and ϕ_0° gives

$$\phi_i^\circ = \frac{(K^{1/n}Y)^i \prod_{i=1}^{i} (n + 1 - i)/i}{1 + \underset{n \text{ terms}}{\sum} [(K^{1/n}Y)^i \prod_{i=1}^{i} (n + 1 - i)/i]} \qquad (1 \leqslant i \leqslant n) \qquad (34)$$

$$\phi_0^\circ = \frac{1}{1 + \underset{n \text{ terms}}{\sum} [(K^{1/n}Y)^i \prod_{i=1}^{i} (n + 1 - i)/i]} \qquad (35)$$

but the denominator of Equations 34 and 35 is a binomial expression. Thus

$$\phi_i^\circ = (K^{1/n}Y)^i \prod_{i=1}^{i} [(n + 1 - i)/i]/(1 + K^{1/n}Y)^i \qquad (36)$$

and

$$\phi_0^\circ = 1/(1 + K^{1/n}Y)^n \qquad (37)$$

Utilizing Equations 36 and 37, and noting the identity

$$n!/[(n - i)!i!] \hat{=} \prod_{1}^{i} (n + 1 - i)/i \qquad (38)$$

and also Equation 5, one can simplify Equation 28 to

$$\Delta G_{t(X)}^{\ominus \text{ mix}} = \Sigma\phi_i^\circ \mu_i^{\ominus \text{ elec}} - \mu_{0(W)}^{\ominus \text{ elec}} + \Sigma(i\phi_i^\circ/n)(\Delta G_{ps}^{\ominus} -$$
$$nRT \ln f_P'/f_W') - RT\Sigma i\phi_i^\circ \ln x_P f_P^\circ/f_P' - RT\Sigma(n - i)\phi_i^\circ$$
$$\times \ln[(1 - x_P)f_W^\circ/f_W'] + RT\Sigma i\phi_i^\circ \ln K^{1/n}Y - nRT \ln (1 + K^{1/n}Y) \quad (39)$$

where all summations are between $i = 0$ and n.

Relation with Experiment

NMR Solute Ion Shift. The NMR chemical shift, a measure of changed magnetic field at the resonating nucleus, is caused by perturbations of the electron cloud around that nucleus. For organic molecules such effects may extend through several chemical bonds, particularly for conjugated systems where the π electron orbitals are fairly polarizable. However, for ionic solutes, the effects of solvent on the solute nuclei chemical shift are unlikely to extend beyond the first layer of solvent. Any given arrangement of solvent persists for only a very short time (10^{-11} sec for alkali metal halides) so the observed chemical shift can be formulated as the weighted average of the chemical shifts arising from all possible arrangements of solvent around the ion (13). Hence, if the ith species has an intrinsic shift of δ_i, the observed shift δ is given by

$$\delta = \sum_{i=1}^{n} \phi_i \delta_i \tag{40}$$

To proceed further it is necessary to make assumptions about how the various δ_i terms are interrelated. It will be assumed (2, 3) that the contribution to the shift from each P or W molecule is additive. The chemical shift δ may be expressed in terms of the shielding constants (14)

$$\delta = \sigma - \sigma_{\text{ref}} \tag{41}$$

where σ is the shielding constant of the nucleus of interest and σ_{ref} that of the reference. Thus

$$\sigma_i = i\sigma_{\text{P}}' + (n - i)\sigma_{\text{W}}' \tag{42}$$

where σ_{P}', σ_{W}' respectively are the contributions from each P or W molecule. If the chemical shift is measured relative to an aqueous solution then

$$\sigma_{\text{ref}} = n\sigma_{\text{W}}' \tag{43}$$

Combining Equations 41, 42, and 43,

$$\delta_i = i(\sigma_{\text{P}}' - \sigma_{\text{W}}') \tag{44}$$

In pure P, $\sigma_{\text{P}} = n\sigma_{\text{P}}'$ and hence

$$\delta_{\text{P}} = n(\sigma_{\text{P}}' - \sigma_{\text{W}}') \tag{45}$$

and thus, from Equations 44 and 45,

$$\delta_i = i\delta_{\text{P}}/n \tag{46}$$

Van Geet has used a similar approach to estimate the chemical shifts of solvated ions (15). He argues that if the repulsive overlap mechanism, shown by Richards et al. (16) to work for ion–ion interactions, also works for ion–solvent interactions, then the repulsive overlap mechanism used by Kondo and Yamashita (17) for alkali metal halide crystals justifies the above assumptions.

Combining Equations 46 and 40 gives

$$\frac{\delta}{\delta_{\text{P}}} = \sum_{1}^{n} i\phi_i/n \tag{47}$$

The quantity $\Sigma i\phi_i$ is the average solvation number of the ion by solvent P and hence the quantity $n - \Sigma i\phi_i$ is the average solvation number of the ion by W. Thus from solute NMR data, the average solvation number of an ion as a function of solvent composition can be obtained. The first attempt to do this was made by Frankel, Stengle, and Langford (18) in 1965.

Ultraviolet Spectral Shifts. Smith and Symons (*19*) in 1957 were the first to propose the use of uv CTTS spectral shifts in mixed solvents to obtain information about preferential solvation of ions. This has been followed up by the Stengle–Langford group (*20, 21*). Although the theoretical interpretation of uv spectral shifts has been the object of many papers, most attention has been directed towards large polar organic solvents. To the knowledge of the present authors, no equation analogous to Equation 47 for NMR shifts has been derived for ionic solutes. However, Mazurenko (*22*), using arguments somewhat related to those presented above, has derived such a relation for a polar organic solvent. It is worthwhile developing this here, using the nomenclature adopted above.

On the basis of an Onsager cavity (*23*) model of dielectrics applied to a polar solute with an intrinsic dipole movement $\mu_r{}^\circ$ in its rth electronic state, Mazurenko gives an equation for the orientational free energy of the solute molecule in a pure polar solvent environment, which can be identified as equivalent to $n\mu_P{}^{\ominus\,\text{chem}}$, thus

$$n\mu_P{}^{\ominus\,\text{chem}} = -[\phi(D_P{}^r) - \phi(n_P{}^2)]\frac{\mu_r{}^2}{2} \tag{48}$$

where $\phi(D^r)$ is a function of the effective relative dielectric permittivity, which is assumed to depend on the number of molecules of polar solvent close to the solute and $\phi(n^2)$ is a function of the refractive index. Further, the shift in uv peak maximum (relative to the same transition in a non-polar solvent Q), corresponding to a species with n molecules of P in its solvation shell is

$$\bar{\nu}_P - \bar{\nu}_Q = -\frac{1}{h}[\phi(D_P{}^r) - \phi(n_P{}^2)]\mu_r\Delta\mu \tag{49}$$

where $\Delta\mu$ is the change in dipole moment between the upper and lower states of the transition. For a species with i molecules of P in its solvation shell and $(n - i)$ molecules of nonpolar Q,

$$i\mu_P{}^{\ominus\,\text{chem}} = -[\phi(D_i{}^r) - \phi(n_i{}^2)]\frac{\mu_r{}^2}{2} \tag{50}$$

and

$$\bar{\nu}_i - \bar{\nu}_Q = \frac{1}{h}[\phi(D_i{}^r) - \phi(n_i{}^2)]\mu_r\Delta\mu \tag{51}$$

assuming with Mazurenko, a linear variation of orientational free energy with the number of polar molecules in the solvation shell.

Substituting Equation 48 into Equation 49, Equation 50 into Equation 51 and combining, gives

$$\bar{\nu}_i - \bar{\nu}_Q = \frac{i}{n}(\bar{\nu}_P - \bar{\nu}_Q) \tag{52}$$

Since

$$\bar{\nu} = \Sigma\phi_i\nu_i$$

then for the shift corresponding to a mixed solvent containing a distribution of solvated species

$$\frac{\bar{\nu} - \bar{\nu}_Q}{\bar{\nu}_P - \bar{\nu}_Q} = \sum_{i=1}^{n} i\phi_i/n \tag{53}$$

Whether this treatment is valid for a dipolar solute in a mixture of polar solvents is doubtful, but it would not appear to be valid for ions without intrinsic dipole moments. Nevertheless, Equation 53, as a ratio of wavelength shifts, has been applied to the analysis of CTTS spectra (19, 20, 21).

Further Thermodynamic Relations

The theory will be developed further using Equation 47 because, Equation 53 being basically similar in form, the resultant equations can be readily obtained if experimental information is available from electronic instead of NMR spectroscopy.

Substituting Equation 36 into Equation 47 gives

$$\frac{\delta}{\delta_P} = \frac{1}{n} \Sigma i\phi_i{}^{\circ}$$

$$= \sum_{n \text{ terms}} \left(\frac{i}{n}\right) (K^{1/n}Y)^i \left(\prod_{i=1}^{i} (n + 1 - i)/i\right) \Big/ (1 + K^{1/n}Y)^n \tag{54}$$

$$= K^{1/n}Y/(1 + K^{1/n}Y) \tag{55}$$

observing again the identity of Equation 38, which identifies the numerator of Equation 54 as a binomial expression in $(n - 1)$. Rearranging Equation 55 gives

$$\Delta G_{ps}{}^{\ominus} = -RT \ln K = -nRT \ln \left[\delta/(\delta_P - \delta)\right]$$
$$- nRT \ln \left[(1 - x_P)/x_P\right] f_W{}^{\circ}/f_P{}^{\circ} \tag{56}$$

From Equations 39, 47, 55, and 56, after considerable simplification, it follows for the free energy of transfer to a mixed solvent that

$$\Delta G_{t(X)}{}^{\ominus \text{ mix}} = \Sigma \phi_i{}^{\circ} \mu_i{}^{\ominus \text{ elec}} - \mu_{0(W)}{}^{\ominus \text{ elec}} - nRT \ln (1 - x_P) f_W{}^{\circ}/f_W{}'$$
$$- nRT \ln \left[\delta_P/(\delta_P - \delta)\right] \tag{57}$$

Rearranging Equation 31 in the form

$$\Delta G_{ps}{}^{\ominus} - nRT \ln f_P{}'/f_W{}' = n(\mu_P{}^{\ominus \text{ chem}} - \mu_W{}^{\ominus \text{ chem}}) - n(\mu_P{}^{\ominus} - \mu_W{}^{\ominus}) \tag{58}$$

shows that, since the right-hand side is solvent composition independent, the left-hand side also should be. Hence, using Equation 56,

$$\Delta G_{ps}{}^{\ominus} - nRT \ln f_P{}'/f_W{}' = - nRT \ln[\delta/(\delta_P - \delta)][(1 - x_P)/x_P]$$
$$- nRT \ln \frac{f_W{}^{\circ}}{f_W{}'} \cdot \frac{f_P{}'}{f_P{}^{\circ}} \tag{59}$$

It will be shown later that for the two solvent systems most thoroughly studied (and others besides), even though these show pronounced deviations from ideality, that the first term on the right-hand side of Equation 59 is a constant independent of solvent composition.   Thus the second term of Equation 59 must be effectively zero, and compensation of activity coefficient factors must occur.

It is pertinent at this stage to inquire the meaning of f_W' and f_P'.   Equations 19 and 20 refer to a free energy change in a given solvent mixture.   Redefining to a new standard state in W denoted by a prime ($'$),

$$\Delta G_i = \mu_i{}^{\ominus'} + RT \ln m_i f_i f_i' - \mu_{i-1}{}^{\ominus} - RT \ln m_{i-1} f_{i-1} f_{i-1}'$$
$$+ \mu_W{}^{\ominus} + RT \ln (1 - x_P) f_W - \mu_P{}^{\ominus} - RT \ln x_P f_P \quad (60)$$

and noting that in proceeding to zero solute concentration f_W/f_P approaches $f_W{}^\circ/f_P{}^\circ$,

$$\Delta G_i{}^{\ominus'} = \mu_i{}^{d\ominus'} - \mu_{i-1}{}^{\ominus'} + \mu_W{}^{\ominus} - \mu_P{}^{\ominus}$$

$$= -RT \ln \frac{m_i}{m_{i-1}} \cdot \frac{f_i'}{f_{i-1}'} - RT \ln \frac{1 - x_P}{x_P} \cdot \frac{f_W{}^\circ}{f_P{}^\circ} = -RT \ln K_i' \quad (61)$$

where f_i, f_{i-1} are activity coefficients for the solvated solute species which go separately to unity at infinite dilution in a given solvent mixture and their ratio is unity by usual Debye–Hückel considerations, and f_i', f_{i-1}' are medium-effect activity coefficients for the solvated ion species.

Thus,

$$K_i' = \frac{m_i}{m_{i-1}} \cdot \frac{1 - x_P}{x_P} \cdot \frac{f_i'}{f_{i-1}'} \cdot \frac{f_W{}^\circ}{f_P{}^\circ} = \frac{m_i}{m_{i-1}} \cdot \frac{1 - x_P}{x_P} \cdot \frac{f_P'}{f_W'} \cdot \frac{f_W{}^\circ}{f_P{}^\circ} \quad (62)$$

if it is assumed consistent with the assumption of Equation 27 that

$$\ln f_i' = (n - i) \ln f_W' + i \ln f_P' \quad (63)$$

Equation 62 for K_i' applies to all solvent mixtures and f_P'/f_W' is identified as synonymous (from Equation 63) with the ratio of medium-effect activity coefficients f_i'/f_{i-1}'.   So,

$$\Delta G_{ps}{}^{\ominus'} = -RT \ln K' = \Sigma \Delta G_i{}^{\ominus'} = \Delta G_{ps}{}^{\ominus} - nRT \ln f_P{}'/f_W'$$
$$= -nRT \ln \left[(\delta/(\delta_P - \delta))][(1 - x_P)/x_P] \right. \quad (64)$$

$\Delta G_{ps}{}^{\ominus'}$ (and K_i') are solvent independent quantities by definition, and the right-hand side of Equation 64 is solvent independent by experimental determination.   Therefore, it remains to be considered whether there is any physical explanation for $f_P'/f_W' = f_P{}^\circ/f_W{}^\circ$.   Although some mirroring of solvent non-ideality in the solvation shell might be expected, all that can be said at this stage is that the exact compensation is surprising.

It is immediately obvious that in Equation 55

$$K^{1/n}Y = (K')^{1/n}y$$

where $y = x_P/(1 - x_P)$. In the interests of simplified nomenclature, $(K')^{1/n}$ will be replaced by $\mathbf{K}$, which is identical with the constant defined by Frankel, Langford and Stengle (20) in their Equations 12 and 13 derived on the basis of analogy between preferential adsorption phenomena and preferential solvation. The experimental results will therefore be discussed in terms of the equations

$$\delta/\delta_p = \mathbf{K}y/(1 + \mathbf{K}y) \qquad (65)$$

$$\Delta G_{ps}^{\ominus'} = -nRT \ln \mathbf{K} = -nRT \ln [(\delta/(\delta_p - \delta))][(1 - x_p)/x_p] \qquad (66)$$

$$\Delta G_{t(X)}^{\ominus \, mix} = \Sigma\phi_i{}^\circ \mu_i{}^{\ominus \, elec} - \mu_{0(W)}{}^{\ominus \, elec} -$$
$$nRT \ln [\delta_p(1 - x_p)/(\delta_p - \delta)] \qquad (67)$$

Comparison with Experiment. DEUTERIUM OXIDE AND WATER MIXTURES. For a solute showing no preferential solvation ($\mathbf{K} = 1$, $\Delta G_{ps}^{\ominus'} = 0$), then $\delta_P = x_P$ from Equation 65, and a plot of δ against x_p is a straight line. Such behavior has been found for the fluoride ion in D_2O–H_2O mixtures (24). The free energy of transfer from Equation 67 for an ion which shows no preferential solvation will be given by

$$\Delta G_{t(X)}^{\ominus \, mix} = \Sigma \phi_i{}^\circ \mu_i{}^{\ominus \, elec} - \mu_{0(W)}{}^{\ominus \, elec}$$

and will be zero for an isodielectric solvent system. Friedman and Krishnan (25) have discussed the available free energy of transfer data for this system. For simple spherical ion salts, the values do not exceed 1 kJ mol^{-1}.

HYDROGEN PEROXIDE AND WATER. In the work of Covington, Lilley, Porthouse, and Newman (2), hydrogen peroxide was chosen as co-solvent because it forms almost isodielectric mixtures with water. Hence the electrical terms in Equation 57 disappear. Measurements of $\Delta G_{t(X)}^{\ominus \, mix}$ were made with ion-selective electrodes (26, 27) in these solvent mixtures to effect a comparison through Equation 57 and NMR measurements with thermodynamic data. Studies were made of ^7Li, ^{23}Na, ^{87}Rb, ^{133}Cs, ^{19}F, and ^{35}Cl shifts in solvents containing up to 85 wt% peroxide. When the solute shift was found to be concentration dependent, values were extrapolated to zero solute concentration (aquamolality) (Figure 1). These infinite dilution shifts (Figure 2) are a measure of ion–solvent interactions only and are free from ion–ion interaction effects. It is not possible to extend measurements to very high peroxide compositions because of solvent instability. It is necessary therefore to make a short extrapolation to obtain a value for δ_p, the shift with respect to the value in water, of the solute resonance in pure peroxide. This can be done conveniently by a rearrangement of Equation 65 to give

$$\frac{1}{\delta} = \frac{1}{\delta_P} \left(1 + \frac{1}{\mathbf{K}y}\right) \qquad (68)$$

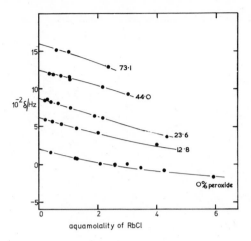

Figure 1.   Extrapolation of ^{87}Rb *chemical shifts for rubidium chloride in hydrogen peroxide solutions as a function of solute aquamolality.   Curves (reading downwards): 73.1, 44.0, 23.6, 12.8 and 0 mole % peroxide*

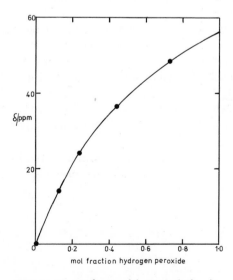

Figure 2.   Infinite dilution shifts for ^{87}Rb *derived from Figure 1*

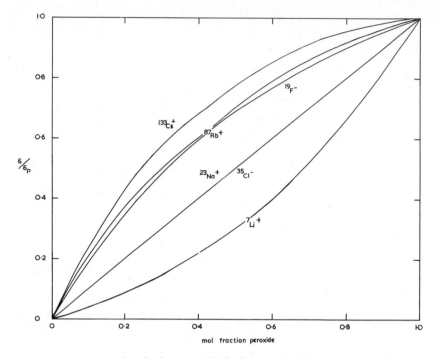

Figure 3.   Normalized chemical shift data (δ/δ_P) for ions in hydrogen peroxide–water mixtures

Thus a plot of $1/\delta$ against $1/y$ will yield $1/\delta_P$ from the intercept and $(1/K\delta_P)$ from the slope.

Normalized plots of δ/δ_P against x_P for all available data for this solvent system are shown in Figure 3.   It is immediately clear from the curvature of the lines in Figure 3 that Rb^+, Cs^+, and F^- are preferentially solvated by peroxide, that Li^+ is preferentially solvated by water, and hardly any preferential solvation exists for Na^+ and Cl^-.

Values are shown in Table I of $\Delta G_{ps}^{\ominus'}/n$.   In order to compare these with emf data of Covington and Thain (27), it is necessary to assume a value of n.   It

Table I.   NMR Solvation Parameters for Hydrogen
Peroxide–Water Mixtures (2)

Ion	δ_p/ppm	$\dfrac{\Delta G_{ps}^{\ominus'}}{n}$ /kJ mol^{-1}
$^7Li^+$	0.50	1.78 ± 0.13
$^{23}Na^+$	14.20	0.06 ± 0.04
$^{87}Rb^+$	56.18	-1.67 ± 0.08
$^{133}Cs^+$	87.75	-2.56 ± 0.01
$^{19}F^-$	25.62	-2.15 ± 0.02
$^{35}Cl^-$	22	0

seems reasonable that n, the average number of solvent molecules around a solute ion, would take values between 4 and 12.   Comparison of values assuming $n = 4$ is given in Figure 4 using Equation 67, where spectroscopic values for both cation and anion have been combined to effect comparison with the thermodynamic values.   The data for lithium fluoride provide a particularly stringent test of the theory because F^- is preferentially solvated by peroxide but Li^+ is solvated by water.   This leads to a subtraction of the ionic contributions and a small negative free energy of transfer as shown in Figure 4.

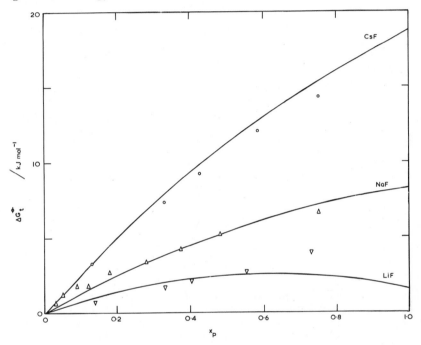

Figure 4.   Free energies of transfer of sodium, lithium and cesium fluoride from water to hydrogen peroxide derived from measurements with ion-selective glass and fluoride electrodes (27)

METHANOL AND WATER.   Methanol and water mixtures have been a popular choice for workers interested in free energies of transfer of ions from water into a mixed solvent.   Such mixtures exhibit a drop in dielectric constant with increasing methanol content.   Hence the electrical term must be estimated in order to compare spectroscopic and thermodynamic quantities.   Feakins and Voice (28) have presented new data and revised earlier data for the alkali metal chlorides.   In advance of carefully determined and extrapolated emf data for fluorides, using the solid state fluoride selective electrode based on lanthanum fluoride, some data of moderate accuracy have been presented (27).   On the

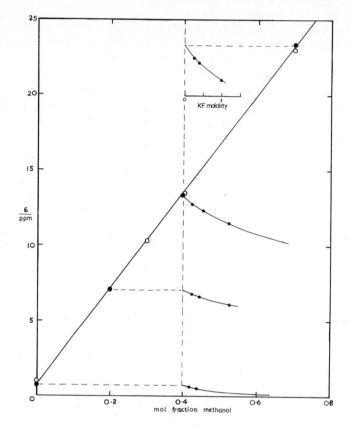

Figure 5. ˙ *Infinite dilution* [19]*F chemical shifts in metha-*
nol–water solutions with respect to a 4.4 molal KF solution
in water. ● *(32),* ○ *(29). Insets show extrapolations*
against molality of KF at various methanol mole fractions.

spectroscopic side, [19]F(KF), [35]Cl(LiCl), [81]Br(LiBr), [133]Cs(CsCl) shift data are
available (*16, 29, 30*) principally from the work of Richards, and these have been
supplemented by studies (*3, 31*) of [23]Na(NaI), [89]Rb(RbI) and confirmatory
measurements on [19]F(KF) shown in Figure 5 (*32*).   Table II shows values of

Table II.   NMR Solvation Parameters for Some Ions in
Methanol–Water Mixtures (*3, 32*)

Ion	δ_p/*ppm*	$\dfrac{\Delta G_{ps}^{\theta'}}{n}$/*kJ mol*[-1]
[23]Na[+]	4.16 ± 0.18	1.28 ± 0.05
[87]Rb[+]	22.9 ± 2.4	0.89 ± 0.05 (32)
[133]Cs[+]	45.92 ± 0.05	0.66 ± 0.06
[19]F[-]	33.0 ± 0.5	−0.15 ± 0.05
[35]Cl[-]	37.0 ± 2.0	0.96 ± 0.06

$\Delta G_{ps}^{\ominus\prime}/n$ derived from these data. To use Equation 67 it is necessary to estimate the electrostatic contributions. A first approximation to these will be given by the Born treatment (8). Thus for a molar standard state,

$$\mu_i^{\circ \text{ elec}} = Lz^2e^2/8\pi r_i\epsilon_0\epsilon_r \qquad (69)$$

where r_i is the radius of the ith solvated species of charge ze, ϵ_r the dielectric constant (relative permittivity), ϵ_0 the permittivity of free space and L, the Avogadro number. If the radii of all solvated species are assumed equal to the radius of the water-solvated species, r_0, then, for the first term in Equation 67,

$$\Sigma\phi_i^{\circ}\mu_i^{\ominus \text{ elec}} - \mu_{0(W)}^{\ominus \text{ elec}} = \frac{Lz^2e^2}{8\pi\epsilon_0 r_0}\left[\frac{1}{\epsilon_r^{\text{mix}}} - \frac{1}{\epsilon_r^{w}}\right] - RT\ln\frac{M\rho_w^{\text{mix}}}{M_w\rho_{\text{mix}}} \qquad (70)$$

where the last term takes into account standard state differences and ρ_{mix}, ρ_W are densities and $M^{\text{mix}} = (1 - x_P)M_W + x_P M_P$, with M_W, M_P being molecular weights of W and P. Previously (3), this last term was overlooked.

The solvation number n remains an adjustable parameter in the comparison of spectroscopic and thermodynamic data. Taking $n = 4$ for Na^+, and $n = 8$ for Cl^- (for which there is strong evidence from x-ray scattering investigations (33, 34) and Gourary and Adrian crystal radii (35), (to which half the O–O distance in ice and solid methanol (1.38 Å) is added for given r_0) allows Equation 70 to be used to calculate the electrostatic contribution for each ion. Such at-

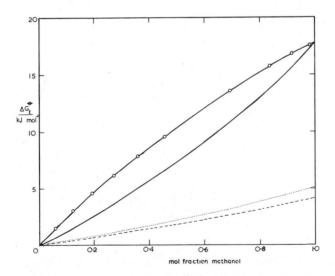

Figure 6. Free energy of transfer of sodium chloride from water to methanol-water mixtures. O emf derived values (28); — predicted values including electrostatic contribution estimated from the Born equation (69) with n = 4 (Na⁺), 8 (Cl⁻); ···· NMR contribution for Na⁺ ion; - - - NMR contribution for Cl⁻ ion

tempts to fit the data for methanol and water mixtures (Figure 6) correlate well
with transfers between the two pure solvents but exhibit the wrong curvature
for transfer to mixtures.   Allowing r_i to vary would not affect the curvature;
reducing r_0 would raise the predicted curve.   The deficiencies of the Born
treatment are well known (8) and alternative approaches to improve on the Born
treatment have been attempted.   Padova (36) has attempted to employ a treatment of Frank (37) to explain all solvent-sorting as arising from electrostatic
contributions.   It is clear that strong electrostatic fields could lead to reinforce-
ment, or reduction, by electrostatic effects of the chemical explanation considered
above.   A modified electrostatic treatment will be presented elsewhere (38), but
as shown in Figure 7 it does not lead to the required curvature in accord with emf
determinations.   The cause may therefore be in what Feakins (39) has called
second order structural effects.

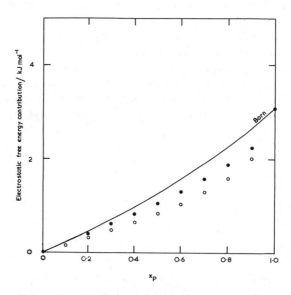

*Figure 7.   Work done (w) in charging a 4 Å ra-
dius ion in methanol–water mixtures.   — Born
equation (69); ● Solvent-sorting treatment assuming ideal mixture (47); ○ Solvent-sorting
treatment with non-ideal mixture (47)*

PROPYLENE CARBONATE (PC) AND WATER.   Data from both spectroscopic
and thermodynamic studies for other solvent systems are sparse and some of it
is of doubtful quality.   For propylene carbonate, Salomon (40) has obtained emf
data using lithium metal and thallium amalgam–thallous chloride or bromide
electrodes.

The only NMR study of PC–water mixtures is a proton resonance study by
Cogley, Butler, and Grunwald (41) who obtained an apparent equilibrium

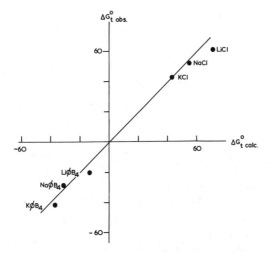

Figure 8. Comparison of observed and calculated values for $\Delta G_t{}^\circ$ (molar standard state) for the transfer of salts from water to propylene carbonate. The electrostatic contribution was estimated from Equation 69 with n = 4 (Li^+, Na^+, K^+), n = 8 (Cl^-, ϕ_4B^-), and the following solvated radii in A: 3.6 (Li^+), 3.9 (Na^+), 4.2 (K^+), 4.3 (Cl^-), and 6.9 (ϕ_4B^-)

constant (essentially a measure of $1/K_n$) for the formation of a monohydrate, obtained by keeping the salt concentration constant and varying the water concentration in PC-rich solutions. Even in the absence of salt, the water shifts are dependent on the water concentration because of dimerization of water, and consequently the salt shifts must be referred relative to the water shift at the appropriate water concentration. Identifying Cogley and co-workers' constant as $n/K[P]$ enables $\Delta G_{ps}{}^{\ominus'}$ to be calculated from Equation 66. A difficulty is the separation of cation and anion effects. Perchlorates and tetraphenylborates showed similar effects which, it was argued, would only happen if the effect of each individually was small. On this basis, a separation was made. Figure 8 shows a comparison of experimental values for $\Delta G_t{}^\circ$ (molar standard state) for some salts derived largely from the work of Salomon (*40*) with calculated values based on the proton resonance data and electrostatic contributions combined for pairs of ions. The agreement is quite remarkable, particularly when it is realized that the values range through 100 kJ mol^{-1}. Whereas Cogley and co-workers (*41*) concluded that the chloride ion was anomalous, this is now seen not to be so.

ACETONITRILE (AN) AND WATER. Free energy of transfer data have been reviewed for this system by Kolthoff and Chantooni (*42*) although they were apparently unaware of some work of Coetzee and Campion (*43*) on lithium salts. Spectroscopic data are available from three sources. Stengle and co-workers (*44*)

Table III.   Comparison of Calculated and Observed $\Delta G_t °$ Values
for AN–Water

Salt	$\Delta G_t °$ (obs)/$kJ\ mol^{-1}$		$\Delta G_t °$ (calc)/$kJ\ mol^{-1}$	
LiCl	71.8		90.6	
NaCl	55.9	61.1	72.6	60.7
NaBr	45.6	50.3	63.8	51.9
NaI	32.7	37.2	54.9	43.0
Lit. ref.	43	42	44, 45	46

have studied [35]Cl, [31]Br, and [127]I chemical shifts in AN–water as well as uv CTTS iodide and bromide bands.   Bloor and Kidd (45) have obtained [23]Na shifts, and Stockton and Martin (46) have used a proton NMR technique similar to that of Cogley and co-workers to obtain apparent association constants for an ion in AN-rich solutions.   Comparison of observed and calculated values of $\Delta G_t °$ is shown in Table III.   Agreement is good in spite of the experimental uncertainties in the data derived from both thermodynamic and spectroscopic sources.   It is encouraging that solvation trends with changing ion size are predicted well.

DIOXANE AND WATER.   Grunwald and co-workers (GBK) (7) used a vapor pressure method to obtain the differential of the free energy of transfer of a solute with respect to solvent mole fraction at 50 wt% dioxane.   On the basis of what has now become known as the large-ion assumption (8), they separated cation and anion effects by equating the free energies of transfer for tetraphenylborate and tetraphenylphosphonium ions.   They concluded that Na^+ was preferentially solvated by dioxane, a surprising result then, but less unexpected now that complexes of the alkali metals with polyethers have been discovered (dioxane

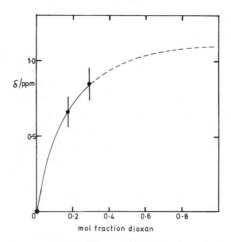

Figure 9.   Infinite dilution [23]*Na shifts in dioxane and water mixtures (47). Vertical bars indicate the experimental uncertainty*

is a cyclic diether).   Figure 9 shows ^{23}Na chemical shifts of NaI in dioxane–water mixtures (47).   Solubility problems restrict the studies to $x_P = 0.3$.   The data are limited and experimental uncertainties are large, but the evidence clearly supports the preferential solvation of Na^+ by dioxane not water.   An approximate value for the chemical contribution to the free energy of transfer can be estimated as -18 ± 9 kJ mol^{-1} (assuming $n = 4$).

NONAQUEOUS SYSTEMS.   A fairly large number of solvents of such purity as to make reasonably precise physicochemical measurements possible are now available (48).   Knowledge of the physical properties of mixtures of these is often lacking except where one component is water.   Gradually, however, more information is being accumulated.   Greenberg and Popov (49) have made the first systematic study of the solvation of Na^+ ions using ^{23}Na shift measurements in all binary mixtures of seven solvents (nitromethane (NM), acetonitrile, hexamethylphosphoramide, dimethylsulfoxide, pyridine (Py), and tetramethylurea).   Results were analyzed in terms of Equation 66 to obtain the free energies of preferential solvation and by the location of the equi- or iso-solvation point (20). This is the point at which both solvents participate equally in the solvation shell. It occurs at the composition at which the chemical shift lies midway between the values for the pure solvents, hence at $\delta/\delta_P = 1/2$ in Equation 65.   However, the δ values do not relate to infinite dilution but to a finite concentration (0.5 M) of sodium tetraphenylborate.   The concentration dependence may be slight, but in view of the low dielectric constant of many of these solvents ion pairs will be present and the dependence will then be non-linear.   The observation that different salts apparently do not extrapolate to the same infinite dilution value (50) causes concern.   Normalized results (δ/δ_P) for acetonitrile as common solvent are shown in Figure 10.   From the values of free energy of preferential solvation, additivity checks are possible.   For example,

$$
\begin{array}{lll}
\text{NM} \longleftarrow \text{AN} & \Delta G_{ps}{}^{\ominus\prime}/n = 4.41 & \pm\ 0.23\ \text{kJ mol}^{-1} \\
\text{Py} \longleftarrow \text{NM} & = -5.51 & \pm\ 0.13\ \text{kJ mol}^{-1} \\
\text{Py} \longleftarrow \text{AN} & = -2.10 & \pm\ 0.47\ \text{kJ mol}^{-1} \\
\hline
\text{Py} \longleftarrow \text{AN(calc)} & = -1.10 & \pm\ 0.36\ \text{kJ mol}^{-1}
\end{array}
$$

Although just outside the estimated errors, this additivity check is encouraging (bearing in mind the reservations mentioned above).   Some systems do not give very good straight line plots with Equation 68.   Hence, additivity checks involving these are suspect.   Further work of the comprehensive, systematic nature of Greenberg and Popov is necessary.

Nonstatistical Distribution of Solvated Species

The introduction of the last term in Equation 27, the statistical factor for the number of ways of arranging i P molecules and $(n - i)$W molecules in the

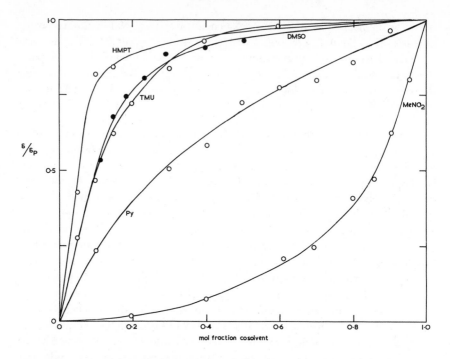

Figure 10.   Normalized ^{23}*Na chemical shifts in some binary solvent mixtures involving transfer from acetonitrile* (49). *DMSO = Dimethylsulfoxide; HMPT = Hexamethylphosphotriamide; TMU = Tetramethylurea; MeNO$_2$ = Nitromethane; Py = Pyridine*

solvation shell, leads to Equation 33 relating K_i values to the constant for the overall process and hence to a statistical distribution of the solvated species, as shown in Figure 11.   However, the process of substituting solvent molecules P for W in the solvation shell may not take place with equal facility for each step. Thus the solvent exchange process may become more or less energetically favorable as the shell becomes richer in P.   In the general case, some simplification is necessary.   The constant for the first step could be assumed to be very much larger than those remaining (9), which may then be related statistically.   A modified model has been adopted (5), in which the free energy change for substitution of one W by one P changes by $RT \ln k$ for each successive step.   A related approach has been used in another connection by Stokes and Robinson (51) and the treatment is also similar to Bjerrum's spreading factor (x) introduced to explain variation in the ratios of successive constants in complex-ion equilibria (12).   It may be noted that $x = k^{-1/2}$; Bjerrum applied x to each step but here k is applied to the second and subsequent steps only.

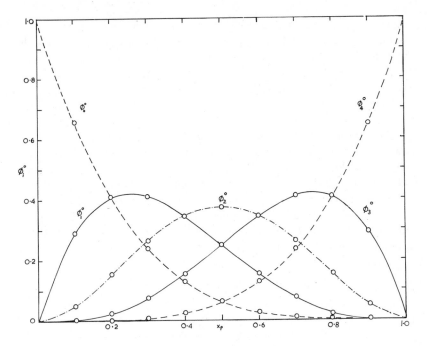

Figure 11.   Statistical distribution of solvated species for $K = 1.0$ *and* n
= 4

The ratio

$$\frac{K_{I+1}}{K_I} = \frac{ki(n - i)}{(i + 1)(n - i + 1)} \tag{71}$$

and for the ith step

$$K_i = \left(\frac{n - i + 1}{i}\right) k^{i-1-(n-1)/2} K^{1/n} \tag{72}$$

Utilizing Equations 47, 23, 24, and 38 gives

$$\frac{\delta}{\delta p} = \frac{Ky \sum_{n \text{ terms}} (Ky)^{i-1}(i/n)k^{i(i-1)/2-(n-1)/2} \prod_1^i (n + 1 - i)/i}{1 + \sum_{n \text{ terms}} (Ky)^i k^{i(i-1)/2-(n-1)/2} \prod_1^i (n + 1 - i)/i} \tag{73}$$

$$= \frac{Ky \sum_{n \text{ terms}} (Ky)^{i-1} k^{i(i-1)/2-(n-1)/2} \cdot (n - 1)!/(n - i)!(i - 1)!}{1 + \sum_{n \text{ terms}} (Ky)^i k^{i(i-1)/2-(n-1)/2} \cdot n!/(n - i)!i!} \tag{74}$$

which does not simplify.

However, for $n = 4$, expanding the summations and selecting values for the parameters K and k, the functional dependence of δ/δ_P on x_P can be calculated. Typical curves are shown in Figure 12 for two arbitrarily selected values of $k = 0.833$ and 1.2 and two values of $K = 0.5$ and 2.0. An illustration of the effect of introducing the factor k into the treatment for $n = 4$ is shown diagrammatically in Figure 13.

Two examples of shift data with nonuniform curvature (unsymmetrical shift data) which can be fitted by this approach are shown in Figures 14 and 15. The study of iodide solvation by uv CTTS and ^{127}I chemical shifts is the only one where both approaches have been used on the same ion (44). The accuracy of the results is questionable but ^{127}I peaks are very broad and measurement of shifts difficult (5). The treatment leading to Equation 53 is not valid for a CTTS transition, and its application is intuitive, but a difference in wave number shifts rather than wavelengths should be plotted. However, the discrepancy resulting is numerically trivial.

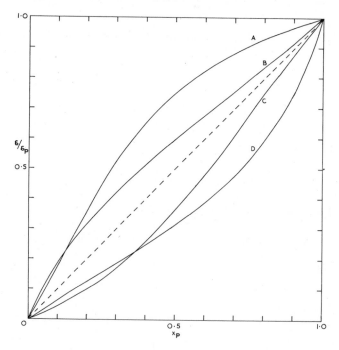

Figure 12.   Effect on normalized chemical shift curves (δ/δ_P) of introduction of factor k. *Curve A: K = 2.0,* k *= 1.2; B: K = 2.0, k = 0.833; C: K = 0.5,* k *= 1.2; D: K = 0.5,* k *= 0.833*

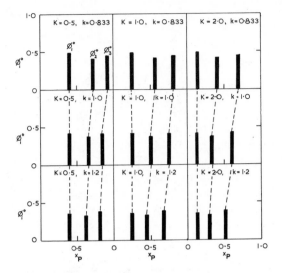

Figure 13.   Schematic representation of the effect of introduction of the factor k *on the positions and maximum obtainable mole fractions for the solvated species* (n = 4), i = 1,2,3.

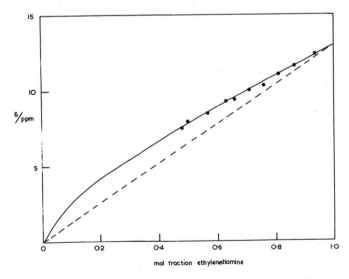

Canadian Journal of Chemistry

Figure 14. ^{23}Na chemical shift data for NaI in H_2O–ethyl-enediamine (en) (45). *— theoretical curve (Equation 74) with* n = 4, **K** = 2.3, k = 0.8. *The asymmetry occurs at the wrong end to account for by bidentate solvate formation by ethylenediamine*

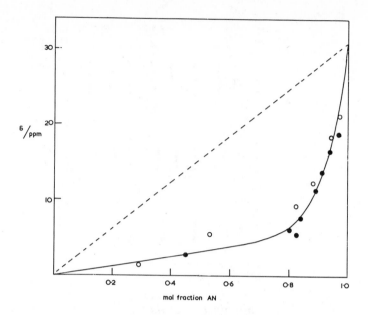

Journal of the American Chemical Society

Figure 15. ^{35}Cl shift data for tetraethylammonium chloride in H_2O-acetonitrile (44). ● 0.225 mol l^{-1}; ○ 0.42 mol l^{-1}; — theoretical curve (Equation 74) with n = 4, K = 0.07, k = 0.9

It is interesting to note that the one sigmoid δ/δ_P curve found for ^{127}I shifts in DMSO–H_2O mixtures by Stengle, Pan and Langford (44) can be fitted (Figure 16) by this treatment (5) with **K** = 2.5, k = 0.5.   In contrast, ^{35}Cl shifts in this solvent mixture analyze well with **K** = 1.0, k = 1.0, i.e., non-preferential solvation of chloride ion (44).

Change of Solvation Number

In the previous section, a possible explanation was advanced for the observation of an unsymmetrical curve joining the ionic nucleus chemical shifts in the pure solvents.   A possible alternative, and physically plausible, explanation is that the solvation number is not constant.   The general case of variable n is intractable, but it is possible to treat the simple case of a change in n from an even integral value in one solvent to half its value in the other pure solvent.   This corresponds to monodentate for bidentate competition for a transition metal ion in solution (52).

Corresponding to Equation 18, the equilibrium for the formation of the ith solvated species $XW_{n-2i}P_i$ (where $0 \leqslant i \leqslant n/2$) can be written

$$XW_{n-2(i-1)}P_{i-1} + P \overset{K_i}{\rightleftharpoons} XW_{n-2i}P_i + 2W \qquad (75)$$

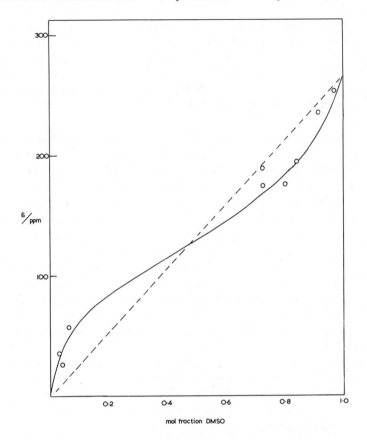

Figure 16. ^{127}I *shift data for NaI in H_2O–DMSO mixtures* (44). — *theoretical curve (Equation 74) with* n = 4, **K** = 2.5, k = 0.5

and Equations 5, 61, and 27 are then modified to

$$\sum_{i=0}^{i=n/2} \phi_i{}^\circ = 1 \tag{76}$$

$$\Delta G_i{}^{\ominus\prime} = -RT \ln K_i{}' = -RT \ln \frac{m_i}{m_{i-1}} - RT \ln \frac{(1 - x_P)^2}{x_P} \cdot \frac{f_W{}^{\circ 2}}{f_P{}^\circ} \tag{77}$$

$$= \mu_i{}^\ominus - \mu_{i-1}{}^\ominus + 2\mu_w{}^\ominus - \mu_P{}^\ominus$$

where $K_i{}'$ takes the values $K_1{}' \cdots K_i{}' \cdots K_{n/2}{}'$. Also,

$$\mu_i{}^{\ominus \, \text{chem}} = (n - 2i)\mu_W{}^{\ominus \, \text{chem}} + i\mu_P{}^{\ominus \, \text{chem}} + (n - 2i)RT \ln f_W{}'$$
$$+ iRT \ln f_P{}' + \text{statistical term.} \tag{78}$$

The form of the statistical term required in Equation 78 is complex in the general case. It should presumably take account of symmetry both of the mixed solvates formed and of the solvating molecules as ligands themselves. In a previous paper (4), an approximate general treatment was attempted by treating pairs of sites in the solvation sphere. Since its limitations were not made fully clear, the problems will be discussed further here.

The statistical term in Equation 78 will be given by the number of ways of arranging $(n - 2i)$ molecules of W and i P bidentate molecules on n sites. If it is assumed that a molecule of P can occupy any pair of sites (this will be true for $n = 4$ but not generally so), then the number of arrangements is $n!/(n - 2i)!i!$ if P is unsymmetrical, with an additional factor of $(2)^i$ to be included in the denominator if P is symmetrical. Introduction of this form into Equation 78 and its subsequent substitution into Equation 77, requires taking the ratio of two such terms for the ith and $(i - 1)$th species leading to $(n - 2i + 2)(n - 2i + 1)/2i$ for symmetrical P compared with $(n + 1 - i)/i$ in Equation 30. The first term arises from the number of ways of adding one end of a P molecule to $n - 2(i - 1)$ sites, and the second term identifies the ways of placing the other end. The term derived previously (4) was $(n - 2i + 2)/2i$, from which it is seen that the model assumed there allowed the other end of the P molecule to be placed in only one way, and that the P molecule must be symmetrical. For $n = 4$, this model would correspond to a tetrahedron with unequal edge lengths. For $n = 6$, or octahedral coordination, the situation is complex because there exist both geometric and optical isomeric solvates (53). For an exact treatment, each case would have to be considered separately. A treatment in terms of symmetry numbers (54, 55) would also lead to the same results.

Equation 23, assuming that P is a symmetrical molecule which can occupy any pair of sites, takes the form

$$\phi_i{}^\circ = \frac{(\mathbf{K}y')^i \prod\limits_{i=1}^{i} (n - 2i + 2)(n - 2i + 1)/2i}{1 + \sum\limits_{n/2 \text{ terms}} (\mathbf{K}y')^i \prod\limits_{i=1}^{i} (n - 2i + 2)(n - 2i + 1)/2i} \qquad (79)$$

where $y' = x_P/x_W{}^2$ and $\mathbf{K} = (K')^{2/n}$.

It is now necessary to relate $\phi_i{}^\circ$ to the chemical shift through Equation 47. This requires an assumption about δ_i, the intrinsic shift of the ith solvation species. It will be assumed (cf. Equation 47) that

$$\delta_i = \sigma_W'(n - 2i) + \sigma_P'i - \sigma_W'n \qquad (80)$$

where σ_W' and σ_P' are shielding constants. When X is solvated by P molecules only

$$\delta_P = \sigma_P' \frac{n}{2} - \sigma_W' n, \text{ so}$$

$$\frac{\delta}{\delta_P} = \frac{2}{n} \sum_i^{n/2} i \phi_i°$$

$$= \frac{\displaystyle\sum_{n/2 \text{ terms}} (\mathbf{K}y')^i \left(\frac{2i}{n}\right) \prod_{i=1}^{i} \frac{(n - 2i + 2)(n - 2i + 1)}{2i}}{1 + \displaystyle\sum_{n/2 \text{ terms}} (\mathbf{K}y')^i \prod_1^{i} \frac{(n - 2i + 2)(n - 2i + 1)}{2i}} \qquad (81)$$

with some reservations about whether Equation 81 is valid if $n > 4$.

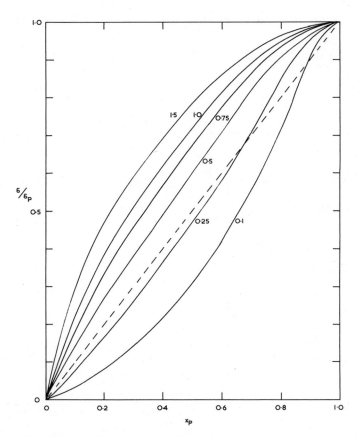

*Figure 17.   Predicted variation of normalized shift (δ/δ_P) from Equation 83 for various values of **K** and P a symmetrical ligand (n = 4)*

Taking $n = 4$ for P symmetrical, Equation 81 gives

$$\frac{\delta}{\delta_P} = \frac{Ky'(1 + Ky')}{1 + 6Ky' + 3(Ky')^2} \qquad (82)$$

The form of variation of δ/δ_P for various values of K is shown in Figure 17. The curves are markedly less asymmetric than those produced using the equation derived previously (4), and none of the data discussed there is of the right form to be fitted with Equation 82.

Again for $n = 4$ and with P an unsymmetrical ligand, gives

$$\frac{\delta}{\delta_P} = \frac{6Ky'(1 + 2Ky')}{1 + 12Ky' + 12(Ky')^2} \qquad (83)$$

and curves generated for various values of K are shown in Figure 18. The effect

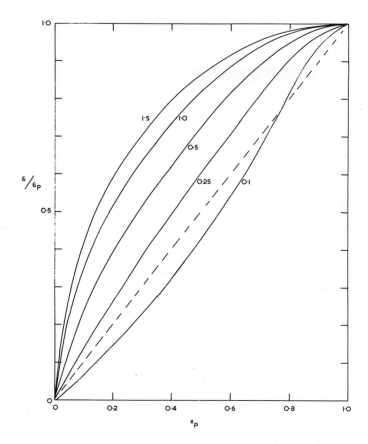

Figure 18.   Predicted variation of normalized shift (δ/δ_P) from Equation 82 for various values of K and P an unsymmetrical ligand $(n = 4)$

of an unsymmetrical P at a given value of **K** is to increase the curvature of the plots in the negative sense.

It would be clearly possible to derive equations similar to Equations 82 and 83 for $n = 6$ and octahedral coordination. No general form like Equation 81 is possible if solvate isomers are considered; account must then be taken of the relations between the K_i values, either from statistical or symmetry number considerations (*12, 55*), where $K = (K')^{1/3}$,

$$K_1 = 24\ K, \quad K_2 = 5\ K, \quad K_3 = \frac{2}{3}\ K \qquad (84)$$

However, the assumption of Equation 80 is now dubious, for the intrinsic shifts of the solvates will depend on their structure and symmetry. Hence the statistical treatment and NMR assumptions are virtually dependent and it is inconsistent to treat them separately.

The previous treatment (*4*), although approximate in the same sense that Flory's (*56*) treatment of polymer solutions was approximate, nevertheless remains useful in indicating possible NMR solute behavior from mono- versus bidentate solvation. In attempting to explain data for both sulfolane and dimethylsulfoxide as bidentate ligands, the fact that only the former is a symmetrical ligand was disregarded (*4*). The most satisfactory explanation of unsymmetrical chemical shift curves is the nonstatistical distribution approach outlined in the preceding section. It is interesting to note that it has been suggested (*7*) that dioxane acted as a bidentate solvating ligand towards potassium. It is unfortunate that the ^{39}K NMR signal is so weak, for this would have been a useful system to study.

Direct Determination of Solvate Fractions

The observation of a single resonance peak in the NMR spectra of ion nuclei in solution indicates that the exchange between separate species, or between ion pairs and such species, is fast. Separate resonance shifts from the solvent coordinated to an ion can be observed in certain cases using ^{1}H or ^{31}P resonance (*57, 58*), if necessary by lowering the temperature to slow the exchange. Fratiello and co-workers (*59, 60*) have used this low temperature technique to study the solvation of ions such as Mg^{2+}, Be^{2+}, and Al^{3+} in aqueous solution to which acetone has been added to lower the freezing point. Green and Sheppard (*61*) and Toma and co-workers (*62*) have shown that if the temperature is lowered even more, the signal from the bound water in the $Mg(ClO_4)_2$–acetone–H_2O system at very low water content splits further and can be resolved into components arising from the differently solvated species. Resolution of the bound acetone signal is much more difficult. From the areas under these resolved peaks it is assumed that the fractions of Mg^{2+} occurring in each mixed solvated form can be determined. These are the quantities ϕ_i defined earlier but with the important difference that they apply not at infinite dilution of the solute ion but at the finite concentration of the experiment. Hence the superscripts° of Equations 23 and 24 are omitted. Covington and Covington (6) have made further experimental analysis of this system at 185°K with the twofold aim of improving the quality of the spectra obtained by using CAT techniques and of checking the inference of Green and Sheppard (61) that the apparent solvation number of Mg^{2+} falls below six at very low water content because of the entry of the perchlorate ion into the solvation shell by the formation of an ion pair. This finding could not be substantiated (6), in agreement with the findings of Toma and co-workers (62). Hence the system can be analyzed by modification of the treatment given earlier in this review applied to the equilibrium represented by Equation 21. Because of the finite Mg^{2+} concentration, it is necessary to distinguish between the stoichiometric (total) mole fractions (T) of the two solvent components which are greater than the actual values of free (F) solvent in the solution because of the amounts bound (B) to the cations. Accordingly it is necessary to rewrite Equation 62 in terms of $y_F = (x_P/x_W)_F$, the ratio of the free mole fractions. Equations 23 and 36, and 24 and 37 are valid for ϕ_i and ϕ_0 (with dropped superscripts) and y_F replaces y.

From mass balance considerations, the total concentration of X is given by

$$[X]_t = [XW_n] \left(1 + \sum_1^n y_F^i \prod_i^n K_i\right) \tag{85}$$

and the bound water concentration by

$$[W]_B = [XW_n] \left(n + \sum_{i=1}^{i=n-1} (n-i)y_F^i \prod_i^{i=n-1} K_i\right) \tag{86}$$

$$= n[XWn]\left[1 + \sum_1^{n-1} \frac{(n-i)}{n}(Ky_F)^i \prod_i^{n-1} \frac{(n+1-i)}{i}\right] \tag{87}$$

using Equation 33 and the definition of $K = (K)^{1/n}$. Using Equation 37 for $\phi_0 = [XW_n]/[X]_T$

$$[X]_B = n[X]_T(1 + Ky_F)^{n-1}/(1 + Ky_F)^n$$

$$= n[X]_T/(1 + Ky_F) \tag{88}$$

Then the ratio of free to bound water is given by

$$F_W = \frac{[W]_F}{[W]_T} = 1 - \frac{nZ}{(1 + Ky_F)} \tag{89}$$

where $Z = [W]_T/[X]_T$, the total number of moles of water per magnesium ion, or rearranging

$$Ky_F = (F_W - 1 + n/Z)/(1 - F_W) \tag{90}$$

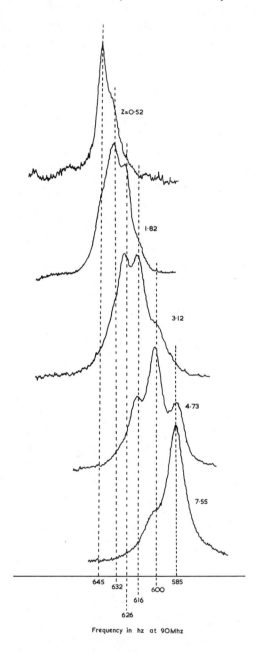

Journal of the Chemical Society, Faraday I

Figure 19.   PMR spectra at 185°K for magnesium perchlorate in acetone–water mixtures at low water content (6) showing mixed solvate species.

Table IV.   Fractions of $[Mg^{2+}]$ in Different Solvated
(Calculated values

Z	$\dfrac{10^2}{Y_T}$	ϕ_0	ϕ_1	ϕ_2
0.26	0.43			
0.52	0.87			
0.80	1.32			
1.33	2.20			0.03 (0.01)
1.82	3.01			0.08 (0.04)
2.35	3.89		0.04 (0.03)	0.13 (0.11)
3.12	5.16		0.11 (0.10)	0.27 (0.24)
4.00	6.62	0.08 (0.08)	0.24 (0.26)	0.34 (0.33)
4.73	7.84	0.19 (0.21)	0.38 (0.37)	0.28 (0.28)
5.51	9.13	0.39 (0.40)	0.43 (0.40)	0.15 (0.17)
6.38	10.58	0.60 (0.61)	0.33 (0.31)	0.06 (0.07)
6.5^a		0.62 (0.62)	0.31 (0.31)	0.07 (0.07)
6.84	11.33	0.72 (0.71)	0.26 (0.25)	0.02 (0.04)
7.55	12.51	0.81 (0.83)	0.19 (0.17)	
9.43	15.64	0.88	0.12	
10.21	16.94	0.93	0.07	
15.62	25.88	1.00		

^{a} Ref. 62.

In the solvent concentration range where separate signals from the mixed solvate species are resolvable, the free water signal is so small as to be undetectable. At higher water concentrations the area under the free water signal and hence F_W in Equation 90 for known values of Z can be obtained.   Taking $n = 6$, the right-hand side is calculable as a fraction of y_T and can be fit in logarithmic form to a quadratic equation, which can be used for extrapolation to give values of Ky_F in the experimentally indeterminate region of F_W.   This enables ϕ_i values to be calculated from Equations 36 and 37.   These calculated values are compared with the direct experimental values obtained from the resolved peak areas in Table IV.   The agreement is within 2% on the average, which is in accord with the accuracy of curve resolution of overlapping peaks (bearing in mind that some spectra contain four components), as shown in Figure 19.

Noting that

$$Y_F = \frac{F_P}{F_W} y_T \tag{91}$$

where $F_P = [P]_F/[P]_T$, which can be obtained from the experimental values of ϕ_i in the region where no bound acetone can be detected in the spectra, it is possible to obtain $K = (1.1 \pm 0.4) \times 10^{-3}$.   With $n = 6$, this leads to $\Delta G_{\theta ps}' = -63 \pm 3$ kJ mol^{-1}.   Some values of ϕ_i derived from Toma and co-workers' experimental results (62) for 185 K and $Z = 1.6$ are included in Table IV, but it has been assumed that the water to magnesium concentration ratio was slightly greater than stated.   Using a 220 MHz instrument, these workers (62) were able to resolve

Species in Mg(ClO₄)₂–Acetone–Water System in parentheses)

ϕ_3	ϕ_4	ϕ_5	ϕ_6
	0.04 (0.02)	0.18 (0.18)	0.78 (0.80)
0.01 (0.01)	0.08 (0.06)	0.32 (0.30)	0.59 (0.63)
0.04 (0.02)	0.15 (0.12)	0.36 (0.37)	0.45 (0.49)
0.12 (0.08)	0.24 (0.24)	0.36 (0.39)	0.25 (0.28)
0.19 (0.16)	0.31 (0.31)	0.30 (0.33)	0.12 (0.15)
0.27 (0.25)	0.32 (0.32)	0.19 (0.22)	0.05 (0.06)
0.32 (0.31)	0.24 (0.23)	0.06 (0.09)	
0.24 (0.23)	0.10 (0.09)		
0.12 (0.11)	0.03 (0.03)		
0.03 (0.04)			

further the spectra involving the ϕ_4 (dihydrate) species and to detect its isomeric forms (*53*) shown below.

The presence of such isomeric solvates may account for the asymmetry noted (*62, 6*) when carrying out the curve resolution of the 90-MHz spectra, but, within the accuracy attainable, the effect is barely significant. There are discrepancies in the values for the intrinsic shifts of the solvated species recorded by all three groups of workers (but differences between successive solvates are more satisfactory). Toma and co-workers calculated (*62*) the theoretically expected and isomeric differences between successive solvates, obtaining reasonable agreement with observed values (Table V).

Similar calculations would be possible with the system Al(ClO₄)₃ in DMSO–H₂O studied by Olander, Marianelli, and Larson (*63*), but the actual experimental results are not available. The Mg²⁺ system in methanol–water has been studied several times (*64, 65*) and appears to be more complex in the interpretation of the spectra obtained. Again detailed systematic measurements at one temperature and a range of solvent compositions are not available.

Table V.   Observed Chemical Shifts of Mixed Solvates of Mg^{2+} in Acetone–Water (δ/ppm)

No. of Moles Water/Mg^{2+}	Reference			
	62	6	61	Calc., 62
6	7.08	7.16	7.12	7.08 (assumed)
5	7.00	7.02	7.03	6.98, 6.93
4	6.89	6.96	6.95	6.84, 6.79
3	6.76	6.84	6.84	6.71, 6.67
2	6.61	6.67	6.66	6.58, 6.54
1	6.45	6.50	6.50	6.45 (assumed)

It is interesting to speculate whether confirmatory experiments on the Mg^{2+} system in acetone–water could be made using ^{25}Mg resonance.   It was concluded (6) that for $K = 10^{-3}$ the distribution of $\phi_i{}^\circ$ values would be that shown in Figure 20 and hence the interesting region where maxima in ϕ_i values occur is experimentally inaccessible because of the extremely low water concentrations (x_W

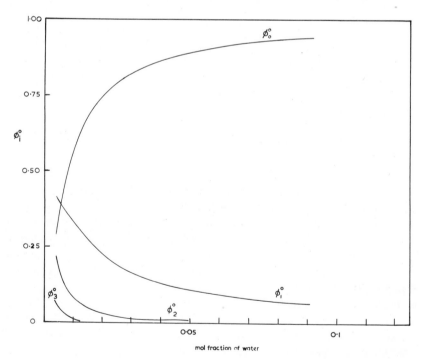

Figure 20.   Predicted variation of $\phi_i{}^\circ$ values with mol fraction of water for $K = 10^{-3}$ in magnesium perchlorate–acetone–water system.

$< 10^{-2}$) which would have to be attained.   The weak signals from ^{25}Mg with an isotopic abundance of 10% makes the experiment even more difficult.

E. L. King and co-workers (*66, 67, 68*) have studied the solvation of Cr^{3+} ions in water–methanol, –ethanol, and –dimethylsulfoxide mixtures.   Species containing up to six bound organic molecules have been separated from each other by cation exchange column procedures.   Analysis of these species yields the ϕ_i values of this section.   If the Cr^{3+} concentration is low in the original solution, these are effectively ϕ_i° values.   Attempts to fit King's data for these three systems yield $K = 0.07(MeOH), 0.03(EtOH)$, and $0.2(DMSO)$.   The last is illustrated in Figure 21.   The ϕ_i values were calculated from Equations 36 and 37 (with y replaced by y_F as necessary) by selecting a value of K by trial and error which most closely fits all the ϕ_i data.   For clarity, only a few experimental points have been shown on Figure 21, the scatter being consistent with the estimated experimental error.   This approach contrasts with King's analysis to obtain equilibrium quotients for the formation of the solvate species making correction for solvate mixture non-ideality.   As a result, the quotients for the successive equilibria were found to be medium dependent.

Discussion and Criticisms of the Present Approach

The basic assumption of the present treatment is that the chemical shift of an ionic nucleus is dependent on its immediate magnetic environment and hence gives a measure of changes in the primary solvation sphere of the ion as the composition of the bulk solvent mixture is changed.   An obvious limitation of the treatment is the need to assume a value for n, the solvation number, if spectroscopic and thermodynamic data are to be compared.   It must be stressed that this solvation number is simply the number of solvent molecules immediately surrounding the ion.   This will not necessarily be the same as the solvation number determined by other methods (*69*) which take into account bonding interactions.   Any configuration about an alkali metal ion will have a very short lifetime, and it is unrealistic to think in terms of rigid solvation shells.   In contrast, some workers consider cesium ions unhydrated in aqueous solution, and the hydration of the chloride ion is often set to zero in order to split into ionic contributions the results from methods which measure hydration numbers for salts.   Solvation numbers determined from radial distribution functions derived from x-ray scattering experiments (*34, 70*) are the most appropriate to the present considerations.   It is unfortunate that there remains some disagreement between the experts about the analysis and interpretation of such data (*71*).   A further distinction should be made between ordered and disordered solvation shells (*46*) depending on the strength of the ion–molecule interaction (*72*).   A trend may occur in a series of increasing ionic radius, but there will certainly be a clear distinction between, for example, a solvated Na^+ and an Al^{3+} ion with octahedrally directed coordination positions for solvating molecules.   Therein lies the problem of whether

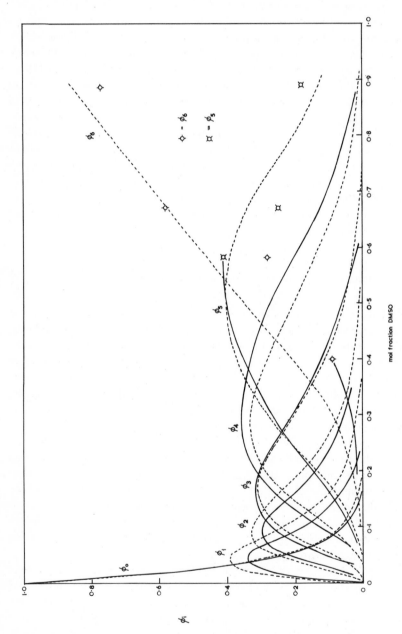

Figure 21. Distribution of solvates in the system Cr^{3+} in H_2O–DMSO (68).
— theoretical curve with $K = 0.2$

to take into account the symmetry of such solvates in a consideration of the likely behavior of ions for monodentate vs. bidentate solvating molecules.

The solvation numbers of ions such as Mg^{2+}, Al^{3+}, and Be^{2+} may be determined by low temperature PMR techniques as mentioned earlier. The solvation number for small spherical ions may be determined in certain circumstances using a titration technique suggested by Van Geet (15). It is based on the competition by water for the solvation sphere of sodium ions in tetrahydrofuran (THF) measured by ^{23}Na shifts. The salt must contain a large anion, which is assumed to be unhydrated during the titration; otherwise a sum of hydration numbers would be determined. The assumptions made by Van Geet are basically those of the present treatment. His apparent constant is for the reverse of the equilibrium of Equation 21 and can be identified as $1/K[P]_F$, where $[P]_F$ is the free THF concentration, effectively constant in the early stages of the titration.

Using Equations 88 and 47 and recognizing that

$$1 - \frac{\delta}{\delta_P} = \frac{1}{n} \Sigma (n - i)\phi_i = \frac{[W]_B}{n[X]_T} \tag{92}$$

where $\Sigma(n - i)\phi_i$ is the average solvation number of W, that is, the number of moles of bound W per mole of solute ion X, then, or alternatively from Equation 65,

$$1 - \frac{\delta}{\delta_P} = \frac{1}{1 + K[P]_F/[W]_F} \tag{93}$$

This can be rearranged, since $Z = [W]_T/[X]_T$, to

$$\frac{Z}{1 - \delta/\delta_P} = n + \frac{K[P]_F}{[X]_T\delta/\delta_P} \tag{94}$$

This is Equation 14 of Van Geet (15) in the present nomenclature. A plot (Figure 22) of the left hand side against $1/\delta$ gives n from the intercept, which can be located with reasonable accuracy if the slope is sufficiently high. For sodium tetraphenylborate, least squares data fitting gave $n = 3.0$ or 3.7 for 0.75 or 0.48 mol/l solutions, respectively. The presence of ion pairs would interfere with this method but it was concluded they were not present in THF. It remains to be seen whether this method can be used for other ions in different solvents.

The second basic assumption of the present treatment concerns the intrinsic chemical shifts or shielding constants of the mixed solvate species. These are proportional to the amounts of co-solvent P which they contain. Some further evidence for the validity of this comes from the studies of Al^{3+} solvates by Delpuech and co-workers (58), who found that when substituting a water molecule by an organic ligand shift changes were approximately additive (3.5 ppm per substitution) as shown in Table VI.

Some of the data analysed using the approach of the present treatment have related to solute chemical shift or uv peak maxima shifts measurements at finite solute concentrations. In order to assume that ion–ion interactions do not in-

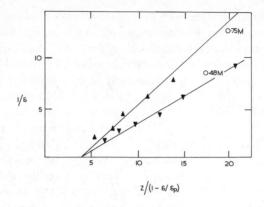

Journal of the American Chemical Society

Figure 22.   Evaluation of n *for* Na^+ *in*
THF–H_2O after Van Geet (15)

Table VI. ^{27}Al Chemical Shifts (ppm) of Al^{3+} Solvates in Nitromethane (58)

No. of Moles Water/Al^{3+}	$P = PO(OMe)_3$	$MePO(OMe)_2$	$HPO(OMe)_2$
6	0	0	0
5	3.7	3.5	3.3
4	6.7	6.8	6.6
3	10.0	10.1	9.1
2	14.0	14.8	14.0
1	17.5	17.5	15.9
0	20.5	20.2	17.7

Journal of the Chemical Society, Chemical Communications

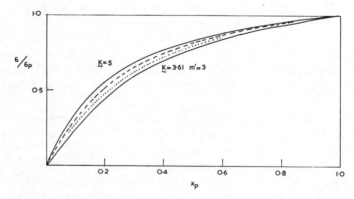

Figure 23.   Effect of solute concentration on the derivation
of **K** *from chemical shift or peak maxima shift data.   upper full*
line m′ = 0 *(infinite dilution),* **K** = 5 *(assumed);* - - - m′ = 1;
····· m′ = 2; *lower full line* m′ = 3 *(the apparent value of* **K** *falls*
to 3.61)

terfere, any solute concentration dependence on the shifts should be checked and, if necessary, extrapolation to zero solute concentration carried out. A second effect of using data relating to finite solute concentrations is that the solvent mole fractions in the bulk solvent may be significantly different from stoichiometric solvent mole fractions because of the quantity of solvent in the ionic solvation spheres. This was discussed in connection with Greenberg and Popov's studies described earlier. Failure to observe this difference especially when **K** is very large or very small can lead to significant errors as illustrated in Figure 23, which was produced by a computer iterative procedure based on the solution of Equation 65 with 95,

$$Y_F = \frac{y_P{}^\circ - \dfrac{m'n}{55.51}(\delta/\delta_P)}{x_W{}^\circ - \dfrac{m'n}{55.51}[1 - (\delta/\delta_P)]} \tag{95}$$

where m' is the aquamolality of the solute ion. The error in $\Delta G_{ps}{}^{\ominus'}$ is about 12% for $m' = 1$ and $\mathbf{K} = 0.5$.

Finally, mention must be made again of the apparent compensation of activity coefficient terms so that the solvent mixture is effectively treated as ideal. The experimental evidence from the constancy of the right hand side of Equation 66 with variation of solvent composition for a variety of solvent mixtures irrespective of the solute is strong. It is supplemented by a simple treatment, ignoring solvent activities, of the directly determined ϕ_i data from the work of King. These lend support to modifications to the theory postulating non-statistical distribution of K_i values or change of solvation number when the right hand side of Equation 64 is not apparently constant. Further light is shed on this problem by treating preferential solvation by the Kirkwood–Buff theory (73) which will be the subject of a paper presented elsewhere (74). There is a discrepancy between the results arising from the choice of standard states in the two treatments, which suggests that a shortcoming of the thermodynamic treatment is the failure to take into account volume changes in the solvation shell as its composition changes. Molecular model considerations suggest that for some systems the volume could change by a factor of up to ten.

Table VII collects the results for all monovalent ion systems for which spectroscopic data are available. Studies of preferential solvation are still at a stage comparable to the establishment of Raoult's and Henry's laws for binary nonelectrolyte solutions. Correlation with thermodynamic data is encouraging for isodielectric solvent systems, but further consideration of the electrostatic terms necessary in the discussion of other systems is required. It is hoped that this present work, which coordinates, correlates, and advances progress made by other workers (7, 18, 19, 20, 45, 46, 61, 62, 66, 67, 68), will stimulate systematic experimental investigations of suitable systems by both spectroscopic and thermodynamic methods.

Table VII. Preferential Solvation[a]

Ion	Solvent mixture	K	k	Ref.
^{7}Li	$H_2O^*-H_2O_2$	0.49	1	2
^{7}Li	$H_2O-DMSO$	0.98	1.38	4
^{7}Li	$DMA-DMSO^*$	1.25	1	4
^{23}Na	$H_2O-H_2O_2$	0.98	1	2
^{23}Na	H_2O^*-MeOH	0.60	1	3
^{23}Na	$AN-H_2O^*$	0.31	0.8	45
^{23}Na	en^*-H_2O	2.3	0.8	45
^{87}Rb	$H_2O-H_2O_2^*$	1.96	1	2
^{87}Rb	H_2O^*-MeOH	0.70	1	32
^{133}Cs	$H_2O-H_2O_2^*$	2.81	1	2
^{133}Cs	$H_2O-DMSO^*$	1.04	1.12	4
^{133}Cs	$H_2O-sulpholane^*$	2.09	1.25	16
^{133}Cs	H_2O^*-MeOH	0.77	1	16
^{19}F	H_2O-D_2O	1.0	1	24
^{19}F	$H_2O-H_2O_2^*$	2.38	1	2
^{19}F	$H_2O-MeOH$	0.96	1	29, 32
^{35}Cl	$H_2O-H_2O_2$	1.0	1	2
^{35}Cl	H_2O^*-MeOH	0.68	1	30
^{35}Cl	H_2O^*-AN	0.07	0.9	44
^{35}Cl	$H_2O-DMSO$	1.0	1	44
^{81}Br	$H_2O-MeOH$	1	1	30
Br$^-$ (CTTS)	$AN-H_2O^*$	0.17	0.75	77
^{127}I	$AN-H_2O^*$			
I$^-$ (CTTS)	$AN-H_2O^*$	0.35	0.75	44
^{127}I	$H_2O-DMSO^*$	2.5	0.5	44
NO_3^- $(n \to \pi^*)$	$DMSO^*-H_2O$	1.68	1.26	1
NO_3^- $(n \to \pi^*)$	$AN-MeOH^*$	0.5	0.7	78

[a] The preferentially solvating component is indicated by an asterisk (*). K is for the formation of a solvated species containing the second mentioned solvent component. CTTS = charge transfer to solvent.

The treatment has wide applicability to coordination chemistry and to other solution phenomena. For example, aspects of it have been applied by Lilley to an explanation of salting-out and salting-in phenomena (75) and to weak interactions in binary nonelectrolyte mixtures in a third solvent (76).

Acknowledgments

This work was supported by grants from Unilever, Ltd., Port Sunlight Research Laboratory, and the Science Research Council. Contributions to this research were made by G. A. Porthouse, I. R. Lantzke, T. H. Lilley, Jennifer M. Thain, and A. D. Covington. Our thanks are given especially to the last two who contributed some of the figures illustrating this review. We are also grateful to J. Clifford, D. G. Hall, W. H. Beck, R. A. Matheson, and R. A. Robinson for valuable and stimulating discussions during the course of the development of the research, to E. L. King (Boulder, Colorado) for drawing our attention to the problems arising from the symmetry of mixed mono-bidentate complexes, to A. I. Popov for letting us see his results prior to publication, and to J. W. Akitt and M. N. S. Hill for help with the experimental NMR aspects.

One of us (A.K.C.) thanks the Royal Society and the University of Newcastle-upon-Tyne for travel grants which made possible the presentation of this paper at the 170th ACS Meeting in Chicago.

Literature Cited

1. Gurney, R. W., "Ionic Processes in Solution," N. Y., McGraw Hill, 1953.
2. Covington, A. K., Lilley, T. H., Newman, K. E., Porthouse, G. A., *J. Chem. Soc. Faraday I* (1973) **69**, 963.
3. Covington, A. K., Newman, K. E., Lilley, T. H., *J. Chem. Soc. Faraday I* (1973), **69**, 973.
4. Covington, A. K., Lantzke, I. R., Thain, J. M., *J. Chem. Soc. Faraday I* (1974) **70**, 1869.
5. Covington, A. K., Thain, J. M., *J. Chem. Soc. Faraday I* (1974) **70**, 1879.
6. Covington, A. D., Covington, A. K., *J. Chem. Soc. Faraday I* (1975) **71**, 831.
7. Grunwald, E., Baughman, G., Kohnstam, G., *J. Am. Chem. Soc.* (1960) **82**, 5801.
8. Popovych, O., *Crit. Rev. Anal. Chem.* (1970) **1**, 73.
9. Kondo, Y., Tokura, N., *Bull. Chem. Soc. Jpn.* (1972) **45**, 818.
10. Schneider, H., Strehlow, H., *Ber. Bunsenges. Phys. Chem.* (1962) **66**, 309.
11. Bjerrum, N., *Z. Phys. Chem.* (1923) **106**, 219.
12. Bjerrum, J., "Metal Ammine Formation in Aqueous Solution: Theory of Reversible Step Reactions," Copenhagen, Haase, 1941.
13. Hertz, H. G., *Angew. Chem. Int. Ed.* (1970) **9**, 124.
14. Akitt, J. W., "N.M.R. and Chemistry," London, Chapman & Hall, 1973, p 17–18.
15. Van Geet, A. L., *J. Am. Chem. Soc.* (1972) **94**, 5583.
16. Halliday, J. D., Richards, R., Sharp, R. R., *Proc. R. Soc.* (1969) **A313**, 45.
17. Kondo, J., Yamashita, J., *J. Phys. Chem. Solids* (1959) **10**, 245.
18. Frankel, L. S., Stengle, T. R., Langford, C. H., *Chem. Commun.* (1965) 393.
19. Smith, M., Symons, M. C. R., *Discuss. Faraday Soc.* (1957) **24**, 206.
20. Frankel, L. S., Langford, C. H., Stengle, T. R., *J. Phys. Chem.* (1970) **74**, 1376.
21. Stengle, T. R., Langford, C. H., *J. Am. Chem. Soc.* (1969) **91**, 4014.
22. Mazurenko, Y. T., *Opt. Spectrosc.* (1972) **33**, 583.
23. Onsager, L., *J. Am. Chem. Soc.* (1936) **58**, 1486.
24. Deverell, C., Schaumburg, K., Bernstein, H. J., *J. Chem. Phys.* (1968) **49**, 1276.
25. Friedman, H. L., Krishnan, C. V., in "Water, a Comprehensive Treatise," F. Franks, Ed., New York, Plenum, 1973, p 83.
26. Covington, A. K., Newman, K. E., Wood, M. *Chem. Commun.* (1972) 1234.
27. Covington, A. K., Thain, J. M., *J. Chem. Soc. Faraday I* (1975) **71**, 78.
28. Feakins, D., Voice, P. J., *J. Chem. Soc. Faraday I* (1972) **68**, 1390.
29. Carrington A., Dravincks, F., Symons, M. C. R., *Mol. Phys.* (1960) **3**, 174.
30. Hall, C., Haller, G. L., Richards, R., *Mol. Phys.* (1969) **16**, 377.
31. Covington, A. K., Newman, K. E., unpublished.
32. Covington, A. K., Thain, J. M., unpublished.
33. Licheri, G., Piccaluga, G., Pinna, G., *Gazz. Chim. Ital.* (1972) **102**, 847.
34. Wertz, D., *J. Solution Chem.* (1972) **1**, 489.
35. Gourary, B. S., Adrian, F. J., *Solid State Phys.* (1960) **10**, 128.
36. Padova, J., *J. Phys. Chem.* (1968) **72**, 796.
37. Frank, H. S., *J. Chem. Phys.* (1955) **23**, 2023.
38. Covington, A. K., Newman, K. E., in preparation.
39. Andrews, A. L., Bennetto, H. P., Feakins, D., Lawrence, K. G., Tomkins, R. P. T., *J. Chem. Soc. A* (1968) 1486.
40. Saloman, M., *J. Phys. Chem.* (1969) **73**, 3299.

41. Cogley, D. R., Butler, J. N., Grunwald, E., *J. Phys. Chem.* (1971) **75**, 1477.
42. Kolthoff, I. M., Chantooni, M. K., *J. Phys. Chem.* (1972) **76**, 2024.
43. Coetzee, J. F., Campion, J. J., *J. Am. Chem. Soc.* (1967) **89**, 2517; Coetzee, J. F., Simon, J. M., Bertozzi, R. J., *Anal. Chem.* (1969) **41**, 766.
44. Stengle, T. R., Pan, Y. C. E., Langford, C. H., *J. Am. Chem. Soc.* (1972) **94**, 9037.
45. Bloor, E. G., Kidd, R. G., *Canad. J. Chem.* (1968) **46**, 3425.
46. Stockton, G. W., Martin, J. S., *J. Am. Chem. Soc.* (1972) **94**, 6921.
47. Covington, A. K., Newman, K. E., unpublished.
48. Covington, A. K., Dickinson, T., Eds., "Physical Chemistry of Organic Solvent Systems," London, Plenum, 1973.
49. Greenberg, M. S., Popov, A. I., *Spectrochim Acta*, (1975) **31A**, 697.
50. Herlem, M., Popov, A. I., *J. Am. Chem. Soc.* (1972) **94**, 1431.
51. Stokes, R. H., Robinson, R. A., *J. Solution Chem.* (1973) **2**, 173.
52. Beck, M. T., "The Chemistry of Complex Equilibria," London, Van Nostrand, 1970, p 262.
53. Toma, F., Villemin, M., Ellenberger, M., Brehamet, L., *16th Ampère Congress* 1970, 1971, pp 317–338.
54. Benson, S. W., *J. Am. Chem. Soc.* (1958) **80**, 5151.
55. King, E. L., private communication, 1975.
56. Flory, P. J., "Principles of Polymer Chemistry," Cornell University Press, Ithaca, N. Y., 1953.
57. Supran, L. D., Sheppard, N., *Chem. Commun.* (1967) 832.
58. Delpuech, J. J., Khaddar, M. H., Peguy, A., Rubini, P., *J. Chem. Soc. Chem. Commun.* (1974) 154.
59. Fratiello, A., Schuster, R. E., *J. Chem. Phys.* (1967) **47**, 1552, 4952.
60. Fratiello, A., Lee, R. E., Schuster, R. E., *J. Phys. Chem.* (1970) **74**, 3726.
61. Green, R. D., Sheppard, N., *J. Chem. Soc. Faraday II* (1972) **68**, 821.
62. Toma, F., Villemin, M., Thiery, J. M., *J. Phys. Chem.* (1973) **77**, 1294.
63. Olander, D. P., Marianelli, R. S., Larson, R. C., *Anal. Chem.* (1969) **41**, 1097.
64. Swinehart, J. H., Taube, H., *J. Chem. Phys.* (1962) **37**, 1579.
65. Nakamura, S., Meiboom, S., *J. Am. Chem. Soc.* (1967) **89**, 1765.
66. Mills, C. C., King, E. L., *J. Am. Chem. Soc.* (1970) **92**, 3017.
67. Kemp, D. W., King, E. L., *J. Am. Chem. Soc.* (1967) **89**, 3433.
68. Scott, L. P., Weeks, T. J., Bracken, D. E., King, E. L., *J. Am. Chem. Soc.* (1969) **91**, 5219.
69. Conway, B. E., Bockris, J. O'M., *Modern Aspects Electro.* (1954) **1**, 47.
70. Narten, A. H., in "Structure of Water and Aqueous Solutions," Luck, W. A. P., Ed., Weinheim, Verlag Chemie, 1974, p 345.
71. Brady, G. W., private communication, 1974.
72. Bernal, J. D., Fowler, R. H., *J. Chem. Phys.* (1933), **1**, 515.
73. Kirkwood, J. G., Buff, F. P., *J. Chem. Phys.* (1951) **19**, 774.
74. Covington, A. K., Newman, K. E., in preparation.
75. Lilley, T. H., to be published.
76. Lilley, T. H., to be published.
77. Blandamer, M. J., Griffiths, T. R., Shields, L., Symons, M. C. R., *Trans. Faraday Soc.* (1964) **60**, 1524.
78. Rotlevi, E., Treinin, A., *J. Phys. Chem.* (1965) **69**, 2645.

RECEIVED July 17, 1975.

The Route to Infinite Dilution

JAN J. SPITZER and H. P. BENNETTO

Department of Chemistry, Queen Elizabeth College (University of London),
Campden Hill, London, W.8

A new theory of electrolyte solutions is described. This theory is based on a Debye–Hückel model and modified to allow for the mutual polarization of ions. From a general solution of the linearized Poisson–Boltzmann equation, an expression is derived for the activity coefficient of a central polarized ion in an ionic atmosphere of non-spherical symmetry that reduces to the Debye–Hückel limiting laws at infinite dilution. A method for the simultaneous charging of an ion and its ionic cloud is developed to allow for ionic polarization. Comparison of the calculated activity coefficients with experimental values shows that the characteristic shapes of the log γ vs. concentration curves are well represented by the theory up to moderately high concentrations. Some consequences in relation to the structure of electrolyte solutions are discussed.

The theory of Debye and Hückel has survived much criticism since the appearance of their celebrated paper (1). This is no doubt because of the simplicity and essential correctness of the limiting laws (2, 3, 4). Nevertheless, many modifications of their treatment have failed to provide a convincing picture of the interionic effects and "structure" in the concentration range of practical importance (5, 6). The work presented here was stimulated by the difficulties of extrapolation encountered in a mixed-solvent emf study (7), and contradicts current trends suggesting that the inadequacy of the DH theory for all but very dilute solutions springs solely from the crudity of the original model. The authors propose a more realistic model that allows the ions to be polarizable and leads to markedly different results.

Modification of the Debye–Hückel Theory

Many authors attribute the limitations of the DH treatment to a failure of the electrostatic approach. We take the view, however, that the noncoulombic aspects have been over-emphasized, and regard with particular caution the inclusion of DH formalism into models that treat the short-range forces explicitly. Some insights have come from various elegant theoretical approaches (8, 9, 10), but these are reviewed elsewhere (2, 3, 6, 11) and will not be dealt with here.

The Structured Ionic Cloud.   We base our treatment on the general solution of the linearized Poisson–Boltzmann equation (LPBE, Equation 1).

$$\nabla^2 \psi = \kappa^2 \psi \tag{1}$$

which is consistent from the electrostatic, thermodynamic, and statistical–mechanical points of view[1]. It is usually written in the radially symmetric form on the assumption that the ionic atmosphere is spherical (2). As shown later, this choice is indeed appropriate for the DH model, but not for ours. The penetrating analysis of Frank and Thompson (16) demonstrated the implausibility of the DH radial solution for an average separation of ions (in water) of about 100 Å; and, following their suggestion that a more satisfactory picture is to be found in a lattice distribution of ions, we show how the concept of a structured ionic cloud can be accommodated within the DH framework.

A well-known result of the DH theory is that the charge dq in the volume element $4\pi r^2 dr$ has a maximum value at a distance κ^{-1} from the central ion (Figure 1a)[2]. The ionic cloud can be *reduced* to a charge on an infinitesimally thin shell placed at a distance $(a + \kappa^{-1})$ from the center of the central ion such that the potential caused by the reduced ionic cloud is

$$\frac{-Q\kappa}{D(1 + \kappa a)},$$

and the total field at $r = a$ is

$$\left(\frac{dV}{dr}\right)_a = -Q/Da^2.$$

A structure may be imposed on the ionic cloud by supposing that dq in the volume element $dv = r^2 \sin\theta \, d\theta \, d\varphi \, dr$ has a finite number, n, of maxima similarly situated at κ^{-1} from the surface of the central ion (Figure 1b). By analogy, this non-radial atmosphere is reducible to a corresponding array of point charges, and this device later enables us to formulate the necessary boundary conditions.

---

[1] The nature of approximations involved in the derivation of the LPBE remains obscure (2), and after the analyses of Fowler (12), Onsager (13), and Kirkwood (14), it appears that no more can be learned about them from a statistical-mechanical argument. Following the early pronouncement of Guggenheim and Fowler (15), supported by other analysis (4), we consider the LPBE the only logical choice for a model in which the ions obey the laws of electrostatics.

[2] All symbols have their usual meaning in the c.g.s. system of units, as given in Ref. 3. The common interpretation that the central ion "sees" its ionic cloud at a distance κ^{-1} away is valid for the point-charge model only. For the DH second approximation the ionic cloud can be reduced to a charge located on a spherical surface at κ^{-1} so as to maintain a constant potential at the surface of the central ion. Therefore, it cannot be replaced by a point charge.

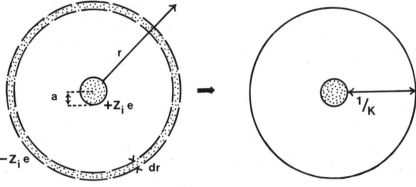

Figure 1a.   Left: a segment of the spherical ionic atmosphere. Right: the reduced ionic cloud of Debye and Hückel.

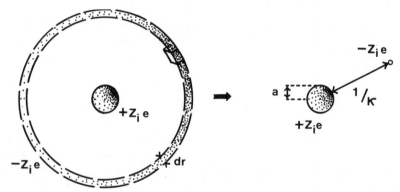

Figure 1b.   Left: the non-spherical ionic atmosphere. The excess charge dq has an absolute maximum in one volume element. Right: the reduced ionic cloud; non-spherical model (n = 1).

Thus the contribution of the structured ionic cloud to the total potential at the surface of the central ion will not be as it is in the DH theory, and because the electrostatic model requires an equipotential surface to be maintained there, a new model is needed. We therefore approximate an ion to a dielectric sphere of radius a, characterized by the dielectric constant of the solvent D, and having a charge Q, residing on an infinitesimally thin conducting surface. This type of model has been exploited by previous workers (17, 18) and may be reconciled with a quantum-mechanical description (18).

The mutual polarization of ions is equivalent to the redistribution of surface charge on the central ion in response to the nonhomogeneous field of the ionic cloud. We need not speculate here on the physical nature of the equipotential surface, except to emphasize that it refers to a solvated species, and one of our central hypotheses is that the self energy of such species is inevitably raised through polarization. This raising of ionic energy is relative to a "ground state" at infinite dilution, and is thus expected to be more pronounced as the concentration increases. The overall effect might be interpreted as a gradual de-solvation of the ions.

Calculations

General Solution of the LPBE. From the preceding discussion it is apparent that a general solution of the LPBE in spherical polar co-ordinates is needed, subject to the usual boundary conditions. Equation 1 can be easily separated into a product solution (*19*) which may be written in the form

$$\psi(r,\theta,\varphi) = r^{-1/2} \sum_{l=0}^{\infty} \sum_{m=0}^{l} K_{l+1/2}(\kappa r) \, P_l^m(\cos \theta) \, [A_{ml} \sin (m\varphi)$$

$$+ \, B_{ml} \cos (m\varphi)] \quad (2)$$

where r, θ, and φ are space coordinates, A_{ml} and B_{ml} are integration constants, $P_l^m (\cos \theta)$ are associated Legendre polynomials, and $K_{l+1/2}(\kappa r)$ are modified Bessel functions (*20*) of the second kind given by Equation 3. This general solution

$$K_{l+1/2}(\kappa r) = \sqrt{\frac{\pi}{2\kappa r}} \exp (-\kappa r) \sum_{\kappa=0}^{l} \frac{(l + k)!}{(l - k)! k!} (2\kappa r)^{-k} \quad (3)$$

satisfies the conditions of continuity and single-valuedness of electric potential and field for a spherical surface, and the condition that the potential vanishes at infinity. A similar solution was obtained in a different context by Scatchard and Kirkwood (*21*).

The conventional viewpoint, which assumes that the ionic atmosphere is spherically symmetric, does not take account of the inevitable effects of ionic polarization. From an analysis of the general solution (*19*), however, it is evident that the ionic atmosphere must be spherically symmetric for nonpolarizable ions, and the DH model is therefore adequate. (Moreover, in very dilute solution polarization effects are negligibly small, and it does not matter whether we choose a polarizable or unpolarizable sphere for our model.) But once we have made the realistic step of conferring a real size on an ion, the ion becomes to some extent polarizable, and the ionic cloud is expected to be nonspherical in any solution of appreciable concentration. Accordingly, we base our treatment on this central hypothesis, that the time-average picture of the ionic solution is best represented with a polarizable ion surrounded by a nonspherical atmosphere. In order to obtain a value for the potential from the general solution of the LPBE we must first consider the boundary conditions at the surface of the central ion.

Boundary Conditions: Solution of the Problem of the Reduced Ionic Cloud. For equipotential conditions the total potential at the surface of the central ion is given from Equation 2 by Equation 4.

$$\psi(a) = a^{-1/2}K_{1/2}(\kappa a)B_{oo} \tag{4}$$

which, considering Equation 3, is seen to be the DH result.   An expression for the electric field on the surface of the central ion can now be obtained with the use of the reduced ionic atmosphere concept.   If the total potential is denoted $V(r,\theta,\varphi)$, the boundary conditions can be written as:

$$\psi(a) = V(a) \tag{5}$$

$$\left(\frac{\partial\psi}{\partial r}\right)_{r=a} = \left(\frac{\partial V}{\partial r}\right)_{r=a} \tag{6}$$

Both these conditions show that the reduced and real ionic atmospheres produce the same effects at the surface of the central ion, and Equation 6 expresses the equality of surface charge distribution which is proportional to the normal component of the electric field.[3]   To write Equations 5 and 6 explicitly and thus determine the integration constants in Equation 2, the classical problem of a charged sphere surrounded by an array of n point charges must be solved.   This is represented in Figure 2, where only one point charge is shown.   The total potential at the observation point $P(r,\theta,\varphi)$ can be written as in Equation 7.

$$V(r,\theta,\varphi) = \frac{Q}{Dr} + V_L(r,\theta,\varphi) + \sum_{i=1}^{n} V^i_{RIA}(r,\theta,\varphi) \tag{7}$$

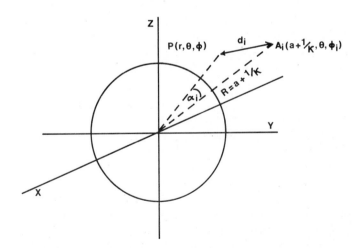

Figure 2.   Coordinate system for the electrostatic problem

[3] Note added in proof: In working out further consequences of the new model, we have found that Conditions 5 and 6 lead to a discontinuity in the potential at $r = a$, the surface of the solvated ion.   We are currently investigating the apparent inconsistency in the formulation of the electrostatic model.

The first term is the potential caused by the spherically distributed charge Q, the second term is the potential caused by redistribution of charge Q in response to the nonhomogeneous field of point charges $-Q/n$, and the third is the ordinary coulombic potential caused by the charges $-Q/n$. The potential $V_L(r,\theta,\varphi)$ is given by the solution of Laplace's equation

$$\nabla^2 V_L = 0 \tag{8}$$

which is:

$$V_L(r,\theta,\varphi) = \sum_{l=0}^{\infty} \sum_{m=0}^{l} r^{-(l+1)} P_m{}^l (\cos\theta) [C_{ml} \sin(m\varphi) + D_{ml} \cos(m\varphi)] \tag{9}$$

where C_{ml} and D_{ml} are integration constants. The potential $V^i{}_{RIA}(r,\theta,\varphi)$ is given by Coulomb's law:

$$V^i{}_{RIA}(r,\theta,\varphi) = -\frac{Q}{nD} \cdot \frac{1}{d_i} \tag{10}$$

With the use of the law of cosines, the function $1/d_i$ can be expanded in Legendre polynomials (22) in the form

$$V_{RIA}{}^i(r,\theta,\varphi) = -\frac{Q}{nD} \sum_{l=0}^{\infty} \left(\frac{r^l}{R^{l+1}}\right) P_l(\cos\alpha_i) \tag{11}$$

which is valid for the region of interest $r \leqslant R = (a + \kappa^{-1})$. When the addition theorem for spherical harmonics (23) is applied, the potential caused by the reduced ionic cloud can be expressed in terms of the coordinates of the points $P(r,\theta,\varphi)$ and $A_i(P_i,\theta_i,\varphi_i)$, where $x_i = \cos\theta_i$:

$$\sum_{i=1}^{n} V_{RIA}{}^i(r,\theta,\varphi) = -\frac{Q}{nD} \sum_{i=1}^{n} \sum_{l=0}^{\infty} \frac{r^l}{R^{l+1}} \Bigg\{ P_l(x) P_l(x_i)$$

$$+ 2 \sum_{m=1}^{l} \frac{(l-m)!}{(l+m)!} P_m{}^l(x) P_m{}^l(x_i) \cos[m(\varphi - \varphi_i)] \Bigg\} \tag{12}$$

In the case of spherical symmetry Equation 12 reduces to the standard DH result

$$\sum_{i=1}^{n} V_{RIA}{}^i(a) = -\frac{Q}{D} \cdot \frac{\kappa}{1 + \kappa a} \tag{13}$$

This result is obtainable directly from electrostatics, and the general solution of the LPBE (Equation 2) determines only the distance at which dq has a maximum value, i.e., $R = (a + \kappa^{-1})$. The distance R gives only the relative maximum, the absolute maximum being arbitrarily specified by the angular coordinates θ_i, φ_i. It will be evident that there are only a few reasonable choices for n, θ_i, and φ_i, depending on the valency of the central ion.

Determination of the Integration Constants A_{ml}, B_{ml}, C_{ml}, D_{ml}. The constants of integration may be determined by a standard technique (22); details are given in Ref. *19*. Conditions 5 and 6 are written explicitly, and multiplied by functions of the type $P_l^m(x) \sin(m\varphi)$, $P_l^m(x) \cos(m\varphi)$ in turn while allowing the indices m and l to vary and so take into account separately the cases where $m = 0, l = 0$; $m = 0, l > 0$; and $m > 0, l > 0$. Each time the multiplication is performed we integrate over the surface of a unit sphere, and on consideration of the orthogonal properties of spherical harmonics we find that Condition 5 determines all the constants C_{ml}, D_{ml} (except D_{oo}) and Condition 6 determines the remaining constants in terms of known C_{ml} and D_{ml}. The final results are given in Table I.

Table I. Constants of Integration for the General Solution of the Linearized Poisson–Boltzmann Equation

$$A_{ml} = \frac{-2Q}{nDa^2 E_{l+\frac{1}{2}}(\kappa a)} \left(\frac{\kappa a}{1 + \kappa a}\right)^{l+1} \left[\frac{(2l + 1)(l - m)!}{(l + m)!}\right] \sum_{i=1}^{n} P_l^m(x_i) \sin(m\phi_i)$$

$$B_{ml} = \frac{-2Q}{nDa^2 E_{l+\frac{1}{2}}(\kappa a)} \left(\frac{\kappa a}{1 + \kappa a}\right)^{l+1} \left[\frac{(2l + 1)(l - m)!}{(l + m)!}\right] \sum_{i=1}^{n} P_l^m(x_i) \cos(m\phi_i)$$

$$B_{oo} = \left(\frac{2\kappa}{\pi}\right)^{\frac{1}{2}} \frac{Q}{D} \frac{\exp(\kappa a)}{1 + \kappa a}$$

$$B_{ol} = \frac{-Q}{nDa^2 E_{l+\frac{1}{2}}(\kappa a)} \left(\frac{\kappa a}{1 + \kappa a}\right)^{l+1} (2l + 1) \sum_{i=1}^{n} P_l(x_i)$$

$$C_{ml} = \frac{2Qa^l}{nD} \frac{(l - m)!}{(l + m)!} \left(\frac{\kappa a}{1 + \kappa a}\right)^{l+1} \sum_{i=1}^{n} P_l^m(x_i) \sin(m\phi_i)$$

$$D_{ml} = \frac{2Qa^l(l - m)!}{nD(l + m)!} \left(\frac{\kappa a}{1 + \kappa a}\right)^{l+1} \sum_{i=1}^{n} P_l^m(x_i) \cos(m\phi_i)$$

$$D_{oo} = 0$$

$$D_{ol} = \frac{Qa^l}{nD} \left(\frac{\kappa a}{1 + \kappa a}\right)^{l+1} \sum_{i=1}^{n} P_l(x_i)$$

The function $E_{l+\frac{1}{2}}(\kappa a)$ is defined by

$$E_{l+\frac{1}{2}}(\kappa a) = \frac{d}{dr} [r^{-\frac{1}{2}} K_{l+\frac{1}{2}}(\kappa r)]_{r=a}$$

The constant D_{oo} vanishes as expected because the potential $V_L(r,\theta,\varphi)$ was defined as due to only the polarization of the central ion; the potential arising from the charge Q itself was taken out as Q/Dr. Thus the "Laplace potential," V_L, results from all the multipoles induced in the surface of the central ion by the structured ionic atmosphere, and vanishes at infinite dilution as required.

The Electrostatic Free Energy

The Charging Process. The calculation of the non-ideal electrostatic energy has sometimes led to difficulties (2), but the use of the LPBE here guarantees that any charging process will give consistent results. A modification of Güntelberg's method is developed below which takes account of the additional effects of the polarization of the ions and their non-spherical distribution, and it is a useful first exercise to consider briefly the DH model in a rigorous manner which recognizes the interactive nature of the central ion and its ionic atmosphere. Thus when the central ion and its ionic atmosphere are charged simultaneously, the total electrostatic energy, W^{DH}, is given by

$$W^{DH} = \frac{1}{2}\frac{Q^2}{Da} + \frac{1}{2}\int_v \rho^{DH}(r)\psi_{ia}{}^{DH}(r)dv + 2W_{int} \qquad (14)$$

The first and second terms on the right are self-energies of the central ion and the ionic atmosphere, and the third contains the interaction energies of the central ion with its ionic cloud and vice versa. According to Green's reciprocal theorem, these energies are equal and are given by

$$W_{int} = \frac{1}{2}Q\psi_{ia}(a) = \frac{1}{2}\int_v \rho^{DH}(r)\psi_{ci}(r)dv \qquad (15)$$

Since the ideal solution is defined by the absence of interactions between ions, the total energy W^{IS} required to charge the central ion and its ionic cloud in an ideal solution is obtained from Equation 14 by setting $W_{int} = 0$:

$$W^{IS} = \frac{1}{2}\frac{Q^2}{Da} + \frac{1}{2}\int_v \rho^{DH}(r)\psi_{ia}{}^{DH}(r)dv \qquad (16)$$

This expression can be formally identified with the chemical potential $\mu_i{}^{IS}$ of the central ion in the ideal solution, where c_i is the ionic concentration on the mole fraction scale:

$$\mu_i{}^{IS} = W^{IS} = \mu_i{}^0 + kT \ln c_i \qquad (17)$$

A similar identification cannot be made for Equation 14 because it contains the interaction energy twice, but on rearranging we derive

$$\mu_i = (W^{DH} - W_{int}) = \frac{1}{2}\frac{Q^2}{Da} + \frac{1}{2}\int_v \rho^{DH}(r)\psi_{ia}{}^{DH}(r)dv + W_{int} \qquad (18)$$

When Equation 16 or Equation 17 is subtracted from Equation 18, we obtain Equation 15, the non-ideal part of the free energy. Thus the self-energies of the central ion and the ionic atmosphere cancel out in the DH model. In the non-spherical model, however, we must expect three contributions to the non-ideal part of the free energy in respect of the DH model: (a) the energy of interaction between the central ion and the ionic cloud will be greater than the DH energy because of the polarization; (b) the internal energy of the central ion will increase because of its polarization; and (c) the self-energy of the ionic cloud will increase because of its structure, which prevents the charge from smoothing out into a spherically symmetric form.

For charging up a central polarizable ion and its ionic cloud we thus derive

$$\mu_i = (W - W_{int}) = \frac{1}{2} \int_s \sigma(a,\theta,\varphi)\psi_{ci}(a,\theta,\varphi)ds$$

$$+ \frac{1}{2} \int_v^s \rho(r,\theta,\varphi)\psi_{ia}(r,\theta,\varphi)dv + W_{int} \quad (19)$$

The chemical potential of the central ion in the ideal solution is again given by Equation 17, and so we obtain

$$kT \ln f_{ci} = \left[\frac{1}{2} \int_s \sigma(a,\theta,\varphi)\psi_{ci}(a,\theta,\varphi)ds - \frac{1}{2}\frac{Q^2}{Da} \right]$$

$$+ \left[\frac{1}{2} \int_v \rho(r,\theta,\varphi)\psi_{ia}(r,\theta,\varphi)dv - \frac{1}{2} \int_v \rho^{DH}(r)\psi_{ia}^{DH}(r)dv \right] + W_{int} \quad (20)$$

We therefore learn that the short-range forces simultaneously affect both the "pure" coulombic interaction energy and the internal energies of the ions themselves, and reach the important conclusion that the nonideality of the solution is not sufficiently defined just by the presence of interactions (unless we also include the interactions of the electrons with the nuclei). That the polarizations should play a part in determining the distribution of ions is a fact which could not be deduced from the original Debye model.

Calculation of the Activity Coefficient. Expression 20 may be rewritten in the form

$$kT \ln f_{ci} = [W_{ci}^{SE} - W_{ci}^{SE(IS)}] + [W_{ia}^{SE} - W_{ia}^{SE(IS)}]$$

$$+ [W_{int} - W_{int}^{DH}] + W_{int}^{DH} \quad (21)$$

where the superscript "SE" denotes "self-energy." All the quantities involved are given quite simply by electrostatics:

$$W_{ci}^{SE} = \frac{1}{2} \int_s \sigma(a,\theta,\varphi) \left[\frac{Q}{Da} + V_L(a,\theta,\varphi) \right] ds \quad (22)$$

$$W_{ci}^{SE(IS)} = \frac{1}{2}\frac{Q^2}{Da} \quad (23)$$

$$W_{ia}^{SE} = \frac{1}{2} \int_v \rho(r,\theta,\varphi)\psi_{ia}(r,\theta,\varphi)dv \quad (24)$$

$$W_{ia}^{SE(DH)} = W_{ia}^{SE(IS)} = \frac{1}{2} \int_v \rho^{DH}(r)\psi_{ia}^{DH}(r)dv \quad (25)$$

$$W_{int} = \frac{1}{2} \int_s \sigma(a,\theta,\varphi)\psi_{ia}(a,\theta,\varphi)ds$$

$$= \frac{1}{2} \int_v \rho(r,\theta,\varphi)\psi_{ci}(a,\theta,\varphi)dv \quad (26)$$

$$W_{int}^{DH} = -\frac{1}{2}\frac{Q^2}{D} \cdot \frac{\kappa}{1 + \kappa a} \tag{27}$$

It is now a matter of straightforward manipulation using the boundary conditions 5 and 6 and the principle of superimposition of potentials to show that the first and third terms in Equation 26 cancel out exactly (19), i.e., the positive increase in internal energy of the central ion is balanced by the negative increment of interaction energy. Hence,

$$kT \ln f_{ci} = [W_{ia}^{SE} - W_{ia}^{SE(DH)}] + W_{int}^{DH} \tag{28}$$

The quantity $W_{ia}^{SE(DH)} = W_{ia}^{SE(IS)}$ is given from the DH theory:

$$W_{ia}^{SE(DH)} = -\frac{1}{4}\frac{Q^2}{D}\frac{\kappa}{(1 + \kappa a)^2} + \frac{1}{2}\frac{Q^2}{D}\frac{\kappa}{1 + \kappa a} \tag{29}$$

The potential $\psi_{ia}(r,x,\varphi)$ is not known explicitly, but Equation 24 can be rewritten as

$$W_{ia}^{SE} = \frac{1}{2}\int_v \rho(r,x,\varphi)\psi(r,x,\varphi)dv - \frac{1}{2}\int_s \sigma(a,x,\varphi)V_{RIA}(a,x,\varphi)ds \tag{30}$$

where the first term is the total energy of the ionic atmosphere (self-energy plus interaction energy) and the second integral is the interaction energy. The surface charge density is given by

$$\sigma(a,x,\varphi) = -\frac{D}{4\pi}\left(\frac{\partial V}{\partial r}\right)_{r=a} = -\frac{D}{4\pi}\left(\frac{\partial \psi}{\partial r}\right)_{r=a} \tag{31}$$

Hence the interaction energy is determined as

$$W_{int} = -\frac{Q^2}{2D}\left(\frac{\kappa}{1 + \kappa a}\right) - \frac{Q^2}{2n^2D}\left(\frac{\kappa}{1 + \kappa a}\right)\sum_{l=1}^{\infty}\left(\frac{\kappa a}{1 + \kappa a}\right)^{2l+1}\left[\sum_{i=1}^{n} P_l(x_i)\right]^2$$
$$-\frac{2Q^2}{n^2D}\left(\frac{\kappa}{1 + \kappa a}\right)\sum_{l=1}^{\infty}\sum_{m=1}^{l}\left(\frac{\kappa a}{1 + \kappa a}\right)^{2l+1}\frac{(l - m)!}{(l + m)!}\left\{\left[\sum_i^n P_l(x_i)\cos(m\varphi_i)\right]^2\right.$$
$$\left. + \left[\sum_i^n P_l(x_i)\sin(m\varphi_i)\right]^2\right\} \tag{32}$$

Here the first term gives the DH result, and all other terms represent the increase of the interaction energy due to multipoles induced on the surface of the central ion. If the ionic atmosphere were spherical, there would be no such induction, and the DH result would be obtained.

The properties of the modified spherical Bessel functions (22) and Equation 33

$$\rho(r,x,\varphi) = -\frac{\kappa^2 D}{4\pi}\psi(r,x,\varphi) \tag{33}$$

are applied to the first term in Equation 30 to give

$$
-W_{ia} = \frac{Q^2}{4D} \frac{\kappa}{(1 + \kappa a)^2} + \frac{Q^2}{2n^2 D} \left(\frac{\kappa}{1 + \kappa a}\right) \sum_{l=1}^{\infty}
$$

$$
\times \frac{(2l + 1) I_{l+1/2}(\kappa a)}{[(l + 1)K_{l-1/2}(\kappa a) + \kappa a K_{l-1/2}(\kappa a)]^2} \times \left(\frac{\kappa g}{1 + \kappa a}\right)^{2l+1} \left[\sum_{i=1}^{n} P_l(x_i)\right]^2
$$

$$
+ \frac{2Q^2}{n^2 D} \frac{\kappa}{1 + \kappa a} \sum_{l=1}^{\infty} \sum_{m=1}^{l} \frac{(2l + 1)I_{l+1/2}(\kappa a)}{[(l + 1)K_{l-1/2}(\kappa a) + \kappa a K_{l-1/2}(\kappa a)]^2} \left(\frac{\kappa a}{1 + \kappa a}\right)^{2l+1}
$$

$$
\times \frac{(l - m)!}{(l + m)!} \left\{\left[\sum_{i=1}^{n} P_l{}^m(x_i) \cos (m\varphi_i)\right]^2 + \left[\sum_{i=1}^{n} P_l{}^m (x_i) \sin (m\varphi_i)\right]^2\right\} \quad (34)
$$

where the integral $I_{l+1/2}(\kappa a)$ is defined by

$$
I_{l+1/2}(\kappa a) = \int_{\kappa a}^{\infty} (\kappa r)K_{l+1/2}{}^2(\kappa r)d(\kappa r) \quad (35)
$$

It is to be expected that this integral is expressible in terms of elementary functions of the type: $\exp (\kappa a)f[\text{polynomial}(\kappa a)]$. Denoting a function $N_{l+1/2}(\kappa a)$ by

$$
N_{l+1/2}(\kappa a) = \frac{(2l + 1) I_{l+1/2}(\kappa a)}{[(l + 1)K_{l-1/2}(\kappa a) + \kappa a K_{l-1/2}(\kappa a)]^2} \quad (36)
$$

we give values for l = 1,2:

$$
N_{l+1/2}(\kappa a) = \frac{3\left(1 + \frac{1}{2} \kappa a\right)}{\left(2 + \frac{2}{\kappa a} + \kappa a\right)^2} \quad (37)
$$

$$
N_{2+1/2}(\kappa a) = \frac{5\left[\frac{1}{2} \kappa a + 3 + \frac{6}{\kappa a} + \frac{3}{(\kappa a)^2}\right]}{\left[\kappa a + 4 + \frac{9}{\kappa a} + \frac{9}{(\kappa a)^2}\right]^2} \quad (38)
$$

Finally, when Equations 34 and 32 are combined, the activity coefficient of the central ion is given by

$$
kT \ln f_i = -\frac{Q^2}{2D} \cdot \frac{\kappa}{1 + \kappa a} + \frac{Q^2}{2D} \cdot \frac{\kappa}{1 + \kappa a} \left\{\sum_{l=1}^{\infty} \sum_{m=0}^{l} \left(\frac{\kappa a}{1 + \kappa a}\right)^{2l+1}\right.
$$

$$
\left. \times \lambda_l{}^m (n,x_i,\varphi_i) \langle 4\rangle [1 - N_{l+1/2}(\kappa a)]\right\} \quad (39)
$$

where

$$\lambda_l^m(n, x_i, \varphi_i) = \frac{(l-m)!}{n^2(l+m)!} \left\{ \left[\sum_{i=1}^{n} P_l^m(x_i) \cos(m\varphi_i) \right]^2 \right.$$
$$\left. + \left[\sum_{i=1}^{n} P_l^m(x_i) \sin(m\varphi_i) \right]^2 \right\} \quad (40)$$

and

$$\langle 4 \rangle = 1 \text{ if } m = 0;$$

$$\langle 4 \rangle = 4 \text{ if } m > 0.$$

For purposes of discussion it is convenient to simplify the above to give Equation 41, in which

$$kT \ln f_i = -\frac{Q^2}{2D} \cdot \frac{\kappa}{1 + \kappa a} \left[1 - \sum_{l=1}^{\infty} \sum_{m=0}^{l} \langle 4 \rangle \pi_l(\kappa a) \lambda_l^m(x_i, \varphi_i) \right] \quad (41)$$

π_l may be called the "polarization term" and λ_l^m is termed a "structure factor," since it defines the ionic cloud through n, x_i, and φ_i.

For 1:1 electrolytes the simplest choice for n is unity (as in Figure 1b) and is shown to be appropriate by comparison with experiment. Thus we have $n = 1$, $x_1 = 1$ ($\cos \theta_1 = 1$, $\theta_1 = 0$), and φ_i can take any value, since $m = 0$ and does not depend on φ. Variants of Equation 39 are easily obtained for other than uni-univalent salts by choosing a structure for the reduced ionic atmosphere in the light of symmetry and chemical intuition. This is illustrated with reference to the divalent ion of a 1:2 electrolyte, where it is reasonable as a first approximation to suppose that the ionic cloud will have two diametrically opposed maxima, each at a distance $1/\kappa$ from the reference ion. It is easy to see that dipoles induced on the central ion by these two charge centers will cancel, as well all higher terms of odd l, but that quadrupolar effects ($l = 2$) and other terms of even l will not. For the structure factor the coordinates of the two maxima in dq are $\theta_1 = 0$ and $\theta_2 = \pi$, while the atmosphere is still symmetrical with respect to the angular coordinate φ. Hence

$$\lambda_l^0(x_i = \pm 1) = \frac{1}{4} \{ [1 + (-1)^l]^2 + 0 \}$$

and $\lambda_1 = 0$, $\lambda_2 = 1$, $\lambda_3 = 0$, $\lambda_4 = 1$ etc.

Results and Discussion

This brief account presents some of the more striking general results of the polarized-ion theory. Since the assessment is incomplete at the time of writing, we deal mainly with activity coefficients of some simple electrolytes, but it will be obvious that there is much scope for application elsewhere. The outline of further developments is, however, intended to be thought-provoking rather than definitive.

Activity coefficients on the molal scale were calculated from Equation 39 by means of a straightforward program containing library sub-routines for evaluation of integrals and modified Bessel functions.

Deviations from the Limiting Law.   In Figure 3 results for 1:1 electrolytes with four different ion-sizes are plotted for hypothetical solutions having $D = 78.54$, $T = 25.0°C$, density = 1.0 and $n = 1$.   The DH limiting-law is also plotted for comparison.   It should be noted that the ionic atmosphere extends outwards from the surface of the central ion, and the parameter a is the mean "effective ionic radius" rather than the "distance of closest approach" of the DH theory.

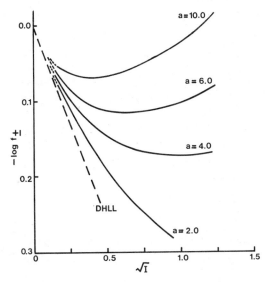

Figure 3.   Activity coefficients calculated from polarized spheres model ($T = 25°C$, $D = 78.54$)

The most notable feature is the prediction of the characteristic shapes and high activity coefficients at high concentrations.   In contrast to the view (*13*) that the deviations arise from some short-range repulsions between ions, the foregoing treatment assigns the positive increments of free energy to the difference between the electrical free energy of an ion in the structured solution and in the ideal solution, which is analogous in symmetry to the DH solution.   The characteristic shapes thus arise from the effects of polarization superimposed on the "hard core" deviations adequately given by the DH treatment for non-polarizable spheres.   The effect of successive terms in the expansion (Equation 39) is illustrated by the figures in Table II.   The terms for $l = 0$ give values of the DH theory (with a), while those for $l = 1, 2, 3 \ldots$ give values relating to the inverse

Table II. Positive Deviations from the Debye–Hückel Limiting-Law
Calculated from the Polarized Spheres Model. Contributions from
Successive Terms

$\Delta \ln f_\pm$ (for $D = 78.54$, $T = 25°C$, $n = 1$)

	Effective Ion Size (Å)					
	2.0		4.0		6.0	
Molality	0.2	0.5	0.2	0.5	0.2	0.5
1 = 0	0.1191	0.2631	0.1941	0.3994	0.2457	0.4828
1	0.0045	0.0166	0.0149	0.0399	0.0240	0.0535
2	0.0003	0.0018	0.0022	0.0100	0.0056	0.0191
3	0.0000	0.0002	0.0003	0.0024	0.0013	0.0069

third, fifth, seventh, etc. powers of the distance r which apparently correspond
to the effects of induced dipoles, quadrupoles, octupoles, etc. For most applications only the first few terms are relevant because the range of validity is at
present restricted to dilute solutions, conveniently defined as the range of concentration for which $\kappa a < 0.5$.[4]

The total potential at the surface of the central ion is the same in our model
as in the DH theory, but the contributions to it are not. The nature of the polarization may be investigated via the quantities V_L or ψ_{ci}, and it is easily shown
that

$$\psi_{ci}(a,\theta,\varphi) = \frac{Q}{Da} + Qa \left(\frac{\kappa a}{1 + \kappa a}\right)^2 \frac{\cos \theta}{Da^2}$$

$$+ 2Qa^2 \left(\frac{\kappa a}{1 + \kappa a}\right)^3 \left(\frac{3 \cos^2 \theta - 1}{2Da^3}\right) + \dots \quad (42)$$

The first term on the right hand side is a pure coulombic potential, while the
induction on the central ion by the structured ionic atmosphere is described by
all the other terms, which disappear at infinite dilution. Thus the second term
is the potential due to a dipole of moment μ given by

$$\mu = Qa \left(\frac{\kappa a}{1 + \kappa a}\right)^2 \quad (43)$$

[4] When the Debye length approaches the effective ionic radius, the form of reduced ionic atmosphere might require modification. The ion nearest to the central ion becomes more and more
responsible for the ionic atmosphere and is increasingly less capable of smearing out into a continuous
charge density. In a further development the reduced atmosphere could be formulated in terms
of a charged sphere rather than a point charge. Such a model might be appropriate for concentrated
solutions and molten hydrates, and would give additional positive contributions to the activity. The
final equation would be expected to contain two ion-size parameters.

Similarly the third term arises from a quadrupole of moment p defined by

$$p = 2Qa^2 \left(\frac{\kappa a}{1 + \kappa a}\right)^3 \tag{44}$$

Figure 4 shows that a cube-root relationship can be recovered from the new model. The cube-root "law" is generally limited to below 0.1 m solutions in water (16), and only one extra term ($l = 1$) is needed in Equation 39 to calculate sufficiently accurate activity coefficients from our theory. In accord with experiment, the theory predicts that the cube-root law must fail both in the DH region where the ion-induced dipole interactions will be negligible, and at higher concentrations, when higher multipoles are induced. In the early stages of this study we had retained a broad view of the possibilities of the quasi-lattice concept (6), but our final conclusion is that the approaches of Frank (5), Bahe (24), and others overestimate the lattice character (25) of the solutions, and that the correct approach is that of Debye and Hückel. It is tentatively suggested, however, that other intriguing empirical "laws" such as the fourth root dependence for solutions

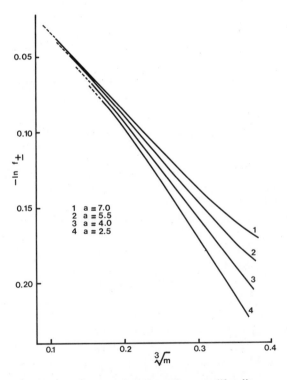

Figure 4. Recovery of the cube-root "law": activity coefficients calculated from polarized sphere model using coulombic and induced dipole terms.

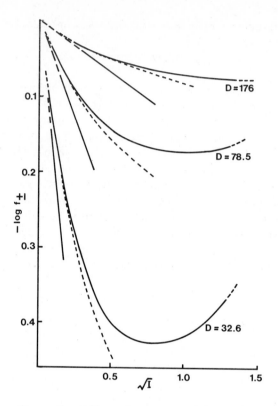

Figure 5.   Effect of variation of dielectric constant

in liquid NH_3 *(26)* might be explained by a superimposition of certain polarization terms, but a more sophisticated treatment would be needed for non-spherical ions.

The effect of changing the dielectric constant is shown for ions of radius 4.0Å in Figure 5, where the values of D correspond to those for the solvents methanol, water, and N-methylpropionamide. The upward turn of $\log \gamma_{\pm}$ at higher concentrations is exaggerated when D is low, which is again in qualitative agreement with some experimental observations *(2)*; the effect is attributed to mutual polarization of ions, which is more effective when the permittivity is low. On an intuitive basis it might have been expected, erroneously, that the polarization would lower the activity coefficients, which could be the reason why ionic polarization has received so little attention in interionic theory.

Multivalent Electrolytes. For 2:1 electrolytes we assume $n = 2$ for the divalent ion and $n = 1$ for the monovalent ion. The latter is a necessary choice which supposes that a monovalent ion "sees" a singly charged ionic cloud represented by a reduced ionic atmosphere with one maximum. This model gives $\log \gamma_{\pm}$ vs. $\sqrt{I}$ plots which are remarkably like the experimental pictures. The form of reduced ionic cloud for higher-valence electrolytes may be deduced from simple symmetry considerations: "triangular" for a trivalent ion, "tetrahedral" for a quadrivalent ion, etc. The same also applies to unsymmetrical electrolytes with a little modification, e.g., for $M_2^{III} X_3^{II}$ the value of n is still three for the M^{III} ion; the second M^{III} ion is assumed to spend equal time associated with each node in the X^{II} atmosphere.

Comparison with Experimental Results for 1:1 Electrolytes. Since precise data are available for HCl and NaCl in water, we choose these electrolytes for detailed analysis, and give summarized results in Table III. For HCl a good match is obtained up to ca. $m = 0.2$ which gives more than twice the range of validity for the DH theory. (Though this is a substantial improvement, it must be admitted that if we look to the fourth place in $\gamma_\pm$, neither the DH second approximation nor the present theory appear to give quite the right shape in the dilute range (*19*).) The results are better for more typical electrolytes like NaCl, where the fit may even extend to above $m = 1.0$. This is probably fortuitous, and a realistic estimate is that the theory usually starts to break down at $\kappa a = 0.5$. Table III also shows some early results for a mixed solvent system (0.50 mole

Table III. Molal Activity Coefficients at 25° C; Comparison of Theory and Experiment

Solvent	Molality	$\gamma_{\pm exp}$	DH Model	Δ^a	Polarized ions model	Δ
Aqueous HCl[b]	0.001	0.966	0.966	0	0.966	0
a = 5.40 Å	0.01	0.905	0.905	0	0.905	0
	0.05	0.830	0.828	−2	0.831	+1
	0.10	0.796	0.786	−11	0.795	−1
	0.20	0.767	0.746	−21	0.762	−5
	0.30	0.756	0.715	−41	0.746	−10
Aqueous NaCl[b]	0.001	0.965	0.965	0	0.965	0
a = 4.14 Å	0.01	0.902	0.902	0	0.902	0
	0.05	0.819	0.817	−2	0.819	0
	0.10	0.778	0.769	−9	0.775	−3
	0.50	0.681	0.645	−36	0.681	0
	1.00	0.657	0.589	−68	0.659	+2
	1.50	0.656	0.554	−102	0.655	−1
NaCl in 0.5 mf	0.0016	0.9202	0.9210	+0.8	0.9210	+0.8
aqueous methanol[c]	0.0144	0.8005	0.8011	+0.6	0.8008	+0.3
a = 3.78 Å	0.0256	0.7554	0.7542	−1.2	0.7547	−0.7
	0.0576	0.6828	0.6795	−3.3	0.6823	−0.5
	0.0784	0.6541	0.6493	−4.8	0.6538	−0.3
	0.1024	0.6290	0.6227	−6.3	0.6293	+0.3

[a] $\Delta = 10^3 (\gamma_\pm \text{ calc} - \gamma_\pm \text{ exp})$. Experimental data from Refs. *2* and *3*.
[b] D = 78.54. Densities taken from International Critical Tables.
[c] D = 49.84. Precise data from cells with transport (*see* J. P. Butler and A. R. Gordon, *J. Am. Chem. Soc.*, (1948) **70**, 2276).

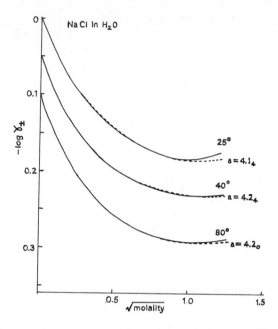

Figure 6.   Fit of theory and experiment at different temperatures for NaCl in water. Solid lines: experimental data from Ref. 2; dotted lines: theory, using values of a as indicated. Lines for 40° and 80° are displaced by 0.05 and 0.1 respectively on the vertical axis.

fraction aqueous methanol), which suggest that the theory is not restricted to the solvent water.

In Figure 6 we show that good theoretical representations are obtained for different temperatures with a small change in the parameter a. On such evidence we might expect to obtain a consistent description of partial molal enthalpies and entropies. However, discussion of these quantities, along with a fuller consideration of solvation aspects, the a parameters, and the dielectric decrements (27, 28), is deferred until a later stage.

Structure, Solvation, and Association in Solutions

From a consideration of the electrostatic free energy alone it is not immediately obvious how the arrangement of ions in the solution as a whole is related to the moving polarized central ion and its structured ionic cloud. It is reasonable to think that the induced multipoles impose restrictions on the mixing of the ions, so that the energy and entropy of configurations described by the structured ionic cloud are lower than in the DH model, as envisaged earlier by Frank and Thompson (16). Such considerations do not lead directly to predictions of the free energy, and the most important conjecture is that although the average arrangement of ions taken over a long time interval is undoubtedly centrosymmetric about a given ion, the structure is nevertheless characterized by the state of polarization for short-lived configurations which may be described by a nonspherical distribution function. Fluctuations in the structure through local changes in enthalpy and entropy are unimportant in themselves, since $\Sigma\delta H$ and $-T\Sigma\delta S$ largely cancel (6), and the free energy depends more critically on an average state of polarization. This reflects the continual strains imposed on the electronic structures of the ions and molecules as they move around in the system. For our model in the present form, importance is attached to the raising in free energy of the ionic atmosphere, since the corresponding effects on the central ion cancel. They need not do so, however, for a more sophisticated version (19).

Solvation Effects. Many previous accounts of the activity coefficients have considered the connections between the solvation of ions and deviations from the DH limiting-laws in a semi-empirical manner, e.g., the Robinson and Stokes equation (3). In the interpretation of results according to our model, the parameter a also relates to the physical reality of a solvated ion, and the effects of polarization on the interionic forces are closely related to the nature of this entity from an electrostatic viewpoint. Without recourse to specific numerical results, we briefly illustrate the usefulness of the model by defining a polarizable cosphere (or primary solvation shell) as that small region within which the solvent responds to the ionic field in nonlinear manner: the solvent outside responds linearly through "mild" Born-type interactions, described adequately with the use of the dielectric constant of the pure solvent. (Our comments here refer largely to activity coefficients in aqueous solution, and we assume complete dissociation of the solute. The polarizability of cations in some solvents, e.g., DMF and acetonitrile, follows a different sequence, and there is probably some ion-association.)

The order of activity coefficients of alkali halogenides (2, 3) in water can now be explained in terms of the properties of polarizable spheres, as follows. The sizes of solvated cations are in the order $H^+ > Li^+ > Na^+ > K^+ > Rb^+ \sim Cs^+$, which is thus also the order of their polarizability (18). The "polarizing power," however, follows the reverse order. The effective sizes of the anions are in the order $F^- > I^- > Br^- > Cl^-$; apparently only F^- is strongly solvated. Remembering that the polarization contribution to the mean ionic activity coefficient is from both kinds of ions, we note that for I^-, Cl^-, and Br^- the order of mean a for cations is $H^+ > Li^+ > Na^+ > K^+ > Rb^+ \sim Cs^+$ and is determined predominantly by the ease of polarization of the solvated cation. For F^- and OH^-, however, the order is reversed, and is determined by the polarizing power of the cations on the large, strongly solvated anions. For H^+, Li^+, Na^+, and K^+ the anionic order is $(F^-) > I^- > Br^- > Cl^-$, as determined by the polarizability of the anions, which therefore give the largest contributions to the mean ionic activity coefficients. However, for Rb^+ and Cs^+ the order is just reversed and is determined by the polarizing power of the anions. In this case the cations will contribute most to the mean activity coefficient. Thus these orders depend on a fine balance between polarizability and polarizing power of both kinds of solvated ions, and will be analyzed in more detail elsewhere.

The explanation given here seems more feasible than that using the concepts of co-sphere overlap (29), and the "structure-making" or "structure-breaking" characteristics of ions (30). This is unsatisfactory because the order shown by the curves does not depend on concentration, and is evident at very low concentrations where the co-spheres would have to be enormous in order to have any overlap. Moreover, the physical nature of the forces arising from such overlap is rather intangible.

A further expectation of the polarized sphere approach is that the labile Gurney co-spheres will change shape in response to the structured ionic cloud. Since the self-energy of the whole entity depends on the solvation and is raised by the polarization, it follows that the ion-solvent interactions will change. The change will become progressively more important as the concentration is increased, and when the co-spheres also start to interpenetrate, the point of breakdown of our model in its present form will probably coincide roughly with the onset of this latter effect. These arguments are important in relation to almost every property of concentrated solutions which depends on ionic sizes, and have been interpreted previously in terms of the structure-breaking or structure-making concepts, e.g., the behavior of osmotic coefficients, upward turn of conductance curves (2), increase in ionic diffusion coefficients (2), incremental decrease in aqueous viscosities (2), etc. The root cause of these effects is the interionic interactions.

Ion Association. One feature of the model is that it requires no vaguely defined "breakdown" of the DH ionic cloud at some critical concentration, but predicts an even gradation of properties over the whole concentration range, or at least up to concentrations where the co-spheres start to overlap.

The structured ionic cloud, which may be compared interestingly with the recent paired-ion descriptions of Stillinger (31) and Fuoss (32), could provide a very reasonable description of the association phenomenon. The original concept of Bjerrum (2, 3, 35) is rather artificial, since in considering the probability of finding an ion near a central ion the potential is assumed to be dominated by the central ion. In our treatment, however, dq is always greatest at κ^{-1} and the sharpening of the maximum in dq at higher concentrations is analogous to the ion-pairing process, and possibly requires no additional description (*see* Footnote 2). Provided that the ions are free to move within a "loose ion-pair" configuration, the DH smoothing process is a reasonable proposition. It is supposed that the activity coefficients will always lie above the limiting-law for the reasons given earlier, while contact ion-pairs may be indicated when the values of $\gamma_{\pm}$ fall below it. This projected picture of ion-pairing is attractively simple and does not appear to be at variance with experimental evidence.

Relation of Electrostatic and Statistical-Mechanical Approaches to Interionic Theory. We believe that the ionic cloud concept is appropriate for the time-averaging superimposition of strong force-fields in an electrolyte solution, but consider it intrinsically different from the statistical averaging process, which deals with the summation of configurations of countable particles and not "smoothed" charges.   Whereas there is general agreement that the two types of description converge for the ideal solution, the repeated failure of attempts to incorporate electrostatic formalism into statistical-mechanical models has been rather sweepingly attributed to the breakdown of the ionic cloud concept.

In seeking to re-establish faith in Debye's intuitive concept, we tend to view that these difficulties derive from the present limitations of statistical mechanics in dealing with what is essentially an electrostatic problem.   Because of the failure to recognize the differences in the two types of averaging process, Debye's critics have illogically attributed the limitations of DH theory to an inadequacy of the LPBE, whereas all the evidence suggests that it is the unlinearized equation which is suspect (*19*).

The LPBE appears to be incapable of providing a description of the interionic effects when straightjacketed by the spherical solution.   Consequently, many investigators have drawn the understandable but incorrect conclusion that because there is something "not quite right" with the LPBE, one must apply some correction such as inclusion of higher terms from the expansion of the exponential in the unlinearized Poisson–Boltzmann equation.   These "corrections" in turn lead to inconsistencies, and to nonsensical conclusions (e.g., "the ionic atmospheres are not additive at higher concentrations").   Such conclusions rest on the validity of the initial assumption that the potential of mean force, $W(r)$, in the Boltzmann equation can be replaced by the electrostatic energy, $z_i e \psi$, which except for the limiting case is inadmissible.   It should be noted, however, that self-consistent treatments of the Poisson–Boltzmann equation have been successfully worked out, and these have been reviewed recently (*35*).   Although we take the view that convincing results have yet to emerge from this direction, we do not reject statistical mechanics as a useful alternative discipline for the treatment of equilibrium properties of electrolyte solutions.

The allowance for polarization in the DH model obviates the need for separation of long-range and short-range attractive forces and for inclusion of additional repulsive interactions.   Belief in the necessity to include some kind of "covolume term" stems from the confused analysis of Onsager (*13*), and is compounded by a misunderstanding of the standard state concept.   Reference to a solvated standard state in which there are no interionic effects can in principle be made at any arbitrary concentration, and the only repulsive or exclusion term required is that described by the DH theory which puts limits on the ionic atmosphere size and hence on the lowering of electrical free energy.   The present work therefore supports the view of Stokes (*34*) that the "covolume term" should not be included in the comparison of statistical-mechanical results with experimental ones.

According to the polarized-ion model, however, the nonideality is not sufficiently described by the coulombic interactions alone, and the chemical state of an ion changes with ionic strength. This poses some problems for the statistical-mechanical approach, since a hypothetical "turning-off" of interactions would leave the ions in a polarized state. In electrostatic terms the energy of a configuration would be given by

$$W = \frac{1}{2} \sum_i^n Q_i U_i \tag{45}$$

where the energy of the i'th ion depends on the charges, shapes, sizes, and positions of all the other spheres, and includes the self-energy.

Concluding Summary

We are forced to reflect that the failure of so many attempts to improve on the DH theory can be attributed to a premature rejection of the DH approach, and a tendency to include extra parameters without proper theoretical foundation. It is surprising that although ionic polarization is emphasized in studies of solvation (36), molten salts (37), and chemistry in general (38), the phenomenon has received little attention in interionic theory. In particular, our attention is drawn to the early work of Fajans and co-workers (39), who first noted the effects of concentration on the ionic molar refractivities of solutions, which were interpreted in terms of a "distorting effect" on the ions. For various reasons the significance of this work has not been appreciated in the field of electrochemistry.

Finally, an important if tentative conclusion emerges concerning the symmetry of distribution functions, which could be relevant to the general theory of polarization and to current ideas about liquid structure. Theoretical analysis in terms of radially symmetric models does not necessarily tell us about the structure of a system, because the short-range forces, by their very nature, have a directional character which is not lost in any averaging process. Is it this averaged directional force which is responsible for "structure" in liquids and solutions?

Acknowledgment

We thank friends and colleagues in several departments of Queen Elizabeth College for help and encouragement, and L. Hough of this department for granting research leave (to J.J.S.).

Literature Cited

1. Debye, P. and Hückel, E., *Phys. Z.* (1923) **24**, 185; "Collected Works of P. J. W. Debye," Fuoss, R. M., Ed., Interscience, New York, 1954.
2. Harned, H. S., Owen, B. B., "Physical Chemistry of Electrolytic Solutions," Reinhold, New York, 1958.
3. Robinson, R. A., Stokes, R. S., "Electrolyte Solutions," Butterworths, London, 1965.
4. Kirkwood, J. G., Poirier, J. C., *J. Phys. Chem.* (1954) **58**, 591; Van Rysselberghe, P., *J. Chem. Phys.* (1933) **1**, 205.

5. Frank, H. S., Thompson, P. T., *J. Chem. Phys.* (1959) **31**, 1086.
6. Bennetto, H. P., *Annu. Rep. Chem. Soc. London* (1973) **70**, 223, and references therein.
7. Bennetto, H. P., Spitzer, J. J., *J. Chem. Soc. Faraday Trans. I.* (1973) **69**, 1491.
8. Mayer, J. E., *J. Chem. Phys.* (1950) **18**, 1426.
9. Friedman, H. L., "Modern Aspects of Electrochemistry," Vol. 6, Bockris, J. O'M., and Conway, B. E., Eds., Butterworths, London, 1971.
10. Rasaiah, J. C., Card, D. N., and Valleau, J. P., *J. Chem. Phys.* (1972) **56**, 3071.
11. Kay, R. L., Ed., "The Physical Chemistry of Aqueous Systems," Plenum, New York, 1973.
12. Powler, R. H., *Proc. Cambridge Philos. Soc.* (1925) **22**, 861; "Statistical Mechanics," Cambridge University Press, 1929, Ch. 13.
13. Onsager, L., *Chem. Rev.* (1933) **13**, 73.
14. Kirkwood, J. G., *J. Chem. Phys.* (1934) **2**, 351.
15. Fowler, R., Guggenheim, E. A., "Statistical Thermodynamics," Cambridge University Press, 1956.
16. Frank, H. S., Thompson, P. T., in "Structure of Electrolytic Solutions," Hamer, W. J., Ed., Wiley, New York, 1959.
17. Schrödinger, E., *2. Phys.* (1921) **4**, 347.
18. Pauling, L., *Proc. R. Soc., London, Ser. A* (1927) **114**, 181.
19. Spitzer, J. J., Ph.D. Thesis, University of London, 1975.
20. Watson, G. N., "Theory of Bessel Functions," Cambridge University Press, 1962.
21. Scatchard, G., Kirkwood, J., *Phys. Z.* (1932) **33**, 297.
22. Byerle, W. E., "Fourier's Series," Ginn & Co., Boston, 1893 (Dover, New York, 1958).
23. Jeans, J., "Mathematical Theory of Electricity and Magnetism," Cambridge University Press, 1948.
24. Bahe, L. W., *J. Phys. Chem.* (1972) **76**, 1062, 1068.
25. Ghosh, J. C., *J. Chem. Soc.* (1918) 113, 449, 707.
26. Baldwin, J., Evans, J., Gill, J. B., *J. Chem. Soc. A* (1971) 3389.
27. Hasted, J. B., "Aqueous Dielectrics," Chapman and Hall, London, 1973.
28. Pottel, R., in "Water: A Comprehensive Treatise," Vol. I, Franks, F., Ed., Plenum Press, New York, 1972, and in "Structure of Water and Aqueous Solutions," Luck, W. A. P., Ed., Verlag Chemie, Weinheim, 1974.
29. Gurney, R. W., "Ionic Processes in Solution," Dover, New York, 1953.
30. Frank, H. S., Wen, W.-Y., *Discuss. Faraday Soc.*, 1957; Frank, H. S., *Z. Phys. Chem. (Leipzig)* (1965) **228**, 364.
31. Stillinger, F. H., Lovett, R., *J. Chem. Phys.*, (1968) **48**, 3858.
32. Fuoss, R. M., *J. Phys. Chem.* (1975) **79**, 525.
33. Justice, J.-C., *J. Chem. Phys.* (1975) **79**, 454.
34. Stokes, R., *J. Chem. Phys.* (1972) **56**, 3382; see also discussions in Ref. *11*, p 40.
35. Owthwaite, C. W., in "Statistical Mechanics," Vol. 2, *Specialist Periodical Report*, The Chemical Society, London, 1975.
36. e.g., Feakins, D., Voice, P. J., *J. Chem. Soc. Faraday Trans. I.* (1972) **68**, 1390.
37. Lumsden, J., "Thermodynamics of Molten Salt Mixtures," Academic Press, New York, 1966, Ch. 4.
38. Pearson, R. G., *Surv. Prog. Chem.* (1969) **5**, 1.
39. Fajans, K., Joos, G., *2. Phys.* (1924) **23**, 1.; Fajans, K., in "Techniques of Chemistry," Vol. 1, "Physical Methods of Chemistry," Weissberger, A., Rossiter, B. W., Eds., Wiley-Interscience, New York, 1971, Pt. 2, 1169.

RECEIVED July 25, 1975.

13

Thermodynamic Study of Hydrobromic Acid in Water–1,2-Dimethoxyethane (Monoglyme) from EMF Measurements

R. N. ROY, E. E. SWENSSON, G. LaCROSS, Jr., and C. W. KRUEGER

Department of Chemistry, Drury College, Springfield, Mo. 65802

Electromotive force measurements of cells of the type:

Pt; H$_2$(g, 1 atm) | HBr(m) in Monoglyme + H$_2$O | AgBr–Ag

at eleven different temperatures ranging from 278.15° to 328.15°K at intervals of 5°K were utilized to evaluate (a) the standard electrode potentials (E°) of the Ag–AgBr electrode in x = 10, 30, and 50 mass percent monoglyme, (b) the mean molal activity coefficients of hydrobromic acid for concentrations ranging from 0.005–0.09 mol kg^{-1}, (c) the relative partial molal enthalpy and heat capacity of HBr in x = 50, and (d) the thermodynamic functions (i.e., ΔG_t^0, ΔH_t^0, and ΔS_t^0) for the transfer of one mole of HBr from the standard state in water to the standard state in x = 10, 30, and 50 mass percent monoglyme. The standard emf was evaluated by using the extended terms of the Debye–Hückel theory with an ion-size parameter of 0.6 nm. The dielectric constants for x = 10, 30, 50, 70, 90, and 100 mass percent monoglyme were measured at the temperatures under investigation. The significance of the results has been discussed in terms of ion-solvent interactions.

The behavior of electrolytes in aqueous organic mixtures, particularly those consisting of dipolar aprotic solvents (1, 2, 3, 4, 5, 6) has long been a subject of considerable importance. Interest in dipolar aprotic solvent–water mixtures arises, in part, from the recent studies of tetrahydrofuran–water mixtures (7), which involved ion-solvation and proton bonding. Because of the scarcity of experimental thermodynamic data concerning the effects of medium changes on the thermodynamic properties of hydrobromic acid, three different compositions of the binary monoglyme–water mixtures were chosen, which contained 10, 30, and 50 mass percent monoglyme, respectively. Other solvent systems of this type are sometimes used in the investigation of the acid–base properties of compounds which are slightly soluble in water, in the spectrophotometric determination of the dissociation constant, pK_a, of m-nitroaniline (8), aniline, and substituted anilines (9), and in studies of chemical kinetics. To meet these requirements, the standard electrode potential of the silver + silver bromide electrode must be known. The silver + silver bromide electrode is highly reproducible (10) and, because of its low solubility, it performs better than the silver + silver chloride electrode, particularly for the determination of the dissociation constants of the nitrogen bases (11).

Measurements of the emf of cells of the type:

$$Pt; H_2(g, 1 \text{ atm}) \,|\, HBr(m) \text{ in monoglyme} + H_2O \,|\, AgBr\text{-}Ag \qquad (I)$$

were made at eleven temperatures extending from 278.15° to 333.15°K and for ten molalities of hydrobromic acid in the nomimal range from 0.01 to 0.1 mol kg^{-1} in 10, 30, and 50 mass percent monoglyme. The standard emf for cells of type I was evaluated by the use of the extended terms (12) of the Debye–Hückel theory. This method of extrapolation is essentially the same as that recently used by Roy, Robinson, and Bates (10) and Roy, Swensson, and LaCross (7). Activity coefficients and the relative partial molal enthalpy and heat capacity of HBr have been derived. The present investigation was undertaken with the intent of evaluating the standard electrode potential of the Ag–AgBr electrode, which will permit the calculation of the dissociation constant of glycine in 50 mass percent monoglyme, and will allow the determination of the standard thermodynamic functions (Gibbs energy, enthalpy, and entropy) for the transfer of one mole of HBr from the aqueous standard state to the standard state in the mixed solvent. The densities and the vapor pressures for $x = 10$, 30, and 50 mass percent monoglyme, and the dielectric constants for $x = 10, 30, 50, 70, 90,$ and 100 percent monoglyme, were measured and are herein reported.

Similar emf measurements on hydrochloric acid solutions in monoglyme–H_2O mixtures containing 8.68, 17.81, 46.52, and 67.03 mass percent monoglyme at 278.15°, 288.15°, 298.15°, and 308.15°K were reported earlier (1), and are shown here for the purposes of comparison of both strong electrolytes. The results of the present study are discussed in terms of hydrogen bonding, as well as preferential solvation of the two solutes (hydrogen ion and bromide ion) by the molecules of the two solvent species, water and monoglyme. Moreover, the results are compared with the similar parallel data (1) for hydrochloric acid in 50 mass percent monoglyme.

Experimental Procedure and Preparation of Solutions

The stock solution, ca. 0.3 mol dm^{-3}, of hydrobromic acid was prepared from a twice-distilled sample of the hydrobromic acid. Its bromide content was determined gravimetrically as silver bromide. Triplicate runs agreed to within ±0.02%. The silver + silver bromide electrode was of the thermal type, prepared by heating twice recrystallized silver bromate (10 mass percent) and silver oxide (90 mass percent) at a temperature of 820°K. The preparation of the silver oxide, the preparation of the hydrogen electrodes, the design of the cell, the purification of the hydrogen gas, and other experimental techniques, have been described earlier (13, 14, 15). The water bath in which the cells were immersed was controlled to within 0.02°K.

Monoglyme (Fisher Certified) was purified in the manner described by Wallace and Mathews (16). The middle fraction of the second distillate was subsequently used in the preparation of the cell solutions. Dry nitrogen was bubbled through the distillation flask during the distillation process. The purity of the middle fractions was verified by gas chromatography. The mixed solvents were prepared by weight dilution methods by diluting the aqueous stock solution of HBr with a known amount of monoglyme and doubly distilled water. Vacuum corrections were applied to all weighings, and weight burets were employed when necessary. Dissolved air was always removed by bubbling purified hydrogen gas into the solutions before the cells, which were fitted with triple saturators, were filled. The molality of the acid in all the solutions reported is correct to within ±0.03% and the monoglyme content of the solutions is accurate to within ±0.02%.

Total vapor pressures (p) for the mixed solvents over the temperature range 278.15°–328.15°K are essential in order to correct the emf data to a hydrogen partial pressure of one atmosphere. The boiling temperature method (1, 17), which utilizes a ballast bulb, a closed-tube manometer, and a triple-neck distillation flask, was employed, and the measurements of the vapor pressures for the mixed solvents under investigation were made in the temperature range of 308°–350°K. Extrapolation and interpolation procedures were based on the linear plots of $\log_{10} p$ as a function of $1/T$, where T is the thermodynamic temperature. As a further verification, the vapor pressures of 50 mass percent methanol–water mixtures were measured and were found to be in good agreement with the literature data (10).

The densities ρ^0 of the mixed solvents required to calculate the parameters A and B of the Debye–Hückel equation were measured with a pycnometer of about 25-cm^3 capacity. Duplicate determinations were always made, and the values agreed to within ±0.005%.

The dielectric constants ϵ, which are also needed to evaluate the Debye–Hückel constants A and B, were measured from 283.15° to 318.15°K at intervals of 5°K, using spectrograde acetone and freshly prepared, doubly distilled water as reference materials. The Janz–McIntyre bridge (18) with Balsbaugh conductivity cell (model 2TN50) having a cell constant of 0.00177 cm^{-1} was used. The temperature of the bath containing the cells was regulated to within 0.02°K. The values of ϵ over the temperature range under investigation were obtained by the method of extrapolation and interpolation, which utilized the straight-line plots of $\log_{10} \epsilon$ vs. $1/T$. The values of the p, ρ^0, ϵ, A, and B are presented in Tables IV, V, and VI for 10, 30, and 50 mass percent monoglyme, respectively, whereas for $x = 70$, 90, and 100, the data for the dielectric constant are included in Table X. The data for the dielectric constant at 298.15°K are plotted in Figure 1 as a function of the mass percent of monoglyme.

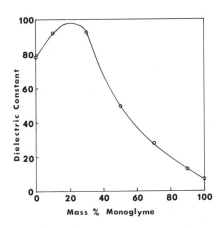

Figure 1. Plot of dielectric constant vs. mass percent monoglyme for 0, 10, 30, 50, 70, 90, and 100 mass percent monoglyme–water mixtures at 298.15°K

The values of the emf were corrected to a partial hydrogen pressure of one atmosphere (101.325 kp$_a$) and are given in Tables I, II, and III (*see* Appendix *for all tables*) for the respective solvent compositions. Each value in the table represents the average of two duplicate cells, prepared for each molality of hydrobromic acid. Typically, the lowest and highest emf values for these duplicate cells differed by no more than 0.14 mV. The emf of the cells was always measured first at 298.15°K, then at descending temperatures to 278.15°K, again at 298.15°K, finally ascending to 328.15°K, and ending with a measurement at 298.15°K once more. The emf data are reliable and stable, as evident from the data at 298.15°K, which were recorded a total of three different times; namely, at the start, the middle, and the end of each temperature run. On the average, these three values agreed to within 0.05 mV. The bias potential of the silver–silver bromide electrode was always within 0.05 mV. Measurements of the emf were made by means of a Leeds and Northrup type K-3 potentiometer, standardized against an Eppley standard cell, and equipped with a Leeds and Northrup D.C. null detector (model 9829), using a sensitivity of 25 μV. The vapor pressures of the solutions were assumed to be the same as those for the pure solvents at the experimental temperatures.

As a final check on the reliability of the data and the performance of the cells, the emf's were measured in 0.01 mol-kg^{-1} aqueous HBr solutions at

298.15°K. The pressure-corrected emf data is 0.31276 V, which is in satisfactory accord with the literature value $(9, 11)$ of 0.31272 V.

Results

The standard emf E° of the cell was determined by means of an extrapolation technique involving a function of the measured emf E (which was measured experimentally), taken to the limit of zero ionic strength I. A linear function of I was observed when the Debye–Hückel equation (in its extended form) (12) was introduced for the activity coefficient of hydrobromic acid over the experimental range of molalities m. With this type of mathematical treatment, the adjustable parameter became a^0, the ion-size parameter, and a slope factor β. This procedure is essentially the same as that used in our earlier determinations $(7, 10)$ although no corrections of E° for ion association were taken into account (ϵ = 49.5 at 298.15°K).

From the Nernst equation for cells of type I, it is obvious that

$$E^0 = E + (2RT/F) \log_e(m\gamma_{\pm}/m^0) \tag{1}$$

in which $m^0 = 1$ mol-kg^{-1}. Furthermore, the extended terms indicated by D (12) can be written in terms of $-\log_{10} \gamma_{\pm}$.

$$- \log_{10} \gamma_{\pm} = Am^{1/2}/(1 + Ba^0m^{1/2}) - \beta m/m^0$$
$$+ \log_{10} (1 + 0.002m \langle M \rangle) - D/\log_e 10 \tag{2}$$

in which A and B are the Debye–Hückel parameters, converted to a molality basis by multiplication by the square root of the solvent density. The values of these constants, which are listed in Tables III, IV, and V, are functions of the thermodynamic temperature T and the dielectric constant ϵ, as well as of the solvent density ρ^0. The mean molar masses $\langle M \rangle$ for x = 10, 30, and 50 mass percent monoglyme are 0.01958, 0.02370, and 0.03003 kg-mol^{-1}, respectively. The contribution arising from the extended terms towards the value of $-\log_{10} \gamma_{\pm}$ in the case of a 1-1 electrolyte may be given by

$$D = 10^{-3}q^3f_3(X) + 10^{-5}q^5f_5(X), \tag{3}$$

in which $q = N_A e^2/\epsilon RTa^0$, where e is the Gaussian charge on the proton, N_A is Avogadro's constant, and $X = \kappa a^0$; that is, $Ba^0m^{1/2}$. To expedite the calculations, the fifth-degree polynomials determined in an earlier investigation (20) were used to obtain $f_3(X)$ and $f_5(X)$:

$$f_3(X) = 0.06024 - 2.682X + 4.935X^2$$
$$- 3.226X^3 + 0.5713X^4 + 0.1005X^5, \tag{4}$$

and

$$f_5(X) = 0.05674 - 3.555X + 18.38X^2 - 32.99X^3 + 25.53X^4 - 7.303X^5 \tag{5}$$

Substitution of the appropriate constants, along with simultaneous combination of these results with those of Equation 3, yield (upon rearrangement)

$$E^{0\prime} = E^0 - 2k\beta m/m^0 = E + 2k \log_{10}(m/m^0) - 2k\{Am^{1/2}/(1 + Ba^0m^{1/2})$$
$$+ \log_{10}(1 + 0.002\langle M\rangle m/m^0) - 0.4343D\} \quad (6)$$

in which k is shown replacing the term $(RT \log_e 10)/F$.

Values of the right-hand side of Equation 6 (that is, those denoting the term $E^{0\prime}$) are expected to be linear in terms of m, when a suitable value for the ion-size parameter is chosen. Several different values for a^0 (e.g., 0.2, 0.4, 0.6, and 0.8 nm) were tested, and the computer calculations showed that the deviations of $E^{0\prime}$ vs. m from linearity were at a minimum when a^0 is equal to 0.6 nm. These results are shown in Table VIII. By means of Equation 6, the intercepts of the extrpolation lines (E^0) can be found and the slopes ($-2k\beta$) computed. The values of the standard emf are summarized in Tables IV, V, and VI for $x = 10, 30$, and 50 mass percent monoglyme, together with the standard deviations of the intercepts (E^0). For $a^0 = 0.6$ nm, the values of β are entered in Table IX.

Table VII summarizes the results of E_N^0 (on the mole fraction basis) and E_c^0 (on the concentration scale), which have been computed from the relations

$$E_N^0 = E^0 - (2RT \log_e 10/F) \log_{10}(\text{kg-mol}^{-1}/\langle M\rangle) \quad (7)$$

and

$$E_c^0 = E^0 + (2RT \log_e 10/F) \log_{10}(\rho^0/\text{g-cm}^{-3}) \quad (8)$$

The mean ionic activity coefficients of hydrobromic acid at round molalities (calculated by means of Equation 2) are summarized in Tables XI, XII, and XIII for $x = 10, 30$, and 50 mass percent monoglyme. Values of $-\log_{10}\gamma_\pm$ at round molalities from 0.005 to 0.1 mol-kg^{-1} were obtained by interpolating a least squares fit to a power series in m which was derived by means of a computer. These values at 298.15°K are compared in Figure 2 with those for hydrochloric acid in the same mixed solvent (1) and that for hydrobromic acid in water (21). The relative partial molal enthalpy $(H_2 - H_2^0)$ can be calculated from the change in the activity coefficient with temperature, but we have used instead the following equations:

$$E/V = a + b(T/K - 298.15) + c(T/K - 298.15)^2 \quad (9)$$

and

$$E^0/V = a' + b'(T/K - 298.15) + c'(T/K - 298.15)^2 \quad (10)$$

The values of the constants in Equations 9 and 10 are not listed, and the experimental values of $(H_2 - H_2^0)$ for $x = 50$ were calculated from the equation:

$$(H_2 - H_2^0)/\text{cal}_{th}\,\text{mol}^{-1} = 23061\{(a' - a) + 298.15(b - b')$$
$$+ (c - c') \times (T/K + 298.15)(T/K - 298.15)\} \quad (11)$$

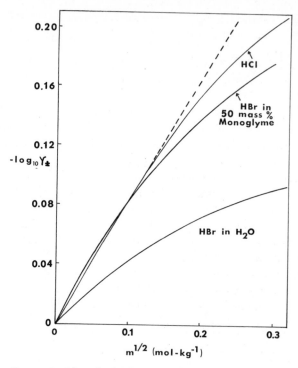

Figure 2. Plot of $-\log_{10} \gamma_\pm$ *vs.* $m^{1/2}$ *for hydrobromic acid in 50 mass percent monoglyme, hydrochloric acid in 50 mass percent monoglyme, and hydrobromic acid in water at 298.15°K. The dashed line represents the Debye–Hückel limiting slope.*

The values of $(H_2 - H_2^0)$ for rounded molalities at 298.15°K were obtained from the plots of $(H_2 - H_2^0)$ against $m^{1/2}$. The smooth curve was drawn through all the experimental points; the average deviation from the smoothed curve was 15 cal$_{th}$ mol^{-1} for 50 mass percent monoglyme. These values are presented in Table XIV and are higher than those found for hydrochloric acid in the same mixed solvents (1). The values of $(H_2 - H_2^0)$ in aqueous medium at 298.15°K are given for direct comparison. The relative partial molal heat capacity $(C_p - C_p^0)$ was calculated by the following formula:

$$(C_p - C_p^0)/\text{cal}_{th} \text{ K}^{-1} \text{ mol}^{-1} = 46122(c - c')T \tag{12}$$

The values for these quantities at $x = 50$ mass percent monoglyme at 298.15°K are summarized in Table XIV and are compared with those for hydrobromic acid in water (21).

Discussion

The values of E_m^0 (on the molal basis) can be expressed by the equations obtained by using the least squares method:

for $x = 10$, $E_m^0 = 0.07137 - 4.608 \times 10^{-4}(t - 25)$
$$- 2.954 \times 10^{-6}(t - 25)^2 \quad (13)$$

for $x = 30$ $E_m^0 = 0.06696 - 6.860 \times 10^{-4}(t - 25)$
$$- 8.259 \times 10^{-6}(t - 25)^2 \quad (14)$$

and, for $x = 50$,

$$E_m^0 = 0.05108 - 1.117 \times 10^{-3}(t - 25) - 1.618 \times 10^{-6}(t - 25)^2 \quad (15)$$

The standard deviations of regression between the experimental data and the values obtained from the above equations are 0.03, 0.03, and 0.12 mV for $x =$ 10, 30, and 50 mass percent monoglyme, respectively.

Some comments should be made in regard to the arbitrary choice of the ion-size parameter, a^0, in the determination of E_m^0. The suitability of the value of the ion-size parameter was based on the standard deviation for regression of the experimental $E^{0\prime}$ in Equation 6 as a function of m using a^0 equal to 0.2, 0.4, 0.6, and 0.8 nm, successively. Table VIII indicates that $a^0 = 0.6$ nm gives the best linear fit, and consequently $a^0 = 0.6$ nm was used at all temperatures. It should be mentioned, however, that the standard deviation of regression is not a reliable guide to the best choice of the ion-size parameter.

We are concerned here with the solvent effects on the equilibrium behavior of acids and bases in dipolar aprotic solvents, such as monoglyme in $x = 10, 30$, and 50 mass percent mixed solvent compositions. The following reaction is of particular interest in such a study:

½ H_2 + AgBr $\rightleftarrows$ HBr (*in 10, 30, and 50 mass percent monoglyme*) + Ag
$$(II)$$

The standard thermodynamic functions for the transfer process:

HBr (*in H_2O*) = HBr (*in respective monoglyme compositions*)

can be derived from the standard emf E^0 of the cell in water (*15*) and in water + 10, + 30, and + 50 mass percent monoglyme, together with the change of E^0 with the temperature. The thermodynamic constants for the transfer from water to the mixed solvents of various compositions can be derived by the relations

$$\Delta G_t^0 = F(E_{N,w}^0 - E_N^0), \quad (16)$$

$$\Delta S_t^0 = -d\Delta G_t^0/dT, \quad (17)$$

and

$$\Delta H_t^0 = \Delta G_t^0 + T\Delta S_t^0, \quad (18)$$

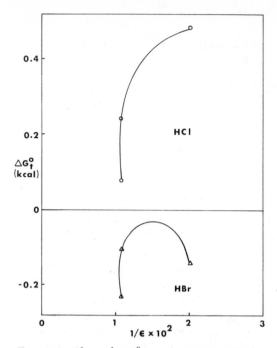

Figure 3. Plots of ΔG_t^0 vs. $1/\epsilon$ at 298.15°K for hydrochloric acid and hydrobromic acid in monoglyme–water mixtures

where the subscript w refers to water and the transfer function is indicated by the subscript t. The values of ΔG_t^0 and related thermodynamic functions are listed in Table XV on the mole fraction basis. The transfer entropy ΔS_t^0 can be obtained by the application of Equation 17 to the temperature variation of E^0 in water as well as in the respective mixed solvent compositions. The thermodynamic quantities refer to transfers between HBr (*in H₂O*) and HBr (*in 10, 30, and 50 mass percent monoglyme*) when each is in its standard state on the mole fraction basis, which is used to eliminate any energy changes due to concentration changes. The Gibbs energy of transfer is an important index of the differences in interactions of the ions (for example, H^+ and Br^-) and the solvent molecules in the two different media. The values of ΔH_t^0 are the same on each scale (i.e., molality and mole fraction), but those of ΔG_t^0 and ΔS_t^0 became larger on the molal scale. For $x = 10$, the value of $\Delta G_t^0 = -7$ cal-mol^{-1}, compared with -106 cal-mol^{-1} on the mole fraction scale. The difference in free energy between the two scales can be calculated by the equation (*16*)

$$2RT \log_e \overline{M}/M_{H_2O} \tag{19}$$

where M_{H_2O} indicates the molecular weight of water. The values of ΔG_t^0 (298.15°K) appear to be negative for $x = 10, 30,$ and 50 mass percent monoglyme.

Thus hydrobromic acid is more strongly stabilized in monoglyme + water mixtures than in water.   Moreover, the negative values of ΔG_t^0 for the experimental mixed solvent compositions support the view that water is less basic than the mixed solvents, if it is assumed that the hydration of a larger bromide ion in aqueous solution is negligible, although our data indicates that the hydration number of chloride in aqueous solution might not be zero (22).

Since anions are much less solvated in dipolar aprotic solvents (23) than in water, the hydrogen ion will be more highly solvated in the mixed solvent because it is preferentially solvated by monoglyme in the monoglyme–water mixtures rather than in the pure aqueous medium.   The selective solvation is an important factor in an understanding of solute–solvent interactions in mixed solvent systems. Unfortunately, the detailed compositions of the primary solvation shell and the secondary mode of solvation (ion–dipole interaction) in mixed solvents are not yet clearly understood.

The error in ΔG_t^0 for all the solvent compositions may be estimated to be within ±8 cal-mol^{-1}.   In Figure 3, it is interesting to note from the plots of ΔG_t^0 against $1/\epsilon$ for HCl (6) and HBr in the monoglyme–water mixtures, that for HBr the curve is parabolic, whereas for HCl it is exponential.

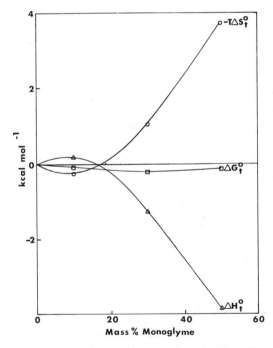

Figure 4. Plots of the various thermodynamic functions of hydrobromic acid vs. mass percent monoglyme in 10, 30, and 50 mass percent monoglyme–water mixtures at 298.15°K

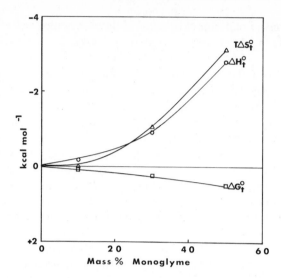

Figure 5. Plots of the various thermodynamic functions of hydrochloric acid vs. mass percent monoglyme in 10, 30, and 50 mass percent monoglyme–water mixtures at 298.15°K

As can be seen from Table XV, the transfer enthalpy and entropy for $x = 10$ appear to be positive at all temperatures. The positive entropy of transfer of HBr from water to monoglyme–water can probably be attributed to a greater structure breaking by HBr in monoglyme–water than in water. Consequently, the degree of solvent orientation is less when $x = 10$ than in aqueous medium. The structure-breaking process by hydrobromic acid when $x = 10$ is endothermic. The negative values of ΔH_t^0 and ΔS_t^0, for $x = 30$ and 50 suggest than when monoglyme is added to a highly polar water molecule, the mixed solvent becomes less associated than pure water itself. Thus, the net amount of order created by both ions, the hydrogen ion and the bromide ion, is more in the mixed solvent than in water. Hydrobromic acid thus "breaks down more structure" (24) in pure water than in monoglyme + water mixtures.

Figures 4 and 5 show the experimental points of ΔG_t^0, ΔH_t^0, and $-T\Delta S_t^0$ for hydrobromic acid, and ΔG_t^0, ΔH_t^0, and $T\Delta S_t^0$ for hydrochloric acid, as a function of mass percent monoglyme at 298.15°K, respectively. All structure-breaking processes, including desolvation of the ions, are endothermic. In Figure 4, ΔH_t^0 shows the endothermic peak in the water-rich region, with a maximum at about $x = 10$. In the monoglyme-rich region, the values of $-T\Delta S_t^0$ are predominant, whereas for ΔG_t^0, the graph is almost linear. The process becomes exothermic before the monoglyme concentration reaches 20 mass percent. As can be seen from Figure 5, the plot of ΔG_t^0 for the transfer of hydrochloric acid from water to monoglyme–water is almost linear and positive, which is just the opposite of that for hydrobromic acid. One reason for this anomalous behavior could be the greater hydrating power of the chloride ion as opposed to the bromide ion. The values of $T\Delta S_t^0$ and ΔH_t^0 show a trend similar to that for hydrobromic acid.

Finally, it is important to note from Figure 1 and Table X that the value of the dielectric constant of pure monoglyme at 298.15°K is 7.20, as compared to the literature value of 7.20 (25). Figure 1 also points out the interesting point that there is a maximum at $x = 20$. This means that the polarity of the mixed solvent at $x = 20$ is enhanced, possibly because of the extensive hydrogen bonding. It may be expected that a monoglyme molecule will induce a water molecule at close proximity and act over a series of water molecules, so that there will be displacements of charges. As a result, the water molecules attached to mono-glyme will be more basic than those in pure water. The following diagrams should provide more insight with regard to the changes in the structural aspects of the solvent.

The seven-membered hydrogen-bonded picture might play an important role in an explanation of the present case, just as the staggered form may also be significant. Similar studies on 1,1-dimethoxymethane–water and 1,3-dim-ethoxypropane–water mixtures are being actively pursued in an effort to sub-stantiate the above-mentioned view.

The activity coefficients of hydrobromic acid in the mixed solvents are lower, as expected, than those in water (20). Hydrobromic acid completely dissociates in the mixed solvents ($\epsilon = 49.5$ at 298.15°K for the 50 mass percent monoglyme) under investigation. Figure 2 clearly indicates that at a particular molality, the stoichiometric activity coefficient of hydrochloric acid is lower than that of hy-drobromic acid in the same mixed solvent, and the heat capacity changes ($C_p - C_p^0$) also suggest that there are no ion-pair formations.

Appendix

Table I. Electromotive Force E of the cell: Pt/H$_2$(g)/HBr(m) in from 278.15°

T/K—273.15 (m/mol kg^{-1})	5	10	15	20	25
0.10000	0.19765	0.19797	0.19859	0.19847	0.19840
0.09000	0.20262	0.20301	0.20364	0.20359	0.20357
0.08000	0.20816	0.20867	0.20932	0.20951	0.20955
0.07000	0.21424	0.21487	0.21562	0.21551	0.21575
0.06000	0.22172	0.22235	0.22325	0.22356	0.22378
0.05000	0.23004	0.23082	0.23167	0.23210	0.23242
0.04000	0.24022	0.24105	0.24228	0.24279	0.24321
0.03000	0.25365	0.25483	0.25609	0.25689	0.25760
0.02000	0.27236	0.27386	0.27530	0.27649	0.27754
0.01000	0.30440	0.30649	0.30847	0.31028	0.31192

Table II. Electromotive Force E of the cell: Pt/H$_2$(g)/HBr(m) in from 278.15°

T/K—273.15 (m/mol kg^{-1})	5	10	15	20	25
0.10000	0.19838	0.19862	0.19856	0.19794	0.19694
0.09000	0.20327	0.20348	0.20347	0.20286	0.20185
0.08000	0.20835	0.20876	0.20899	0.20820	0.20729
0.07000	0.21440	0.21478	0.21494	0.21445	0.21357
0.06000	0.22136	0.22192	0.22212	0.22167	0.22086
0.05000	0.22953	0.23017	0.23037	0.23016	0.22940
0.04000	0.23957	0.24033	0.24068	0.24060	0.24005
0.03000	0.25255	0.25358	0.25415	0.25424	0.25403
0.02000	0.27088	0.27219	0.27315	0.27346	0.27369
0.01000	0.30253	0.30441	0.30592	0.30708	0.30779

**Water + 10 Mass Percent Monoglyme/AgBr/Ag at Temperatures *T*
to 328.15°K**

30	35	40	45	50	55
0.19820	0.19795	0.19747	0.19724	0.19632	0.19585
0.20343	0.20323	0.20280	0.20260	0.20174	0.20107
0.20949	0.20928	0.20879	0.20863	0.20763	0.20723
0.21573	0.21559	0.21551	0.21513	0.21470	0.21438
0.22392	0.22359	0.22341	0.22335	0.22277	0.22236
0.23266	0.23278	0.23272	0.23276	0.23231	0.23197
0.24368	0.24415	0.24397	0.24422	0.24419	0.24387
0.25809	0.25864	0.25897	0.25912	0.25917	0.25937
0.27849	0.27929	0.27993	0.28060	0.28088	0.28131
0.31345	0.31483	0.31606	0.31727	0.31817	0.31915

**Water + 30 Mass Percent Monoglyme/AgBr/Ag at Temperatures *T*
to 328.15°K**

30	35	40	45	50	55
0.19551	0.19323	0.19170	0.18916	0.18619	0.18351
0.20048	0.19832	0.19657	0.19413	0.19122	0.18858
0.20604	0.20394	0.20237	0.19977	0.19694	0.19425
0.21244	0.21039	0.20887	0.20637	0.20363	0.20121
0.21966	0.21798	0.21633	0.21407	0.21133	0.20873
0.22844	0.22683	0.22541	0.22303	0.22057	0.21791
0.23926	0.23784	0.23651	0.23437	0.23199	0.22944
0.25343	0.25220	0.25095	0.24903	0.24681	0.24442
0.27343	0.27238	0.27156	0.26997	0.26799	0.26584
0.30807	0.30786	0.30735	0.30642	0.30508	0.30330

Table III.   Electromotive Force E of the cell: Pt/H$_2$(g)/HBr(m) in
from 278.15°

T/K−273.15 (m/mol kg^{-1})	5	10	15	20	25
0.10000	0.20289	0.19993	0.19707	0.19320	0.19130
0.09000	0.20729	0.20447	0.20162	0.19782	0.19587
0.08000	0.21222	0.20946	0.20663	0.20299	0.20110
0.07000	0.21836	0.21578	0.21311	0.20958	0.20762
0.06000	0.22442	0.22206	0.21953	0.21609	0.21402
0.05000	0.23232	0.22991	0.22750	0.22452	0.22239
0.04000	0.24195	0.23969	0.23744	0.23462	0.23264
0.03000	0.25436	0.25256	0.25029	0.24784	0.24592
0.02000	0.27212	0.27057	0.26865	0.26638	0.26467
0.01000	0.30262	0.30150	0.30030	0.29875	0.29738

Table IV.   Dielectric Constant ϵ, Vapor Pressure p, and Density ρ^0
Standard emf (E°, molal scale) and its Standard Deviation $\sigma(E°)$

$\dfrac{T}{K}$−273.15	ϵ	p (Torr)	ρ^0 (g cm^{-3})
5	87.0	16.8	0.9969
10	88.4	22.6	0.9960
15	89.7	30.3	0.9951
20	90.9	39.9	0.9938
25	92.3	52.5	0.9922
30	93.5	67.7	0.9904
35	94.7	86.8	0.9885
40	96.1	111.4	0.9864
45	97.1	139.9	0.9840
50	98.4	176.6	0.9816
55	99.6	217.5	0.9791

^{a}Torr = (101.325/760)kP$_a$.

Water + 50 Mass Percent Monoglyme/AgBr/Ag at Temperatures T to 328.15°K

30	35	40	45	50	55
0.18916	0.18726	0.18402	0.17980	0.17763	0.17970
0.19363	0.19166	0.18855	0.18443	0.18214	0.18372
0.19888	0.19686	0.19370	0.18971	0.18744	0.18868
0.20538	0.20334	0.20029	0.19654	0.19415	0.19486
0.21182	0.20977	0.20679	0.20320	0.20079	0.20113
0.22013	0.21811	0.21521	0.21183	0.20939	0.20929
0.23038	0.22844	0.22562	0.22255	0.22006	0.21950
0.24371	0.24188	0.23924	0.23641	0.23404	0.23314
0.26272	0.26108	0.25870	0.25635	0.25408	0.25297
0.29582	0.29463	0.29283	0.29095	0.28920	0.28797

of the Mixed Solvent, Debye–Hückel Parameters A and B, and the in Water + 10 Mass Percent Monoglyme from 278.15° to 328.15°K[a]

A ($kg^{1/2}\ mol^{-1/2}$)	B ($kg^{1/2}\ mol^{-1/2}\ cm^{-1}$)	E^0 (V)	$\sigma(E^0)$ (mV)
0.4839	0.3228	0.07939	0.06
0.4598	0.3172	0.07759	0.07
0.4383	0.3121	0.07574	0.07
0.4180	0.3071	0.07364	0.12
0.3981	0.3020	0.07137	0.11
0.3805	0.2973	0.06898	0.11
0.3637	0.2927	0.06647	0.08
0.3473	0.2880	0.06375	0.05
0.3332	0.2838	0.06097	0.08
0.3188	0.2794	0.05804	0.06
0.3056	0.2753	0.05503	0.06

Table V. Dielectric Constant ϵ, Vapor Pressure p, and Density ρ^0
Standard emf (E^0, Molal Basis) and its Standard Deviation $\sigma(E^0)$

$\dfrac{T}{K} - 273.15$	ϵ	$\dfrac{p}{Torr}$	$\dfrac{\rho^0}{g\,cm^{-3}}$
5	87.1	15.8	0.9967
10	88.8	24.3	0.9940
15	90.2	33.1	0.9911
20	91.7	43.9	0.9882
25	93.1	58.0	0.9852
30	94.2	75.6	0.9819
35	95.8	97.1	0.9788
40	97.1	125.7	0.9754
45	98.4	160.9	0.9719
50	99.8	203.0	0.9683
55	101.1	253.9	0.9646

[a] Torr = $(101.325/760)k\mathrm{P_a}$.

Table VI. Dielectric Constant ϵ, Vapor Pressure p, and Density ρ^0
Standard emf (E^0, Molal Basis) and its Standard Deviation $\sigma(E^0)$

$\dfrac{T}{K} - 273.15$	ϵ	p (Torr)	ρ^0 ($g\,cm^{-3}$)
5	52.7	58.6	0.9843
10	51.7	73.0	0.9804
15	51.2	90.0	0.9761
20	50.3	110.0	0.9715
25	49.5	133.4	0.9673
30	48.8	161.7	0.9629
35	48.4	193.5	0.9584
40	47.6	231.6	0.9540
45	47.0	275.3	0.9494
50	46.4	323.1	0.9447
55	45.8	379.9	0.9400

[a] Torr = $(101.325/760)k\mathrm{P_a}$.

of the Mixed Solvent, Debye–Hückel Parameters A and B, and the
in Water + 30 Mass Percent Monoglyme from 278.15° to 328.15°K[a]

A	B	E^0	$\sigma(E^0)$
$kg^{1/2}\ mol^{1/2}$	$kg^{1/2}\ mol^{-1/2}\ cm^{-1}$	V	mV
0.4832	0.3226	0.07737	0.06
0.4566	0.3163	0.07542	0.07
0.4335	0.3106	0.07299	0.08
0.4119	0.3050	0.07016	0.05
0.3916	0.2996	0.06696	0.03
0.3748	0.2949	0.06336	0.04
0.3561	0.2896	0.05925	0.05
0.3398	0.2848	0.05487	0.05
0.3248	0.2802	0.04991	0.03
0.3101	0.2756	0.04465	0.04
0.2965	0.2712	0.03896	0.09

of the Mixed Solvent, Debye–Hückel Parameters A and B, and the
in Water + 50 Mass Percent Monoglyme from 278.15° to 328.15°K[a]

A	B	E^0	$\sigma(E^0)$
$(kg^{1/2}\ mol^{1/2})$	$(kg^{1/2}\ mol^{-1/2}\ cm^{-1})$	(V)	(mV)
1.0194	0.4120	0.07271	0.05
1.0206	0.4116	0.06754	0.06
1.0071	0.4092	0.06221	0.05
1.0053	0.4083	0.05660	0.06
0.9997	0.4070	0.05103	0.03
0.9938	0.4056	0.04527	0.01
0.9807	0.4031	0.03993	0.02
0.9799	0.4024	0.03393	0.03
0.9726	0.4008	0.02813	0.03
0.9650	0.3991	0.02210	0.01
0.9592	0.3976	0.01631	0.04

Table VII.   Standard emf's $E°$ of the Cell in

$\dfrac{T}{K}-273.15$	x = 10		x = 30
	E_c°/V	E_N°/V	E_c°/V
5	0.07924	−0.10916	0.07721
10	0.07739	−0.11435	0.07513
15	0.07550	−0.11958	0.07255
20	0.07333	−0.12507	0.06956
25	0.07097	−0.13073	0.06619
30	0.06848	−0.13651	0.06241
35	0.06586	−0.14241	0.05811
40	0.06301	−0.14852	0.05353
45	0.06009	−0.15469	0.04835
50	0.05701	−0.16101	0.04286
55	0.05385	−0.16741	0.03692

[a] Reference 15.

Table VIII.   Values of the Standard Potentials E_m° (on the Molal Scale) and the Corresponding Variations of the Ion-Size Parameter, $a°$, with the Standard Deviations of $\sigma(E_m^\circ)/mV$ for x = 10, 30, and 50 Mass Percent Monoglyme–Water Mixtures at 298.15°K

$a°/nm$	x = 10		x = 30		x = 50	
	E_m°/V	$\sigma(E_m^\circ)/mV$	E_m°/V	$\sigma(E_m^\circ)/mV$	E_m°/mV	$\sigma(E_m^\circ)/mV$
0.2	0.07094	0.13	0.06654	0.08	0.04827	0.36
0.4	0.07118	0.11	0.06678	0.05	0.05023	0.12
0.6	0.07137	0.11	0.06696	0.03	0.05103	0.03
0.8	0.07154	0.11	0.06713	0.04	0.05167	0.08

Table IX.   Values of the Parameter β of Equation 6 for Mixtures of Monoglyme with Water (x = Mass Percent Monoglyme)

$\dfrac{T}{K}-273.15$	x = 10	x = 30	x = 50
5	0.256	0.00308	0.0784
10	0.212	0.0499	0.0897
15	0.142	0.113	0.0594
20	0.120	0.145	0.115
25	0.0773	0.176	0.00914
30	0.0407	0.207	0.125
35	0.00631	0.214	0.229
40	0.0320	0.292	0.263
45	0.0875	0.334	0.206
50	0.104	0.371	0.332
55	0.151	0.467	0.765

Monoglyme–Water Mixtures (x = Mass Percent Monoglyme)

$x = 30$	$x = 50$		$x = 0^a$
E_N°/V	E_c°/V	E_N°/V	$E_N^\circ, w/V$
−0.10202	0.07195	−0.09534	−0.11292
−0.10720	0.06657	−0.10353	−0.11827
−0.11285	0.06101	−0.11188	−0.12373
−0.11891	0.05514	−0.12051	−0.12941
−0.12533	0.04932	−0.12910	−0.13532
−0.13215	0.04329	−0.13788	−0.14127
−0.13949	0.03767	−0.14624	−0.14745
−0.14709	0.03139	−0.15526	−0.15365
−0.15528	0.02528	−0.16409	−0.16009
−0.16377	0.01893	−0.17314	−0.16664
−0.17268	0.01281	−0.18195	−0.17341

Table X. Dielectric Constant ϵ in Water + 70 Mass Percent Monoglyme, Water + 90 Mass Percent Monoglyme, and in Pure Monoglyme from 278.15° to 328.15°K

$\dfrac{T}{K} - 273.15$	ϵ		
	$x = 70$	$x = 90$	$x = 100$
5	33.41	14.83	8.00
10	32.14	14.38	7.73
15	30.23	14.01	7.58
20	29.59	13.60	7.37
25	28.50	13.20	7.20
30	27.42	12.85	7.03
35	26.44	12.60	6.85
40	25.47	12.28	6.71
45	24.81	11.94	6.55
50	23.93	11.62	6.41
55	23.10	11.36	6.27

Table XI.   Activity Coefficients $\gamma_\pm$ of Hydrobromic Acid in 10 Mass Percent Monoglyme

$T/K-273.15$	$\gamma_\pm$					
$(m/mol\ kg^{-1})$	5	15	25	35	45	55
0.005	0.935	0.940	0.945	0.948	0.951	0.954
0.01	0.915	0.921	0.926	0.931	0.934	0.938
0.02	0.893	0.898	0.904	0.909	0.912	0.915
0.03	0.879	0.883	0.890	0.894	0.896	0.899
0.04	0.870	0.873	0.879	0.882	0.883	0.886
0.05	0.863	0.865	0.870	0.873	0.872	0.874
0.06	0.858	0.858	0.863	0.865	0.863	0.864
0.07	0.855	0.853	0.857	0.858	0.855	0.855
0.08	0.852	0.849	0.852	0.852	0.847	0.846
0.09	0.850	0.845	0.847	0.846	0.840	0.838

Table XII.   Activity Coefficients $\gamma_\pm$ of Hydrobromic Acid in 30 Mass Percent Monoglyme

$T/K-273.15$	$\gamma_\pm$					
$(m/mol\ kg)$	5	15	25	35	45	55
0.005	0.933	0.938	0.943	0.947	0.950	0.952
0.01	0.910	0.916	0.922	0.927	0.930	0.932
0.02	0.882	0.888	0.895	0.901	0.903	0.904
0.03	0.864	0.869	0.875	0.882	0.882	0.881
0.04	0.849	0.854	0.860	0.866	0.865	0.862
0.05	0.838	0.841	0.846	0.853	0.850	0.845
0.06	0.828	0.830	0.835	0.841	0.836	0.829
0.07	0.819	0.820	0.824	0.830	0.823	0.815
0.08	0.812	0.810	0.814	0.819	0.811	0.801
0.09	0.805	0.802	0.805	0.810	0.800	0.787

Table XIII.   Activity Coefficients $\gamma_\pm$ of Hydrobromic Acid in 50 Mass Percent Monoglyme

	$\gamma_\pm$					
$T/K-273.15$ $(m/mol\ kg^{-1})$	5	15	25	35	45	55
0.005	0.867	0.868	0.868	0.868	0.870	0.865
0.01	0.827	0.829	0.828	0.827	0.829	0.820
0.02	0.782	0.783	0.782	0.777	0.779	0.761
0.03	0.753	0.754	0.752	0.744	0.746	0.720
0.04	0.732	0.733	0.730	0.718	0.721	0.688
0.05	0.716	0.716	0.712	0.698	0.701	0.660
0.06	0.702	0.703	0.698	0.680	0.684	0.636
0.07	0.691	0.691	0.685	0.665	0.669	0.614
0.08	0.682	0.682	0.674	0.651	0.656	0.594
0.09	0.673	0.673	0.665	0.639	0.644	0.576

Table XIV.   Relative Partial Molal Enthalpy $(H_2 - H_2°)$ and Relative Partial Molal Heat Capacity $(C_p - C_p°)$ of HBr in 50 Mass Percent Aqueous Monoglyme and in Water at 298.15°K $(cal_{th} = 4.184J)$

$m(HBr)/mol\ kg^{-1}$	0.02	0.03	0.04	0.05	0.06	0.07	0.08	0.09	0.1
	$(H_2 - H_2°)/cal_{th}\ mol^{-1}$								
50 Mass percent monoglyme	52	142	225	297	368	445	508	580	675
Water[a]	85	—	—	124	—	—	—	—	163
	$(C_p - C_p°)/cal_{th}\ K^{-1}\ mol^{-1}$								
50 Mass percent monoglyme	2.3	3.8	5.7	8	10.1	12.4	15	18	20.6
Water[a]	1.5	—	—	2.2	—	—	—	—	2.9

[a] Reference 20.

Table XV.   Standard Thermodynamic Functions (Mole Fraction Basis) for the Transfer of HBr from Water to $x = 10$, 30, and 50 Mass Percent Monoglyme[a]

$T/K - 273.15$	$x = 10$			$x = 30$			$x = 50$		
	ΔG_t° $(cal_{th}$ $mol^{-1})$	ΔH_t° $(cal_{th}$ $mol^{-1})$	ΔS_t° $(cal_{th}$ $mol^{-1})$	ΔG_t° $(cal_{th}$ $mol^{-1})$	ΔH_t° $(cal_{th}$ $mol^{-1})$	ΔS_t° $(cal_{th}$ $mol^{-1})$	ΔG_t° $(cal_{th}$ $mol^{-1})$	ΔH_t° $(cal_{th}$ $mol^{-1})$	ΔS_t° $(cal_{th}$ $mol^{-1})$
5	− 87	140	0.81	−252	+ 103	+ 1.3	−405	−4261	−13.9
15	− 96	156	0.87	−251	− 572	− 1.1	−273	−4073	−13.2
25	−106	172	0.93	−230	−1275	− 3.5	−143	−3874	−12.5
35	−116	189	0.99	−184	−1999	− 5.9	− 28	−3676	−11.8
45	−125	209	1.05	−111	−2745	− 8.3	+ 92	−3459	−11.2
55	−138	224	1.10	− 17	−3518	−10.7	+197	−3245	−10.5

[a] $cal_{th} = 4.184J$.

Acknowledgment

All calculations described above were performed with the help of IBM 370-165 computer at Southwest Missouri State University.   The authors are indebted to R. G. Bates for providing the facility for measurements of the dielectric constants in his laboratory, as well as to A. Chatterjee and J. J. Gibbons for their comments and constructive criticisms of this paper.

Literature Cited

1. Johnson, D. A., Sen, B., *J. Chem. Eng. Data* (1968) **13**, 376.
2. Khoo, K. H., *J. Chem. Soc.*, A (1971) 2932.
3. Roy, R. N., Sen, B., *J. Chem. Eng. Data* (1967) **12**, 584.
4. Roy, R. N., Sen, B., *J. Chem. Eng. Data* (1968) **13**, 79.
5. Roy, R. N., Vernon, W., Gibbons, J. J., Bothwell, A. L. M., *J. Chem. Soc.*, A, (1971) 3589.
6. Roy, R. N., Bothwell, A. L. M., *J. Chem. Eng. Data* (1971) **16**, 347.
7. Roy, R. N., Swensson, E. E., LaCross, G., *J. Chem. Thermodynamics* (1975) **7**, 4566.
8. Roy, R. N., White, T., Gibbons, J. J., to be submitted for publication.
9. Reynaud, R., *Bull. Soc. Chim. France* (1967) 4597.
10. Roy, R. N., Robinson, R. A., Bates, R. G., *J. Chem. Thermodynamics* (1973) **5**, 559.
11. Roy, R. N., Robinson, R. A., Bates, R. G., *J. Am. Chem. Soc.* (1973) **95**, 8231.
12. Gronwall, T. H., LaMer, V. K., Sandved, K., *Z. Physik* (1928) **29**, 358.
13. Bates, R. G., "Determination of pH." 2nd. Ed. Wiley, New York, 1973, Ch. 8, 10.
14. Gary, R., Bates, R. G., Robinson, R. A., *J. Phys. Chem.* (1964) **68**, 1186.
15. Hetzer, H. B., Robinson, R. A., Bates, R. G., *J. Phys. Chem.* (1962) **66**, 1423.
16. Robinson, R. A., Stokes, R. H., "Electrolyte Solutions," 2nd Ed., Butterworths, London, 1959, pp 31, 352.
16. Wallace, W. J., Mathews, A. L., *J. Chem. Eng. Data* (1963) **8**, 496.
17. Shoemaker, P. D., Garland, C. W., "Experiments in Physical Chemistry," McGraw-Hill, New York, 1966, p 162.
18. Janz, G. J., McIntyre, J. D. E., *J. Electrochem. Soc.* (1961) **108**, 272.
19. Harned, H. S., Keston, A. S., Donelson, J. G., *J. Am. Chem. Soc.* (1936) **58**, 989.
20. Harned, H. S., Owen, B. B., "The Physical Chemistry of Electrolytic Solutions," 3rd Ed. Reinhold, New York, 1958, p 727.
20. Pool, K. H., Bates, R. G., *J. Chem. Thermodynamics*, (1969) **1**, 21.
22. Bates, R. G., Staples, B. R., Robinson, R. A., *Anal. Chem.* (1970) **42**, 867.
23. Parker, A. J., *Quart. Rev.* London (1962) **16**, 163.
24. Feakins, D., "Physico-Chemical Processes in Mixed Aqueous Solvents," Franks, F., Ed., American Elsevier Publishing Co., New York, 1967, p 71.
25. Carvajal, C., Tölle, K. J., Smid, J., Szwarc, M., *J. Am. Chem. Soc.* (1965) **87**, 5548.

RECEIVED July 10, 1975.

14

Electrolytic Conductance of Lithium Bromide in Acetone and Acetone–Bromosuccinic Acid Solutions

CHARLES W. JONES[1] and CLARENCE M. CUNNINGHAM

Oklahoma State University, Stillwater, Okla. 74074

The Fuoss–Onsager–Skinner equation satisfactorily describes the electrolytic conductance of lithium bromide in acetone. Values of $198.1 \pm 0.9 \; \Omega^{-1} \, cm^2 \, eq^{-1}$ and $(3.3 \pm 0.1) \times 10^3$ are established for Λ_0 and K_A, respectively, at 25°C; furthermore, a value of 2.53 Å is obtained for the sum of the ionic radii (a). When bromosuccinic acid is added to 10^{-5} N lithium bromide in acetone, there is a decrease in the specific conductance of lithium bromide rather than the increase that is observed at higher concentrations. As the concentration of bromosuccinic acid is increased, the values obtained for Λ_0 and K_A decrease, while those for a increase when the bromosuccinic acid and acetone are considered to constitute a mixed solvent. These results do not permit any simple explanation. When bromosuccinic acid and acetone are considered a mixed solvent, the Fuoss–Onsager–Skinner theory does not describe the system.

This study was undertaken to determine whether or not the electrolytic conductance of the lithium bromide–bromosuccinic acid–acetone system can be described by the Fuoss–Onsager–Skinner equation (FOS equation)—Equation 2—by treating the system as lithium bromide in a mixed solvent, and to establish values for Λ_0 and K_A for lithium bromide in anhydrous acetone with the same equation. The equation requires knowledge of the concentration and corresponding equivalent conductance along with the dielectric constant and viscosity of the solvent and the temperature; that is,

$$F(c, \Lambda, D, \eta, T) = 0 \qquad (1)$$

The essential data were compiled from both the experimental portion of this study

<hr>

[1] Current address: Lake Superior State College, Sault Ste. Marie, Mich. 49783

and the work of Cunningham and co-workers. The experimental portion consisted of measuring the electrical resistance, at 25°C, of solutions of varying amounts of lithium bromide in acetone, lithium bromide in bromosuccinic acid and acetone, and lithium bromide in dimethyl bromosuccinate and acetone.

This is an extremely complicated system for such a study, inasmuch as it is a three-component system consisting of an ionophore (lithium bromide) and an ionogen (bromosuccinic acid) in a smenogenic solvent (acetone). Further, the solvent has a high affinity for water and a comparatively high vapor pressure at 25°C.

In 1962 Fuoss and Onsager began a revision of their treatment of the conductance of symmetrical electrolytes. In their first paper they considered the potential of total force; in the second, the relaxation field; in the third, electrophoresis; and in the fourth, the hydrodynamic and osmotic terms in the relaxation field ($1,2,3,4$). In 1965 Fuoss, Onsager, and Skinner (5) combined the results of the four papers and formulated a general conductance equation:

$$\Lambda = \Lambda_0 - Sc^{1/2}\gamma^{1/2} + E'c\gamma\ln(\tau^2\gamma) + Lc\gamma - K_Ac\gamma f_\pm^2\Lambda \tag{2}$$

where Λ is the equivalent conductance, Λ_0 is the equivalent conductance at infinite dilution, c is the normal concentration, γ is the fraction of electrolyte existing as free ions, K_A is the association equilibrium constant, $f_\pm$ is the mean ionic activity coefficient,

$$S = \alpha\Lambda_0 + \beta \tag{3}$$

$$E' = E_1'\Lambda_0 - E_2' \tag{4}$$

$$\tau = (6E_1'c)^{1/2} \tag{5}$$

and

$$L = 3.202E_1'\Lambda_0 - 3.420E_2' + \alpha\beta + 2E_1'\Lambda_0(2b^{-1} + 2b^{-2} - b^{-3}) \\ + 44b^{-1}E_2'/3 - 2E'\ln b \tag{6}$$

Substitution of numerical values for physical constants yields the following equations for 1:1 electrolytes, in terms of the sum of the ionic radii, a (in angstroms); the dielectric constant, D; the absolute temperature, T (in Kelvin); and the viscosity, η (in poise):

$$\alpha = 8.205 \times 10^5 (DT)^{-3/2} \tag{7}$$

$$\beta = 82.49\eta^{-1}(DT)^{-1/2} \tag{8}$$

$$b = 1.671 \times 10^{-3}(aDT)^{-1} \tag{9}$$

$$K_A = 2.523 \times 10^{21}a^3\exp(b) \tag{10}$$

$$E_1' = 2.943 \times 10^{12}(DT)^{-3} \tag{11}$$

$$E_2' = 4.333 \times 10^7\eta^{-1}(DT)^{-2} \tag{12}$$

and $$f_\pm = \exp[-8.404 \times 10^6 (c\gamma)^{1/2}(DT)^{-3/2}] \tag{13}$$

These equations may be used in conjunction with Equations 3, 4, 5, and 6 for evaluating Equation 2.

Olson and Cunningham (6) found that the specific conductance of $0.01m$ lithium bromide in acetone was increased by 30% when sufficient bromosuccinic acid, was added to make the solution $0.2m$ with respect to the acid. When dimethyl bromosuccinate was added in lieu of bromosuccinic acid, the specific conductance was diminished by 6%; and when lithium perchlorate was substituted for lithium bromide, the specific conductance decreased linearly as bromosuccinic acid was added. These observations motivated Cunningham and his co-workers to continue work in the field.

Bjornson (7) investigated the electrolytic conductance of systems consisting of lithium halides and some carboxylic acids in acetone to determine the effect of the addition of successive increments of acid. He found that such additions were usually accompanied by an anomalous rise in specific conductance. By means of a Fuoss (8, 9) plot he obtained values of 2.13×10^{-4} and $196.0 \ \Omega^{-1} \ \text{cm}^2$ eq^{-1} for K_D (the reciprocal of K_A) and Λ_0, respectively, for lithium bromide in acetone. He suggested that in the lithium bromide–bromosuccinic acid–acetone system, bromosuccinic acid is a stronger acid than hydrogen bromide, and hydrogen bromide would be formed. Bailey (10) applied the same method to the hydrogen bromide–acetone system, and derived values of 1×10^{-6} and 110–120 $\Omega^{-1} \ \text{cm}^2 \ \text{eq}^{-1}$ for K_D and Λ_0, respectively, for hydrogen bromide in acetone. Muller (11) measured viscosities, densities, and dielectric constants of solutions composed of lithium bromide and some carboxylic acids in acetone. Mahan (12) also made density measurements on solutions of lithium bromide in acetone.

A value for the equivalent conductance at infinite dilution for lithium bromide in acetone was first calculated in 1905 by Dutoit and Levier (13) for 18°C: 166 $\Omega^{-1} \ \text{cm}^2 \ \text{eq}^{-1}$. A graphical method involving Ostwald's dilution law ($\Lambda^{-1} = \Lambda_0^{-1} + c\Lambda/K_D\Lambda_0^2$), applied to their data in 1913 by Kraus and Bray (14), produced values of 5.7×10^{-4} for K_D and 165 $\Omega^{-1} \ \text{cm}^2 \ \text{eq}^{-1}$ for Λ_0. Deviations from the mass action law (nonlinearity in the graph) become appreciable at concentrations of ca. $10^{-3}N$. Both groups pointed out that measurements in acetone are liable to error from several sources, including the presence of solvent impurities and exposure to light. A solvent correction of 21% was applied to their most dilute solution.

In 1910 Serkov (15) determined the conductance of several salts (including lithium bromide) at 25°C in water, methanol, ethanol, acetone, and binary mixtures of these solvents, reporting a value of 144 $\Omega^{-1} \ \text{cm}^2 \ \text{eq}^{-1}$ for Λ_0 for lithium bromide in acetone. He found that, unlike the other mixtures, acetone solutions exhibit no parallelism between conductance and fluidity, and concluded that when the surveyed ionophores are dissolved in acetone, the complexity of the solvates formed increases as Λ_0 for the ionophores decreases.

In 1939 Dippy, Jenkins, and Page (16) found that the phoreogram for lithium bromide in acetone at 25°C contains an inflection point, and they were unable to get Λ_0 by extrapolation.   Inspection of their phoreogram indicates, however, that Λ_0 is nearer Serkov's value than that of Kraus and Bray.   They noted that although different batches of acetone had different specific conductances, the data points of the phoreogram lay uniformly on a smooth curve.   This they considered evidence of the adequacy of the solvent correction employed.

Reynolds and Kraus (17) obtained conductance for 14 salts in acetone at 25°C, and used the Fuoss method to calculate their equivalent conductances at infinite dilution.   Among the salts were tetra-*n*-butylammonium fluorotriphenylborate, tetra-*n*-butylammonium picrate, lithium picrate, and tetra-*n*-butylammonium bromide.   They then derived ionic equivalent conductances at infinite dilution by the method of Fowler (18) using tetra-*n*-butylammonium fluorotriphenylborate as the reference electrolyte and obtained a value of 188.7 Ω^{-1} cm^2 eq^{-1} for Λ_0 for lithium bromide.

In 1953 Olson and Konecny (19) studied the conductance of lithium bromide in acetone–water mixtures at 25°C and 35°C.   They calculated K_D and Λ_0 in the acetone-rich solvents by the Fuoss method and Λ_0 in the water-rich solvents by extrapolation of the phoreogram.   They found that as the water content increases: K_D increases, Λ_0 decreases but then undergoes an increase, and *a* increases from slightly less than the sum of the crystal ionic radii to the sum of the radii of the fully hydrated ions.   Extrapolation of their data for Λ_0 to zero water content is not reliable because of the large concave upward negative slope; however, it would appear to lead to a value of about 220 Ω^{-1} cm^2 eq^{-1}.   Similar extrapolations of values for K_D and *a* yield 2.0×10^{-4} and 2.2 Å, respectively.

Two years later Nash and Monk (20) also measured conductances at 25°C using aqueous acetone (12.5 wt % water) as the solvent.   For K_D they obtained values of 1×10^{-3} and 6×10^{-3} for lithium bromide and hydrogen bromide, respectively, by the Davies (21) method and 101.1 Ω^{-1} cm^2 eq^{-1} and 117.1 Ω^{-1} cm^2 eq^{-1} for Λ_0 for lithium bromide and hydrogen bromide, respectively, by the Fuoss method.

A more recent study using dry acetone and acetone–water mixtures was reported by Nilsson and Beronius (22).   For acetone containing 0.005% by weight they found Λ_0, K_A, and *a* to be 195.0 Ω^{-1} cm^2 eq^{-1}, 4202, and 9.3 Å, respectively. All three of these parameters were found to decrease as the concentration of water was increased.   They suggest that the change in these parameters is a result of the strong solvation of the salt by the water.   Nilsson (23) conducted similar studies on acetone and methanol mixtures and noted similar results.

Experimental

Reagents.   Several batches of very nearly anhydrous acetone were prepared by the method of Howard and Pike (24).   A two-dm^3 borosilicate flask containing about 1.5 dm^3 of acetone (Fisher certified ACS) and 200–250g of $\frac{1}{16}$-in. synthetic zeolite pellets (Linde Type 5A molecular sieve) was stoppered and stored in a dark cabinet for a minimum of two days, during which time the contents were swirled and mixed intermittently.   It was connected to a Pyrex fractionating distillation column of approximately 50 theoretical plates and distilled with a 1:1 reflux ratio, the first 250-cm^3 fraction being discarded.   One batch of dry acetone had a specific conductance κ of 3×10^{-9} Ω^{-1} cm^{-1} and the others 2×10^{-8} Ω^{-1} cm^{-1}.   The one with the lowest specific conductance was prepared from acetone which had remained in contact with molecular sieves for several months.   The specific conductance of the distilled acetone increased upon standing even in the borosilicate flask.   For this reason all solvents were prepared from freshly distilled acetone.

The mass spectrum of the bromosuccinic acid (K & K Laboratories, Inc.), a snow-white powder which melted smoothly in the range of 160°–165°C, showed no peak corresponding to the parent compound.   No impurities could be identified; in particular, there were no peaks corresponding to fragments containing two bromine atoms.   The mass spectrum for bromosuccinic acid was not found in the literature, but that of the prepared acid was analogous to the one for succinic acid, e.g., no parent peak (25).

Dimethyl bromosuccinate was prepared from bromosuccinic acid by the diazomethane method (26) using the procedure of Eisenbraun, Morris, and Adolphen (27).   It was distilled under vacuum (0.08–0.1 Torr) at 45°–49°C to yield a clear colorless oil.   Thin layer chromatography with benzene as the solvent on SiO$_2$ yielded a symmetrical single spot, indicating either a pure compound or no separation with this particular solvent.   Its mass spectrum had a very small peak corresponding to the parent compound, but none to a dibromo compound. The mass spectrum for dimethyl bromosuccinate was not found in the literature, but that for dimethyl succinate also has a small peak corresponding to the parent compound (25).

Reagent grade anhydrous lithium bromide powder (Matheson, Coleman and Bell) was used after drying in a vacuum oven at 100°–120°C.   Potassium chloride (Fisher certified ACS) was dried at 110°C and used to calibrate the conductance cell.   Deionized water from the laboratory supply was piped directly into a Pyrex glass still (Corning model AG-1a) and distilled into a polyethylene vessel, where it was kept until needed.   The specific conductance of the water was 1×10^{-6} Ω^{-1} cm^{-1}.

Apparatus.   All electrical resistances were measured with an electrolytic conductivity bridge (Leeds and Northrup model 4666) which was constructed according to specifications set forth by Jones (28) and described by Dike (29). The audio-frequency source was a General Radio Co. type 1311-A audio oscillator used with the frequency regulated at 1000 Hz and the output at about 5 V.   The detector circuit consisted of a high-gain low-noise tuned amplifier and null detector (General Radio Co. type 1232-A) and an oscilloscope (Heathkit model O-11);

the input and output transformers were Leeds and Northrup models 019200 and 019201, respectively.

The conductance cell employed was used in an oil bath maintained at 25.00 ± 0.01°C. It was a dilution cell modified from the design of Shedlovsky (*30*), who incorporated the recommendations of Jones and Bollinger (*31*). The cell bulb was almost spherical and the electrodes were vertical parallel circular plates ca. one mm apart, with their centers aligned on an axis perpendicular to the planes of the plates. The diameter of each electrode was ca. four cm. The bulb had a glass tube leading from the top, the tube having a Teflon stopcock so as to eliminate the need for stopcock grease. (Stopcock grease contamination was a major source of trouble prior to use of the Teflon.) The dilution bulb—100 cm^3—had an outer 24/40 standard tapered joint, the top of which was above the top of the cell bulb and fitted with a stopper. All the glass (Pyrex) in the conductance cell was one piece. A 0.01000 demal standard potassium chloride solution, as defined by Jones and Bradshaw (*32*), was used to calibrate the cell. A correction was made for the specific conductance of the water, and the deviation in the measured resistance for the standard solution was taken into account. A value of 0.017363 ± 0.000007 cm^{-1} was obtained for the cell constant.

Procedure. Equation 1 indicates that it is necessary to determine the concentration, resistance, dielectric constant, viscosity, and temperature of the system. These data were acquired for five different solvent systems. A series of measurements, in which the concentration of lithium bromide was varied from about $10^{-5}N$ to $10^{-3}N$, was made on each system. The solvents used were acetone (I), 0.02063m bromosuccinic acid in acetone (II), 0.05009m bromosuccinic acid in acetone (III), 0.09958m bromosuccinic acid in acetone (IV), and 0.05047m dimethyl bromosuccinate in acetone(V). Each solvent was used to prepare stock solutions of 10^{-2} and $10^{-3}m$ lithium bromide. All mixed solvents and solutions were prepared in the dry box.

For each series of measurements about 50 g of solvent was transferred quantitatively in the dry box to the cell by pouring it into the dilution bulb; this was the minimum amount required to fill the cell bulb. The cell was removed from the dry box, placed in the oil bath, and connected to the bridge. Time was allowed for the attainment of thermal equilibrium; then at least three resistance measurements were made at five-min intervals, and the average value was calculated. The cell was removed from the bath and returned to the dry box. Dilute stock solution was quantitatively added to the cell by means of a weighing buret. The contents of the cell were carefully mixed, and the resistance of the solution was measured as before. The procedure just described was repeated several times with the dilute stock solution and then with the concentrated stock solution. About ten concentrations with a hundredfold range were obtained. A portion of the final solution in the cell (the most concentrated solution) was removed, and the infrared spectrum taken; no absorption band indicative of traces of water was observed at 3600 cm^{-1}. It was necessary to obtain the densities of the solution in order to calculate the lithium bromide normality. In all cases the density of the lithium bromide solution was assumed to be the same as the density of the solvent. The densities of the solvents were calculated from data obtained by Muller (11), who found that the density of a solution, ρ, of m molal bromosuccinic acid in acetone is given by the equation

$$\rho = \rho_0(1 + 0.129m) \qquad (14)$$

in which ρ_0 is the density of acetone. The molality and normality of each solution were then calculated.

The viscosities of the acetone–bromosuccinic acid mixed solvents were derived from the Jones–Dole (33) equation and data acquired by Muller, who used the special viscometer described by Tuan and Fuoss (34). The values used for the viscosities (in poise) of solvents I–V were 3.02×10^{-3}, 3.05×10^{-3}, 3.08×10^{-3}, 3.13×10^{-3}, and 3.02×10^{-3}, respectively. The literature value for the dielectric constant of acetone, 20.7, was used as the dielectric constant for each solvent. This is justified because at the highest concentration of bromosuccinic acid its mole fraction is less than 0.004.

Results

The experimental results are summarized for each series (the series numbers correspond to the respective solvent numbers) in Table I. The first column is simply for reference, and the second is the normality of the lithium bromide. The third column gives the experimental equivalent conductance as calculated from the corrected specific conductances.

A computer program for the solution of the FOS equation, which is a modification of the method of Fuoss, Onsager, and Skinner (5), was written in Fortran IV and executed on an IBM System 360/50 (Operating System—H Level) computer. The program uses the method of Wentworth (35) for least-squares

Table I. Experimental Results and Functions Calculated From FOS Equation

Point	$c \times 10^5$, N	Λ (exp)	γ	S Term	E Term	L Term	K_A Term	Λ (cal)	$\delta\Lambda$
Series I		(pure acetone, $\kappa = 1.825 \times 10^{-8}\ \Omega^{-1}\ cm^{-1}$)							
1	1.262	187.0	0.957	2.37	−0.18	−0.20	7.06	188.33	1.32
2	2.650	180.4	0.930	3.39	−0.33	−0.42	13.54	180.47	0.08
3	5.160	170.3	0.886	4.62	−0.55	−0.77	22.99	169.22	−1.08
4	8.644	158.6	0.833	5.80	−0.79	−1.22	32.75	157.59	−1.06
5	13.40	146.4	0.777	6.97	−1.06	−1.76	42.43	145.92	−0.51
6	19.56	134.7	0.723	8.12	−1.35	−2.39	51.49	134.79	0.05
7	30.83	119.6	0.653	9.69	−1.77	−3.40	62.52	120.77	1.08
8	53.84	101.6	0.569	11.95	−2.43	−5.17	76.24	102.35	0.75
9	92.96	84.72	0.491	14.58	−3.23	−7.71	88.56	84.07	−0.66

Table I.   (*continued*)

Point	$c \times 10^5$, (N)	Λ (exp)	γ	S Term	E Term	L Term	K_A Term	Λ (cal)	$\delta\Lambda$
Series II	(0.02063 m bromosuccinic acid, κ = 7.701 $\times$ 10^{-7} Ω^{-1} cm^{-1})								
1	0.7508	154.6	0.964	1.66	−0.09	−0.14	1.05	159.25	4.65
2	1.469	157.2	0.986	2.35	−0.16	−0.29	2.09	157.30	0.14
3	3.729	154.4	0.981	3.74	−0.36	−0.72	4.99	152.38	−1.98
4	5.777	150.8	0.967	4.62	−0.51	−1.11	7.27	148.69	−2.15
5	8.633	146.3	0.948	5.60	−0.69	−1.62	10.05	144.24	−2.06
6	11.26	142.5	0.932	6.34	−0.84	−2.08	12.30	140.65	−1.85
7	22.56	128.3	0.865	8.64	−1.37	−3.86	19.29	129.04	0.78
8	34.13	118.6	0.820	10.35	−1.79	−5.54	24.40	120.11	1.55
9	55.52	106.1	0.766	12.76	−2.43	−8.41	31.00	107.60	1.47
10	96.95	91.18	0.706	16.19	−3.37	−13.55	38.97	90.13	−1.05
Series III	(0.05009 m bromosuccinic acid, κ = 1.408 $\times$ 10^{-6} Ω^{-1} cm^{-1})								
1	0.6024	140.9	0.993	1.42	−0.07	−0.06	0.40	141.51	0.59
2	1.214	139.3	0.986	2.02	−0.12	−0.13	0.79	140.42	1.17
3	3.354	137.5	0.986	3.35	−0.28	−0.35	2.06	137.43	−0.07
4	5.445	135.7	0.981	4.26	−0.42	−0.56	3.19	135.04	−0.61
5	8.243	133.2	0.973	5.22	−0.59	−0.84	4.57	132.25	−0.94
6	10.88	131.0	0.965	5.97	−0.73	−1.10	5.76	129.92	−1.05
7	22.59	121.6	0.923	8.41	−1.24	−2.18	9.87	121.76	0.20
8	36.04	114.1	0.892	10.45	−1.71	−3.36	13.45	114.51	0.36
9	59.29	104.3	0.849	13.07	−2.35	−5.26	17.80	104.98	0.69
10	100.2	92.79	0.805	16.54	−3.24	−8.42	22.87	92.41	−0.38
Series IV	(0.09958 m bromosuccinic acid, κ = 2.590 $\times$ 10^{-6} Ω^{-1} cm^{-1})								
1	0.6932	132.0	0.984	1.47	−0.07	−0.07	0.17	133.97	1.95
2	1.333	133.3	0.999	2.05	−0.12	−0.13	0.32	133.11	−0.17
3	2.663	131.0	0.989	2.88	−0.22	−0.26	0.61	131.76	0.80
4	5.030	130.5	0.997	3.98	−0.37	−0.50	1.13	129.76	−0.79
5	8.122	128.8	0.996	5.06	−0.55	−0.81	1.74	127.59	−1.26
6	10.97	127.4	0.994	5.87	−0.70	−1.09	2.26	125.82	−1.55
7	23.68	119.9	0.969	8.51	−1.25	−2.30	4.12	119.56	−0.33
8	37.53	113.3	0.945	10.59	−1.72	−3.56	5.65	114.22	0.88
9	61.24	105.8	0.926	13.38	−2.40	−5.69	7.73	106.55	0.75
10	105.8	95.38	0.905	17.39	−3.37	−9.61	10.40	94.97	−0.41
Series V	(0.05047 m dimethyl bromosuccinate, κ = 1.911 $\times$ 10^{-7} Ω^{-1} cm^{-1})								
1	0.9796	171.8	0.968	2.01	−0.13	−0.54	2.21	175.20	3.42
2	1.642	171.7	0.973	2.61	−0.20	−0.90	3.67	172.70	1.04
3	2.573	170.7	0.975	3.27	−0.30	−1.42	5.62	169.48	−1.25
4	4.155	166.6	0.962	4.12	−0.44	−2.26	8.55	164.72	−1.93
5	6.549	159.6	0.934	5.10	−0.62	−3.46	12.20	158.70	−0.87
6	9.2	156	0.924	6.01	−0.81	−4.81	16.14	152.31	−3.34
7	39.0	111.7	0.742	11.10	−2.08	−16.38	34.46	116.07	4.35
8	66.4	95.5	0.696	14.03	−2.92	−26.18	43.54	93.41	−2.05

computation. The details of the program are available from the authors. Table I also gives the numerical values of the terms in the theoretical equation. Each row shows the values calculated for a given point.

The constants of the FOS equation are given in Table II for each series. Values of b were calculated from K_A and a combination of Equations 9 and 10 set up for iteration in the program as follows:

$$b_n = \ln \left[\frac{K_A}{2.523 \times 10^{21}} \left(\frac{DT}{16.71 \times 10^{-4}} \right)^3 \right] - 3\ln (b_{n-1}) \qquad (15)$$

Values of a, expressed in Å, obtained from b and Equation 9 are entered in the last column.

Table II. Calculated Constants of FOS Equation for Lithium Bromide

Series	Λ_0	L	K_A	b	a
I	198.14 ± 0.90	$-16\,905 \pm 8357$	3320 ± 145	11.53	2.35
II	162.20 ± 1.62	$-19\,781 \pm 15\,040$	982 ± 254	9.83	2.75
III	143.47 ± 0.50	$-10\,443 \pm 4876$	500 ± 84	8.84	3.06
IV	133.74 ± 0.70	$-10\,036 \pm 6340$	195 ± 111	7.34	3.69
V	180.09 ± 2.54	$-56\,567 \pm 39\,330$	1433 ± 609	10.37	2.61

Discussion

Lithium Bromide in Acetone. It can be seen from Table I that the solvent correction for the specific conductance of lithium bromide in Series I is negligible throughout, ranging from 0.025 to 0.76%. The relative standard deviations of the calculated constants (the standard deviation of the constant divided by the value of the constant) can be computed from data in the first row of Table II; they are 0.0045, 0.49, and 0.044 for Λ_0, L, and K_A, respectively. The standard deviation for L is especially large; however, this great uncertainty in L is not uncommon for smenogenic solvents. Fuoss, Onsager, and Skinner pointed out that "no useful information can be obtained from the ion-pair term at high dielectric constants nor from the linear term at low" (5). It should be noted that L and K_A are not independent variables. If Equation 2 is to be considered a theoretical equation, it should be solved with two adjustable parameters by making K_A and Λ_0 the adjustable parameters in solutions of low dielectric constant and L and Λ_0 the adjustable parameters if the dielectric constant of the solvent is high. Of course a better fit to experimental data is obtained by making K_A, L, and Λ_0 adjustable parameters. In acetone this changes the values of K_A and Λ_0 very little; and since others have made three-parameter solutions for this equation, the same procedure was followed in this work.

Table I shows that γ decreases as the concentration increases, which is expected and illustrated in Figure 1. All the terms listed in the table diminish the

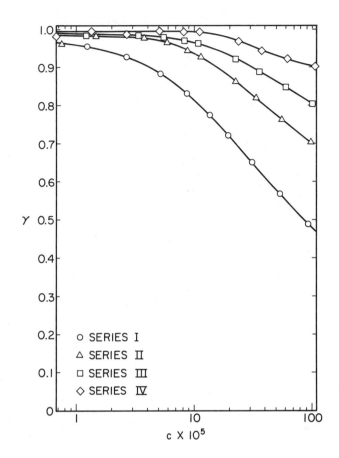

Figure 1. Fraction of lithium bromide existing as free ions in Solvents I–IV as a function of lithium bromide concentration

equivalent conductance from Λ_0, and the magnitude of the reduction for each term increases with concentration. The K_A term makes the largest contribution to the decrease throughout the entire concentration range and the E term the smallest. At the lower concentrations the E term and L term are of the same magnitude, but at the higher concentrations the L term is about double the E term. The S term is greater than the L term at all concentrations. The calculated and experimental values for Λ are in good agreement as is shown by $\delta\Lambda$ in Table I and the phoreogram in Figure 2.

For comparison, the data of Bjornson (7) and Dutoit and Levier (13) were compiled and run through the program. Table III gives the calculated constants of the FOS equation for the Bjornson data, as well as the reported values of K_A and a for Nilsson and Beronius (22). The value for Λ_0 from Bjornson's data is in excellent agreement with the value he obtained from the same data using

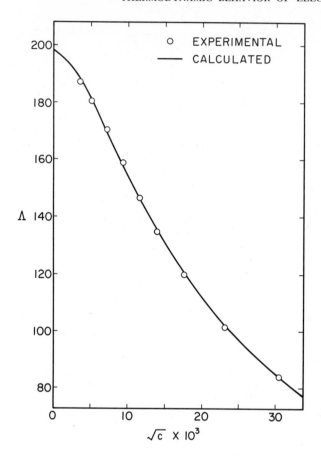

*Figure 2. Phoreogram of lithium bromide in acetone
(solvent I)*

the Fuoss method. The relative standard deviation of L is smaller for Series I than for the Bjornson data. The value obtained for L with Bjornson's data gives a positive contribution from the L term in the FOS equation, whereas our data produce a negative contribution. If, however, L is calculated from Equation 6 with the value procured for b from the experimental K_A and Equation 10, a negative value for L is obtained for both. The relative standard deviation of K_A reported from the data of Nilsson and Beronius is indeed very small. In fact,

Table III.   Constants of FOS Equation for Lithium Bromide

Data Source	Λ_0	L	K_A	b	a
Nilsson and Beronius	195.0 ± 0.02		4202 ± 0.6		9.3
Bjornson	196.92 ± 4.17	$7420 \pm 43\,880$	4755 ± 782	12.0	2.25

it is smaller than might be expected from the deviation reported for the values for Λ. However, it should be noted that their experimental data points are at concentrations that are an order of magnitude higher than those of Bjornson and Series I. This might account for the fact that their Λ_0 and K_A are lower than those reported in Table II for Series I. It can be argued that the accuracy of the points at very low concentrations is not good, but the theory certainly would not be expected to apply as well at higher concentrations.

It appears that by and large the FOS equation does satisfactorily describe the electrolytic conductance of the lithium bromide–acetone system. However, scrutiny of the last column of Table I shows that the change in sign for $\delta\Lambda$ for Series I may not be random. In particular, the most dilute solution has a positive and larger value for $\delta\Lambda$ than for any other. In the range 5–$10 \times 10^{-5}N$ lithium bromide, $\delta\Lambda$ is negative; it becomes positive as the concentration is increased further, but negative again for the most concentrated solution. This is shown in the phoreogram in Figure 2 and in itself would not be significant. However, the data of Bjornson and of Dutoit and Levier show the same trend. In fact, Kraus and Bray (*14*) rejected the point corresponding to the first row in Table I. In all cases the experimental phoreogram has a greater curvature and inflection than the calculated phoreogram. The calculated phoreogram is high for the very dilute solution and low for the more concentrated solutions; it has a less negative slope at the point of inflection. This same pattern is observed for Series II–V as shown in Table I. Of course, these systems contain another component, but nonetheless the pattern exists. These results suggest that phenomena are occurring which are not accounted for by the Fuoss–Onsager–Skinner theory. For example, adsorption of ions at the electrodes would introduce a larger error for the very dilute solutions.

Lithium Bromide–Bromosuccinic Acid–Acetone System. The specific conductance of bromosuccinic acid in acetone at the concentrations used in this research is about two orders of magnitude greater than that of acetone, and approaches the order of magnitude of conductance of very dilute solutions of lithium bromide in acetone. A plot of the corrected specific conductance of bromosuccinic acid in acetone as a function of its concentration is shown in Figure 3. This agrees with the results of Bjornson (*7*), who found the specific conductance of $0.02m$ and $0.2m$ bromosuccinic acid in acetone to be ca. $8 \times 10^{-7}\ \Omega^{-1}$ cm^{-1} and $4 \times 10^{-6}\ \Omega^{-1}\ cm^{-1}$, respectively. The solvent correction for the specific conductance of lithium bromide in the bromosuccinic acid–acetone mixed solvents is no longer negligible throughout the entire concentration ranges. Table I shows that, as expected, the corrections become more significant with decreasing concentration of lithium bromide and increasing concentration of bromosuccinic acid. The corrected specific conductance of lithium bromide as a function of normality for Series I–IV is shown in Figures 4 and 5. Figure 4 shows that in the dilute solutions the increase in specific conductance with increasing concentration of lithium bromide is diminished with augmentation of bromosuccinic acid.

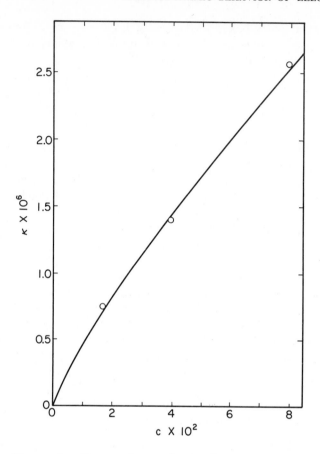

*Figure 3. Corrected specific conductance of bro-
mosuccinic acid in acetone as a function of bromosuccinic acid concentration*

Figure 5 indicates this holds true up to a concentration of about 2–3 $\times$ $10^{-4}N$
lithium bromide, whereas above about 5 $\times$ $10^{-4}N$ the specific conductance is
enhanced by the addition of bromosuccinic acid.   On the basis of the divergence
of the curves at these concentrations, it is reasonable to assume that if the concentration of lithium bromide is further increased, the enhancement of the
specific conductance from the addition of the bromosuccinic acid should be still
greater.   This is consistent with the results of Bjornson, who measured the specific
conductance of solutions containing $0.01m$ lithium bromide in acetone and
varying amounts of bromosuccinic acid.   He found the specific conductance of
the solution to be 3.0 $\times$ 10^{-4} Ω^{-1} cm^{-1} in the absence of bromosuccinic acid; for
$0.02m$, $0.05m$, $0.1m$, and $0.2m$ bromosuccinic acid in the solution, the specific
conductances were (Ω^{-1} cm^{-1}) 3.4 $\times$ 10^{-4}, 4.0 $\times$ 10^{-4}, 4.4 $\times$ 10^{-4}, and 4.6 $\times$ 10^{-4},

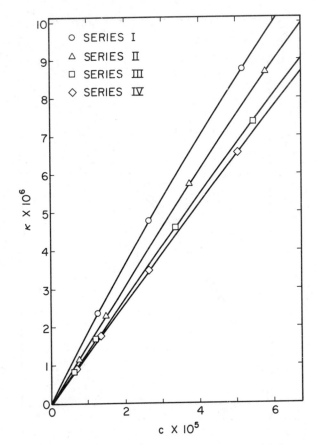

Figure 4. Corrected specific conductance of lithium bromide as a function of lithium bromide concentration for Series I–IV

respectively. These results are also consistent with the observations of Olson and Cunningham.

It is to be noted from Table II that, as in Series I, the standard deviation of Λ_0 is small, that of L is large, and that of K_A is intermediate for Series II–IV. These large standard deviations can be rationalized as in the preceding section. The table shows that Λ_0 and K_A decrease systematically while a increases with increasing bromosuccinic acid concentration.

Table I shows that γ decreases with increasing concentration of lithium bromide for each series, but that the decrease becomes smaller as the bromosuccinic acid concentration gets larger. The effect of lithium bromide and bromosuccinic acid concentration on γ is also demonstrated in Figure 1. All the terms listed in Table I diminish the equivalent conductance from Λ_0 for Series II–IV and the magnitude of the reduction for each term increases with increasing

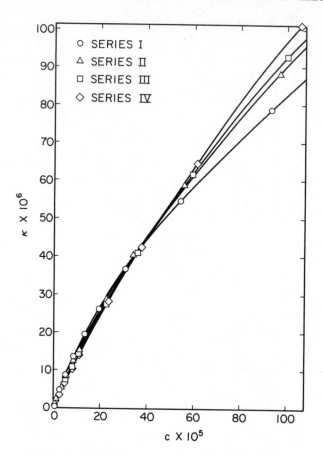

*Figure 5. Corrected specific conductance of lithium
bromide as a function of lithium bromide concentration
for Series I–IV*

lithium bromide concentration, just as in Series I.   The contributions of the E
terms and L terms are analagous to Series I; that is, the E terms make the smallest
contribution to the decrease throughout the entire concentration ranges, and at
the lower concentrations the E and L terms are of the same magnitude,
but at the higher concentrations the L terms are two to three times the E terms.
For series II and III the K_A terms make the largest contribution to the reduction
in equivalent conductance at higher concentrations, but, unlike the case in Series
I, at lower concentrations the S terms are the largest contributors.   The S term
makes the largest contribution to the decrease in equivalent conductance at all
concentrations in Series IV.   As previously mentioned, the same trends are noted
in $\delta\Lambda$ for series II–IV as in Series I, but are even more pronounced.   There is
general agreement between the experimental and calculated equivalent con-

ductances, but the agreement is not as good as in Series I. The calculated phoreograms for Series I–IV are shown in Figure 6. In each series the first derivative is negative throughout and there is a point of inflection, the second derivative being negative at low concentrations and positive at high ones.

If the system behaved ideally, the specific conductances should be additive. Figure 7 shows the specific conductance of the solution corrected (by subtraction) for the specific conductances of the acetone and lithium bromide for various fixed amounts of lithium bromide as a function of bromosuccinic acid concentration. Inasmuch as this should be equal to the equivalent conductance of bromosuccinic acid, if there were no interaction among the conducting species all four curves should coincide with the curve for no lithium bromide. Clearly, some type of interaction must occur.

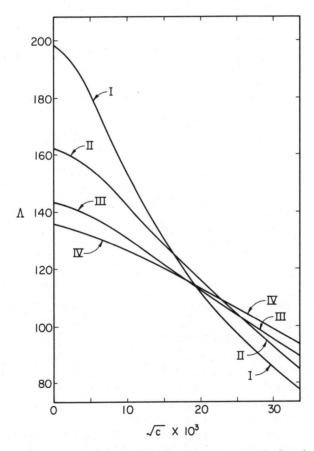

Figure 6. Lithium bromide phoreogram calculated from the FOS equation and experimental data for solvents I–IV

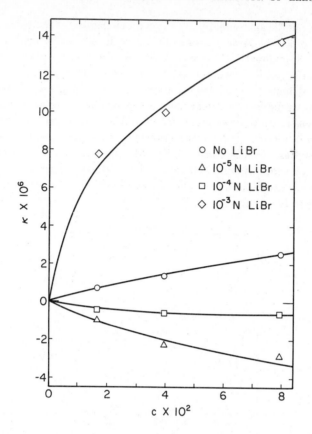

Figure 7. Specific conductance of solution minus specific conductance of acetone and lithium bromide as a function of bromosuccinic acid concentration for various fixed amounts of lithium bromide

Bjornson worked at sufficiently high concentration ($0.01m$ lithium bromide) to observe only the increase in specific conductance due to the addition of bromosuccinic acid to a lithium bromide–acetone solution. Lithium bromide is an ionophore and in acetone exists as lithium ions and bromide ions (conductors) in equilibrium with associated lithium bromide ion pairs (nonconductors), while bromosuccinic acid is an ionogen which exists in acetone as bromosuccinic acid molecules (nonconductors) in equilibrium with hydrogen ions and bromosuccinate ions (conductors). In order to explain the anomalous increase in specific conductance, Bjornson proposed that when bromosuccinic acid is added to the lithium bromide–acetone solution, bromide ion from the lithium bromide combines with the hydrogen ion from the bromosuccinic acid and forms molecular hydrogen bromide (a nonconductor). This would result in a decrease in concentration of these ions; however, as bromide ions and bromosuccinate ions were removed, less lithium bromide would be associated and more bromosuccinic acid would be dissociated.   The final result after establishment of equilibrium among lithium ions, bromide ions, hydrogen ions, bromosuccinate ions, bromosuccinic acid molecules, hydrogen bromide molecules, and lithium bromide ion pairs would be a net increase in conducting species and therefore an increase in specific conductance.

Bjornson also measured the specific conductance of a solution of $0.01m$ lithium bromide in acetone with various amounts of dimethyl bromosuccinate added and found a slight linear decrease in specific conductance with addition of dimethyl bromosuccinate.   These results, along with those of Olson and Cunningham, lent support to Bjornson's postulate, in that when the acidic hydrogens of bromosuccinic acid were replaced with methyl groups, or the bromide ions of lithium bromide were replaced with perchlorate ions, the increase in specific conductance was not observed.

Series V consisted of runs in which lithium bromide was added to a fixed amount of dimethyl bromosuccinate in acetone.   Table I shows that the solvent correction is greater than for Series I, but less than for Series II–IV.   The specific conductance of lithium bromide in dimethyl bromosuccinate–acetone is only slightly less than in acetone.   This is in contrast to Series II–IV.   Table II shows that both Λ_0 and K_A are less than for Series I but greater than for Series II–IV. Table I indicates that for Series V the trends in each column are the same as for Series I.   The results of Series V are in agreement with those of Bjornson and those of Olson and Cunningham.

The decrease in K_A and Λ_0 with increase in concentration of bromosuccinic acid observed in this work is consistent with the trend observed by Nilsson and Beronius (22) for water and acetone and by Nilsson (23) for methanol and acetone.   However, the a values obtained from these data do not match too well with those obtained for the water–acetone and water–methanol systems, but it should be noted that these a values are not calculated in the same manner.   With acetone and bromosuccinic acid the change in the a values is just the reverse of what was observed for the other systems as the concentration of the second solvent is increased.

In any event, it can be concluded that in acetone there are strong associations between bromosuccinic acid and lithium bromide ions.   The largest concentration of bromosuccinic acid studied (Series IV) is approximately equal to the lowest concentration studied for the acetone–water and acetone–methanol systems.   As expected, the data show the association with the acid to be much greater than with either the methanol or the water.   From the K_A and a values it is evident that the association between the salt and dimethyl bromosuccinate is much less than the association of the salt with bromosuccinic acid but greater than the salt–acetone association.   In view of this, it is concluded that the association between the bromide ion and the second solvent accounts for the change in K_A. Knowledge of the precise nature of the association will have to await further investigations.

Acknowledgment

This work was supported in part by the National Science Foundation through a faculty fellowship and terminal traineeship.

Literature Cited

1. Fuoss, R. M., Onsager, L., *J. Phys. Chem.* (1962) **66**, 1722.
2. Fuoss, R. M., Onsager, L., *J. Phys. Chem.* (1963) **67**, 621.
3. Fuoss, R. M., Onsager, L., *J. Phys. Chem.* (1963) **67**, 628.
4. Fuoss, R. M., Onsager, L., *J. Phys. Chem.* (1964) **68**, 1.
5. Fuoss, R. M., Onsager, L., Skinner, J. F., *J. Phys. Chem.* (1965) **69**, 2581.
6. Olson, A. R., Cunningham, C. M., University of California, 1948, unpublished observations.
7. Bjornson, G., M.S. Thesis, Oklahoma State University, Stillwater, Okla., 1960.
8. Fuoss, R. M., *J. Amer. Chem. Soc.* (1935) **57**, 488.
9. Fuoss, R. M., *J. Amer. Chem. Soc.* (1935) **57**, 2604.
10. Bailey, T. E., M.S. Thesis, Oklahoma State University, Stillwater, Okla., 1962.
11. Muller, J. H., unpublished observations, 1964.
12. Mahan, K., unpublished observations, 1963.
13. Dutoit, P., Levier, A., *J. Chem. Phys.* (1905) **3**, 435, from Ref. *14*.
14. Kraus, C. A., Bray, W. C., *J. Amer. Chem. Soc.* (1913) **35**, 1315.
15. Serkov, S., *J. Russ. Phys. Chem. Soc.* (1909) **40**, 399; **41**, 1, from *Chem. Abstr.* (1910),
    4, 1921, through *Chem. Zentr.* (1909), **I**, 1452.
16. Dippy, J. F. J., Jenkins, H. O., Page, J. E., *J. Chem. Soc.* (1939) 1386.
17. Reynolds, M. B., Kraus, C. A., *J. Amer. Chem. Soc.* (1948) **70**, 1709.
18. Fowler, D. L., Kraus, C. A., *J. Amer. Chem. Soc.* (1940) **62**, 2237.
19. Olson, A. R., Konecny, J., *J. Amer. Chem. Soc.* (1953) **75**, 5801.
20. Nash, G. R., Monk, C. B., *J. Chem. Soc.* (1955) 1899.
21. Davies, C. W., *Trans. Faraday Soc.* (1927) **23**, 351.
22. Nilsson, A. M., Beronius, P., *Z. Phys. Chem.* (1972) **79**, 83.
23. Nilsson, A. M., *Acta Chem. Scand.* (1973) **27**, 2722.
24. Howard, K. S., Pike, F. P., *J. Phys. Chem.* (1959) **63**, 311.
25. "Atlas of Mass Spectral Data" Stenhagen, E., Abrahamsson, S. and McLafferty, F. W., Eds. Interscience Publishers, New York, 1969.
26. Moore, J. A., Reed, D. E., *Org. Syn.* (1961) **41**, 16.
27. Eisenbraun, E. J., Morris, R. N., Adolphen, G., *J. Chem. Educ.* (1970) **47**, 710.
28. Jones, G., Josephs, R. C., *J. Amer. Chem. Soc.* (1928) **50**, 1049.
29. Dike, P. H., *Rev. Sci. Instrum.* (1931) **2**, 379.
30. Shedlovsky, T., *J. Amer. Chem. Soc.* (1932) **54**, 1411.
31. Jones, G., Bollinger, G. M., *J. Amer. Chem. Soc.* (1931) **53**, 411.
32. Jones, G., Bradshaw, B. C., *J. Amer. Chem. Soc.* (1933) **55**, 1780.
33. Jones, G., Dole, M., *J. Amer. Chem. Soc.* (1929) **51**, 2950.
34. Tuan, D. F., Fuoss, R. M., *J. Phys. Chem.* (1963) **67**, 1343.
35. Wentworth, W. E., *J. Chem. Educ.* (1965) **42**, 96.

RECEIVED July 15, 1975.

15

A Potentiometric Method for Determination of the Thermodynamics of Ionization Reactions in Partially Aqueous Solvents

CHARANAI C. PANICHAJAKUL and EARL M. WOOLLEY

Department of Chemistry, Brigham Young University, Provo, Utah 84602

A potentiometric method for determination of ionization constants for weak acids and bases in mixed solvents and for determination of solubility product constants in mixed solvents is described. The method utilizes glass electrodes, is rapid and convenient, and gives results in agreement with corresponding values from the literature. After describing the experimental details of the method, we present results of its application to three types of ionization equilibria. These results include a study of the thermodynamics of ionization of acetic acid, benzoic acid, phenol, water, and silver chloride in aqueous mixtures of acetone, tetrahydrofuran, and ethanol. The solvent compositions in these studies were varied from 0 to ca. 70 mass % nonaqueous component, and measurements were made at several temperatures between 10° and 40°C.

Ionization reactions have been investigated *(1)* by a variety of methods that lead to reasonably accurate values of equilibrium constants over rather wide ranges of temperatures, pressures, and dissolved salt concentrations. However, the status of measurements leading to ionization constants in aqueous organic mixed solvents has not been developed nearly so well, in spite of the excellent work of Harned, Grunwald, Bates, and others *(1–10)*. Experimental methods have been difficult and those methods that utilize the hydrogen electrode can be applied only to systems in which there are no complicating reduction reactions.

We have recently devised a rapid and convenient method for determination of the ionization constant for water in mixed aqueous organic solvents *(11–16)*. The method utilizes glass electrodes and gives results in satisfactory agreement with earlier work.

As a result of our continuing interest in ionization equilibria in mixed solvent systems (*11–20*), we have now devised a convenient and rapid method for determining ionization constants for weak acids (*16*) and solubility product constants for certain slightly soluble salts in aqueous organic mixed solvents. Two features of the method are significantly different from earlier methods (*1–10*). First, provision is made to determine glass electrode responses. Second, the experimental procedure of diluting aqueous electrolyte solutions with a nonaqueous solvent component eliminates the need for most of the solution preparation and handling associated with the earlier methods. In this paper we describe the details of our method and report on its application to the determination of the ionization constants for acetic acid, benzoic acid, phenol, water, and silver chloride in mixtures of water with ethanol, tetrahydrofuran, and acetone. The mixtures containing ethanol and acetone were studied at $10°$, $15°$, $20°$, $25°$, $30°$, $35°$, and $40°C$, and the mixtures containing tetrahydrofuran were studied at $15°$, $25°$, and $35°C$.

Method and Calculations

In our investigations we describe the ionization reactions in solvent S as follows:

$$H_2O(S) = H^+(S) + OH^-(S); \quad \Delta G_w°, \Delta H_w°, \Delta S_w° \tag{1}$$

$$K_w = a_H a_{OH}/a_w = C_H C_{OH}(y\pm)^2/1 \tag{2}$$

$$HA(S) = H^+(S) + A^-(S); \quad \Delta G_a°, \Delta H_a°, \Delta S_a° \tag{3}$$

$$K_a = a_H a_A/a_{HA} = C_H C_A(y\pm)^2/C_{HA}y_{HA} \tag{4}$$

$$MX(c) = M^+(S) + X^-(S); \quad \Delta G_s°, \Delta H_s°, \Delta S_s° \tag{5}$$

$$K_s = a_M a_X/a_{MX} = C_M C_X(y\pm)^2/1 \tag{6}$$

In Equations 2, 4, and 6, a_i represents thermodynamic activities based on molar concentrations C_i of the species indicated, $y\pm$ represents mean ionic activity coefficients, y_{HA} is the activity coefficient of HA(S) molecules, and the activity of water is chosen to be one in all solvents. Consequently values of K, $\Delta G°$, and $\Delta S°$ are based on these choices regarding standard states.

This study is based on measurement of the potentials of the cells represented by

glass electrode|soln A:

$$HCl(C_1), KNO_3(C_2), \text{ in solvent } S|AgCl, Ag \quad (A)$$

glass electrode|soln B:

$$KOH(C_3), KCl(C_4), \text{ in solvent } S|AgCl, Ag \quad (B)$$

glass electrode|soln C:

$$HA(C_5), KOH(C_6), KCl(C_7), \text{ in solvent } S|AgCl, Ag \quad (C)$$

glass electrode|soln D:

$$AgNO_3(C_8), HNO_3(C_9), \text{ in solvent } S|AgCl, Ag \quad (D)$$

In cells A–D, C_1–C_9 represent total formal analytical concentrations.
A general expression for the potentials of these cells is given by

$$E = k_1 + k_2 \log (a_H a_{Cl}) \quad (7)$$

and specific equations for cells A–D are given in terms of the definitions in
Equations 2, 4, and 6 of K_w, K_a, and K_s as

$$E_A = k_1 + k_2 \log (C_1)^2 + k_2 \log (y\pm)_A{}^2 \quad (8)$$

$$E_B = k_1 + k_2 \log (K_w C_4/C_3) \quad (9)$$

$$E_C = k_1 + k_2 \log [(C_H)_C C_7] + k_2 \log (y\pm)_C{}^2 \quad (10)$$

$$E_D = k_1 + k_2 \log (K_s C_9/C_8) \quad (11)$$

As shown previously (*11–16*), combination of Equations 8 and 9 gives Equation
12 when the solvent composition is the same in cells A and B.

$$pK_w = (E_A - E_B)/k_2 - \log [(C_1)^2 C_3 (y\pm)_A{}^2/C_4] \quad (12)$$

Equation 12 is used first to determine electrode responses k_2, and then to determine pK_w values, as described previously (*11–16*). Equation 13 is used to estimate $y\pm$ values.

$$\log (y\pm) = -\frac{1.825 \times 10^6 (d/D^3 T^3)^{1/2} I^{1/2}}{1 + 2.298 \times 10^2 (d/DT)^{1/2} I^{1/2}} \quad (13)$$

Density and dielectric data were obtained from the literature (*23*) or from experimental measurements.

Combination of Equations 8 and 10 leads to Equation 14, which can be used
in conjunction with the "buffer ratio" defined by Equation 15 in obtaining values
of K_a as shown previously (*16, 24, 25*).

$$\log (C_H)_C = (E_C - E_A)/k_2 + \log (C_1{}^2/C_7) + \log [(y\pm)_A{}^2/(y\pm)_C{}^2] \quad (14)$$

$$C_A/C_{HA} = (C_6 + C_H - C_{OH})/(C_5 - C_6 - C_H$$
$$+ C_{OH}) \sim C_6/(C_5 - C_6) \quad (15)$$

The value of k_2 to be used in Equation 14 may be taken as that value calculated from Equation 12, or one could calculate a value of k_2 from Equation
14 from a known value of K_a in pure water and from measured values of E_A and
E_C in pure water in conjunction with Equations 2, 4, 13, and 15.

Values of K_h for the hydrolysis reaction

$$A^-(S) + H_2O(S) = HA(S) + OH^-(S) \tag{16}$$

can be obtained from known values of K_w and K_a by using Equation 17.

$$K_h = a_{HA}a_{OH}/a_A a_w = K_w/K_a = K_b \tag{17}$$

Combination of Equations 8 and 11 leads to Equation 18 when the solvent composition is the same in cells A and D.

$$pK_s = (E_A - E_D)/k_2 - \log [(C_1)^2 C_8 (y\pm)_A^2/C_9 a_{MX}] \tag{18}$$

Values of K_s can be obtained from measured E_A and E_D values in any solvent S when the analytical concentrations C_1, C_8, and C_9 are known, when a_{MX} is known, and where $y\pm$ is calculated from Equation 13. In this work, a_{MX} is taken to be unity.

The value of k_2 used in Equation 18 may either be taken as that value calculated from Equation 12 or one could calculate a value of k_2 from Equation 18 from a known value of K_s in pure water and from measured values of E_A and E_D in pure water in conjunction with Equation 13.

The assumptions made in the use of these methods to obtain the above ionization constants in mixed solvents have been summarized (16). The fact that the pK values calculated using the above assumptions are in good agreement with those values reported in the literature is an indication that any errors resulting from these assumptions are probably relatively small.

The calculation of $\Delta H°$ and $\Delta S°$ values from the pK–temperature data in each solvent mixture was performed by the nonempirical method of Clarke and Glew (26) as simplified by Bolton (27). In this method the thermodynamic parameters are considered to be continuous, well-behaved functions of temperature, and their values are expressed as perturbations of their values at some reference temperature θ by a Taylor's series expansion. The basic equation is:

$$R \ln K = - \frac{\Delta G_\theta°}{\theta} + \frac{\Delta H_\theta°}{\theta} t_1 + \Delta C_{p,\theta}° t_2 + \frac{\theta}{2} \left(\frac{\partial \Delta C_p°}{\partial T} \right)_\theta t_3 + \cdots \tag{19}$$

where the thermodynamic parameters are the regression coefficients and the terms t_i are the temperature dependent variables. All equilibrium constants were converted to the molality scale prior to the above analysis (21, 22, 24). Values of K, $\Delta G°$, and $\Delta S°$ were then converted back to the molarity scale and are expressed on that basis in the "Results" section.

Experimental

Potential measurements were made on cells A, B, C, and D with a Model E436 Metrohm Potentiograph recording potentiometric titrator. The sensitivity was set to 50-mV full scale so that potentials were readable to 0.1 mV. The glass electrodes used were the Fisher 13-639-1 and the Coleman 3-472 wide pH range electrodes. Silver–silver chloride electrodes were prepared from Beckman 39261 Silver Billet electrodes by electrolysis in chloride solution (5).

Experimental measurements were made by immersing a pair of the electrodes in a 15-ml portion of purely aqueous solution A, B, C, or D and allowing the potentials to stabilize. When the potential became stable, a portion of the nonaqueous cosolvent was added to the solution in the cell and the potential was again recorded. This procedure was continued until 50 ml of the cosolvent had been added. The temperature of the cells was kept constant to within ±0.05°C of the reported temperatures throughout the experiments. The potential measurement–cosolvent addition experiments were performed at least twice with each combination of glass electrodes and silver–silver chloride electrodes on at least two independently prepared solutions.

Total ionic strengths of solutions in the cells were varied from about $0.005M$ to ca. $0.02M$. The concentrations of solutions in cell C were made so that the buffer ratio in Equation 15 always had a value between 0.4 and 0.6. The nonaqueous cosolvents used in this study were Reagent Grade or better, and they were tested to be sure that they were free from significant quantities of potentially interfering substances such as halide ions, acids, and bases. Densities of tetrahydrofuran–water mixtures were determined pycnometrically at 15°C and at 35°C.

Results and Discussion

Values of k_2/T calculated from Equations 12, 14, and 18 ranged from 0.1960 to 0.1980 mV/°K for different electrode combinations in different solutions at different temperatures, compared to $2.303 R/F = 0.1984$ mV/°K.

Data for a typical series of measurements on cells A, B, C, and D are given in Table I. In Table I are also given the auxiliary data and the results (pK_a, pK_h, pK_w, and pK_s values) for this series of measurements.

Each pK value obtained in this work is the average result of at least two independent series of measurements using different combinations of electrodes and different solutions in the cells. All these replicate measurements led to pK values which have average deviations of <0.03 in solvent mixtures containing up to 50 mass percent organic component and <0.05 in solutions of higher organic content. In most cases, these average deviations were less than 0.02 and 0.03 in these two regions, respectively.

Values of $\Delta H_\theta{}^\circ$ and $\Delta S_\theta{}^\circ = (\Delta H_\theta{}^\circ - \Delta G_\theta{}^\circ)/\theta$ ($\theta = 298.15°K$) obtained by Equation 19 had statistical uncertainties (26, 27) that were typically 300–500 cal and 1–2 cal/°K, respectively. Standard deviations in the pK–temperature correlation as analyzed according to Equation 19 with terms involving $\Delta C_{p,\theta}{}^\circ$ and higher order terms did not decrease significantly from the two-parameter values of the standard deviations of between 0.01 and 0.04. More accurate pK

Table I. Data from a Typical Series of Measurements with Silver Chloride and Acetic Acid in Water–Ethanol Mixtures at 25°C[a]

wt% EtOH	D	d (g/ml)	E_A (mV)	$-E_B$ (mV)	E_C (mV)	$-E_D$ (mV)	pK_w	pK_a	pK_h	pK_s
0.00	78.5	0.997	193.6	386.9	36.6	138.1	14.00	4.75	9.25	9.73
6.16[•]	70.7	0.988	194.5	388.4	34.6	141.2	14.12	4.85	9.27	9.89
11.6	64.1	0.979	195.3	390.2	32.1	144.1	14.24	4.95	9.29	10.03
16.5	58.9	0.972	195.5	392.0	29.1	146.9	14.34	5.04	9.30	10.15
20.8	54.0	0.966	195.6	393.4	25.8	149.5	14.43	5.14	9.29	10.26
28.3	48.5	0.952	194.6	395.9	18.8	155.5	14.58	5.31	9.27	10.47
34.4	43.6	0.942	193.4	397.6	12.1	160.9	14.69	5.47	9.22	10.65
44.1	38.3	0.923	192.0	399.4	3.0	169.0	14.88	5.71	9.17	10.94
51.2	35.4	0.910	191.3	400.3	−4.0	175.0	15.03	5.89	9.14	11.17
56.8	33.3	0.895	190.9	400.9	−10.0	179.6	15.15	6.05	9.10	11.37
61.2	31.6	0.884	190.9	401.1	−15.5	183.4	15.27	6.20	9.07	11.54
64.8	30.5	0.876	190.9	401.3	−19.8	186.7	15.36	6.32	9.04	11.69
67.8	29.4	0.868	191.0	401.3	−23.4	189.4	15.45	6.43	9.02	11.83
72.4	28.5	0.858	192.3	401.4	−29.5	192.9	15.61	6.62	8.99	12.05

[a] 15.00 ml of solution in each cell, with $C_1 = C_2 = C_3 = C_4 = C_6 = C_7 = C_8 = C_9 = 0.01000$ and $C_5 = 0.01912$; $k_2 = 58.76$ mV from Equation 12, $k_2 = 58.82$ mV from Equation 14, and $k_2 = 58.97$ mV from Equation 18.

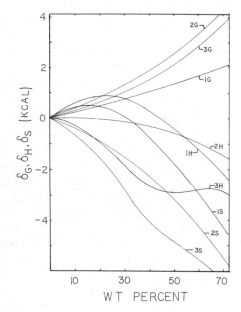

Figure 1. Thermodynamics for $H_2O(S) = H^+(S) + OH^-(S)$ vs. wt% organic solvent component. Equations 1, 2, 20–22: 1 = water–ethanol, 2 = water–acetone, 3 = water–tetrahydrofuran; G = δG, H = δH, S = δS.

measurements are necessary to determine $\Delta C_p°$ values (21, 22, 24, 26, 27). We estimate that the "total" uncertainties in our reported $\Delta G°$, $\Delta H°$, and $\Delta S°$ values are typically 50 cal, 500 cal, and 2 cal/°K, respectively.

The results of all our pK determinations are summarized graphically in Figures 1–8.   In these plots we show our results as

$$\delta G = \Delta G°(S) - \Delta G°(H_2O) \tag{20}$$

$$\delta H = \Delta H°(S) - \Delta H°(H_2O) \tag{21}$$

$$\delta S = T\Delta S°(S) - T\Delta S°(H_2O) \tag{22}$$

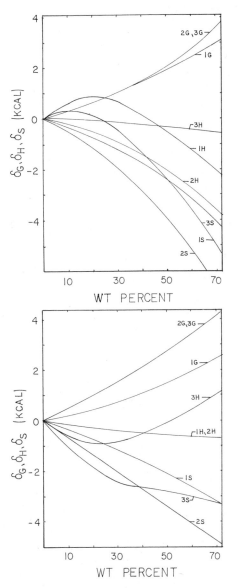

Figure 2.   Thermodynamics for $AgCl(c) = Ag^+(S) + Cl^-(S)$ vs. *wt % organic solvent component. Equations 5, 6, 20–22:* 1 = water–ethanol, 2 = water–acetone, 3 = water–tetrahydrofuran; G = δG, H = δH, S = δS.

Figure 3.   Thermodynamics for $HAc(S) = H^+(S) + Ac^-(S)$ *where Ac = acetate vs. wt % organic solvent component.   Equations 3, 4, 20–22:* 1 = water–ethanol, 2 = water–acetone, 3 = water–tetrahydrofuran; G = δG, H = δH, S = δS.

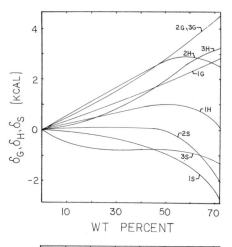

Figure 4. Thermodynamics for HBz(S) = H⁺(S) + Bz⁻(S) where Bz = benzoate vs. wt% organic solvent component. Equations 3, 4, 20-22: 1 = water-ethanol, 2 = water-acetone, 3 = water-tetra-hydrofuran; G = δG, H = δH, S = δS.

Figure 5. Thermodynamics for HPh(S) = H⁺(S) + Ph⁻(S) where Ph = phenolate vs. wt% organic solvent component. Equations 3, 4, 20-22: 1 = water-ethanol, 2 = water-acetone, 3 = water-tetra-hydrofuran; G = δG, H = δH, S = δS.

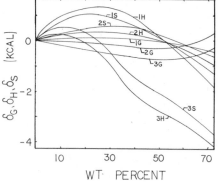

Figure 6. Thermodynamics for Ac⁻(S) + H₂O(S) = HAc(S) + OH⁻(S) where Ac = acetate vs. wt% organic solvent component. Equations 16, 17, 20-22: 1 = water-ethanol, 2 = water-acetone, 3 = water-tetrahydrofuran; G = δG, H = δH, S = δS.

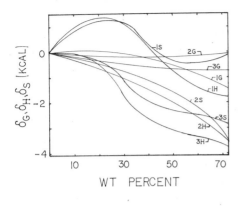

Figure 7. Thermodynamics for Bz⁻(S) + H₂O(S) = HBz(S) + OH⁻(S) where Bz = benzoate vs. wt % organic solvent component. Equations 16, 17, 20–22: 1 = water–ethanol, 2 = water–acetone, 3 = water–tetrahydrofuran; G = δG, H = δH, S = δS.

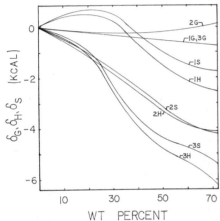

Figure 8. Thermodynamics for Ph⁻(S) + H₂O(S) = HPh(S) + OH⁻(S) where Ph = phenolate vs. wt % organic solvent component. Equations 16, 17, 20–22: 1 = water–ethanol, 2 = water–acetone, 3 = water–tetrahydrofuran; G = δG, H = δH, S = δS.

In Equations 20–22, the S and H_2O refer to values in solvent S and pure water, respectively. Values of $\Delta G°$, $\Delta H°$, and $\Delta S°$ for all five ionization reactions (Equations 1, 3, and 5) in pure water solvent are given in Table II. Values of $\Delta G°$, $\Delta H°$, and $\Delta S°$ for the three hydrolysis reactions (Equation 16) in pure water can be obtained by combination of the values for ionization of water and for ionization of each of the acids.

Values of pK_w at 25°C are in good agreement with our previous results obtained by the same method (*11, 12*). The values in ethanol–water mixtures also agree well with those obtained by Gutbezahl and Grunwald (*29*). Values of pK_a for acetic acid and benzoic acid in ethanol–water mixtures at 25°C are also in excellent agreement with those values reported previously (*1, 2, 4, 7, 16*), with maximum deviations of less than 0.08.

Our $\Delta H_w°$ values are in fair agreement with earlier calorimetric values (*30*). Our values are more positive by ca. one kcal at the highest ethanol compositions.

Table II. Thermodynamic Parameters at 25°C for Reactions 1, 3, and 5 in Pure Water as Solvent

Reaction	$\Delta G^{\circ a}$ (kcal)	ΔH° (kcal)	$\Delta S^{\circ a}$ (cal/°K)	Ref.
$H_2O = H^+ + OH^-$	19.11	13.34	-19.3_5	21
$AgCl(c) = Ag^+ + Cl^-$	13.31	15.65	7.8_5	28
$HAc = H^+ + Ac^{-b}$	6.48	-0.10	-22.0_7	21
$HBz = H^+ + Bz^{-b}$	5.73	0.10	-18.8_8	21
$HPh = H^+ + Ph$ ^{b}	13.61	5.48	-27.2_7	21

[a] Standard states of all ions and acid solutes based on molarities. Standard states of water and silver chloride are chosen so that their activities are unity.
[b] Ac is acetate, Bz is benzoate, and Ph is phenolate.

Our reported pK_s values in ethanol–water and in acetone–water mixtures at 25°C are in fair agreement with previously reported values (18, 31). Deviations are less than 0.08 in all cases where there is more than 50% water present.

One can interpret the trends in δH and δS in Figures 1–8 in terms of solvation of the reactants or products. Negative values of δH are a possible indication that the products in the reaction are more strongly solvated in solvent S than they are in water. Negative δH values also might indicate that reactants in the reaction are less strongly solvated in solvent S than in water. Conversely, positive values of δH are a possible indication that products are less solvated in solvent S than in water or that reactants are more solvated in solvent S than in water. Another possible factor in the interpretation of δH vs. solvent composition data is the possibility of large contributions of solvent–solvent interactions.

Similar complicated possibilities exist in interpreting δS data, for one must consider the total entropy change for the reaction, including reactants, products, and solvent.

One simplifying approach would be to ascribe all effects to the interaction of the ions with each other. If this approach were valid, we would expect to see the values of δG, δH, and δS nearly zero in all solvents for isoelectronic processes, such as the hydrolysis reactions in Figures 6–8. The fact that the trends are generally somewhat smaller in magnitude for these hydrolysis reactions than for the ionization reactions in Figures 1–5 lends some support to this approach.

A specific example of some of the possible interpretations of the data is as follows. From Figure 8 we see that δH and δS for hydrolysis of phenolate ion in acetone–water mixtures decrease in rather regular fashion. We might interpret this behavior in one or a combination of the following ways. (a) Phenol is more strongly solvated in tetrahydrofuran–water mixtures than in water. (b) Hydroxide ion is more strongly solvated in tetrahydrofuran–water mixtures than in water. (c) Water is more strongly solvated in water than in tetrahydrofuran–water mixtures. (d) Phenolate ion is more strongly solvated in water than in tetrahydrofuran–water mixtures.   (e) Phenol and/or hydroxide ion tend to cause tetrahydrofuran–water mixtures to be more structured than they tend to cause water to be structured.   (f) Phenolate ion and/or water tend to cause water to be more structured than they tend to cause tetrahydrofuran–water mixtures to be structured.   Of these options, (a) and (c) appear to be most reasonable, and (b) and (e) appear to be least reasonable in this case.

Literature Cited

1. Harned, H. S., Owen, B. B., "The Physical Chemistry of Electrolyte Solutions," 3rd ed., Reinhold, New York, 1958.
2. Grunwald, E., Berkowitz, B. J., *J. Am. Chem. Soc.* (1951) **73**, 4939.
3. Grunwald, E., *J. Am. Chem. Soc.* (1951) **73**, 4934.
4. Frohliger, J. O., Gartska, R. A., Irwin, H. W., Steward, O. W., *Anal. Chem.* (1968) **40**, 1408.
5. Bates, R. G., "Determination of pH, Theory and Practice," 2nd ed., Wiley, New York, 1973.
6. Schindler, P., Robinson, R. A., Bates, R. G., *J. Res. Nat. Bur. Stand.* (1968) **72A**, 141.
7. Spivey, H. O., Shedlovsky, T., *J. Phys. Chem.* (1967) **71**, 2171.
8. Salomon, M., *J. Electrochem. Soc.* (1974) **121**, 1584.
9. Feakins, D., Lawrence, K. G., Voice, P. J., Willmott, A. R., *J. Chem. Soc.* (A) (1970) 837.
10. Kratohvil, J., Težak, B., *Arh. Kem.* (1954) **26**, 243.
11. Woolley, E. M., Hurkot, D. G., Hepler, L. G., *J. Phys. Chem.* (1970) **74**, 3908.
12. Woolley, E. M., Tomkins, J., Hepler, L. G., *J. Solution Chem.* (1972) **1**, 341.
13. Woolley, E. M., Hepler, L. G., *Anal. Chem.* (1972) **44**, 1520.
14. Woolley, E. M., George, R. E., *J. Solution Chem.* (1974) **3**, 119.
15. George, R. E., Woolley, E. M., *J. Solution Chem.* (1972) **1**, 279.
16. Panichajakul, C. C., Woolley, E. M., *Anal. Chem.* (1975) **47**, 1860.
17. Matsui, T., Hepler, L. G., Woolley, E. M., *Can. J. Chem.* (1974) **52**, 1910.
18. Anderson, K. P., Butler, E. A., Woolley, E. M., *J. Phys. Chem.* (1971) **75**, 93.
19. Anderson, K. P., Butler, E. A., Woolley, E. M., *J. Phys. Chem.* (1973) **77**, 2564.
20. Anderson, K. P., Butler, E. A., Woolley, E. M., *J. Phys. Chem.* (1974) **78**, 2244.
21. Larson, J. W., Hepler, L. G., in "Solute-Solvent Interactions," J. F. Coetzee and C. D. Ritchie, Eds., Marcel Dekker, New York, 1969.
22. Hepler, L. G., Woolley, E. M., in "Water—A Comprehensive Treatise," F. Franks, Ed., Vol. 3, Plenum, New York, 1973.
23. Timmermans, J., "The Physico-Chemical Constants of Binary Systems in Condensed Solutions," Vol. 4, Interscience, New York, 1960.
24. King, E. J., "Acid-Base Equilibria," Pergamon, Oxford, England, 1965.
25. Butler, J. N., "Ionic Equilibrium, A Mathematical Approach," Addison-Wesley, Reading, Mass., 1964.
26. Clarke, E. C. W., Glew, D. N., *Trans. Faraday Soc.* (1966) **62**, 539.
27. Bolton, P. D., *J. Chem. Educ.* (1970) **47**, 638.
28. Wagman, D. D., Evans, W. H., Parker, V. B., Halow, I., Bailey, S. M., Schumm, R. H., *U.S. Natl. Bur. Std. Techn. Notes* **270-3** (1968); **270-4** (1969).
29. Gutbezahl, B., Grunwald, E., *J. Am. Chem. Soc.* (1953) **75**, 565.
30. Bertrand, G. L., Millero, F. J., Wu, C. H., Hepler, L. G., *J. Phys. Chem.* (1966) **70**, 699.
31. Anderson, K. P., Butler, E. A., Anderson, D. R., Woolley, E. M., *J. Phys. Chem.* (1967) **71**, 3566.

RECEIVED July 10, 1975.   This work supported in part by the Brigham Young University Research Division.

16

Heat Capacities and Volumes of Transfer of Electrolytes from Water to Mixed Aqueous Solvents

JACQUES E. DESNOYERS, OSAMU KIYOHARA, and GÉRALD PERRON

Department of Chemistry, Université de Sherbrooke, Sherbrooke, Quebec, Canada J1K 2R1

LÉVON AVÉDIKIAN

Laboratoire d'Etudes des Interactions Solutés-solvants, Université de Clermont, 63170 Aubière, France

The densities and volumetric specific heats of some alkali halides and tetraalkylammonium bromides were undertaken in mixed aqueous solutions at 25°C using a flow digital densimeter and a flow microcalorimeter. The organic cosolvents used were urea, p-dioxane, piperadine, morpholine, acetone, dimethylsulfoxide, tert-butanol, and to a lesser extent acetamide, tetrahydropyran, and piperazine. The electrolyte concentration was kept at 0.1m in all cases, while the cosolvent concentration was varied when possible up to 40 wt %. From the corresponding data in pure water, the volumes and heat capacities of transfer of the electrolytes from water to the mixed solvents were determined. The converse transfer functions of the nonelectrolyte (cosolvent) at 0.4m from water to the aqueous NaCl solutions were also determined. These transfer functions can be interpreted in terms of pair and higher order interactions between the electrolytes and the cosolvent.

In recent years we have undertaken a systematic investigation of the volumes and heat capacities of transfer of alkali halides and tetraalkylammonium bromides from water to mixed aqueous solvents (*1–6*). These properties are important because, when combined with enthalpies and free energies, they can be used to calculate the temperature and pressure dependences of various equilibrium properties of electrolytes in mixed solvents. Since the properties of electrolytes in mixed aqueous solvents are closely related to the corresponding properties of the nonelectrolyte in an electrolyte solution, information on the factors affecting salting out and salting in can be obtained from the thermodynamics of electrolytes in mixed aqueous solvents. Finally, different properties measure different proportions of the various interactions between components, and in an aqueous solvent properties such as heat capacities are much more sensitive to structural changes in the medium than are free energies.

In a ternary system the property we are seeking is the standard function (Y) of transfer of an electrolyte (E) from water (W) to a mixed solvent of a nonelectrolyte in water (W + N). This function is defined by

$$\Delta Y_E^0 (W \rightarrow W + N) = \overline{Y}_E^0 (W + N) - \overline{Y}_E^0 (W) \tag{1}$$

where $\overline{Y}_e^0$ is the standard partial molal quantity of the electrolyte in water and in the mixed solvent. The converse transfer functions of the nonelectrolyte from water to an electrolyte solution are given by

$$\Delta Y_N^0 (W \rightarrow W + E) = \overline{Y}_N^0 (W + E) - \overline{Y}_N^0 (W) \tag{2}$$

With most properties (enthalpies, volumes, heat capacities, etc.) the standard state is infinite dilution. It is sometimes possible to obtain directly the function near infinite dilution. For example, enthalpies of solution can be measured in solution where the final concentration is of the order of 10^{-3} molar. With properties such as volumes and heat capacities this is more difficult, and, to get standard values, it is usually necessary to measure apparent molal quantities ϕ_Y at various concentrations and extrapolate to infinite dilution ($\phi_Y^0 = \overline{Y}^0$). Fortunately, it turns out that, at least with volumes and heat capacities, the transfer functions $\Delta Y_E (W \rightarrow W + N)$ do not vary significantly with the electrolyte concentration as long as this concentration is relatively low (3). With most of the systems investigated, the transfer functions were calculated from apparent molal quantities at $0.1m$ and assumed to be equivalent to the standard values.

The cosolvents chosen for this study were urea (U), acetone (ACT), dimethylsulfoxide (DMSO), *p*-dioxane (D), piperidine (PD), morpholine (M), *tert*-butanol (TBA), and to a lesser extent acetamide (ACM). The study of the binary system was also extended to piperazine (PZ) and tetrahydropyran (THP). This choice of cosolvents is sufficiently varied to allow an examination of the various factors which influence the transfer functions.

As typical electrolytes we have taken LiCl, NaCl, Me$_4$NBr, and Bu$_4$NBr; NaCl is the most studied alkali halide, Bu$_4$NBr is a well-known hydrophobic electrolyte, Li$^+$ is more solvated than Na$^+$ (coulombic hydration) but its structural hydration is usually smaller, and Me$_4$NBr is a weak structure breaker. A comparison of these electrolytes should therefore give us some idea of the various interactions between different electrolytes and mixed solvents.

Experimental

Rather precise apparent molal data are required to see the cosolvent dependence of the transfer functions. With volumes, a precision in densities of

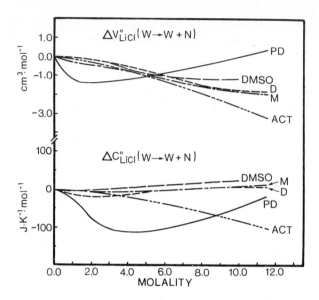

Figure 1. Transfer functions of 0.1m LiCl from water to aqueous mixed solvents at 25°C

a few ppm is required. Differences in densities of this precision are readily available by various buoyancy or dilatometric techniques. A flow digital densimeter (7) is about as sensitive but has the advantage of being rapid (10 min per measurement) and requires only a small quantity of liquid (generally less than 10 cm³).

Standard heat capacities of transfer can be derived from the temperature dependence of standard enthalpies of solution (8). While this technique can give general trends in the transfer functions from water to mixed solvents (9), it is not always sufficiently precise to detect the differences between similar cosolvents, and the technique is rather laborious. Direct measurements of the difference between heat capacities per unit volume of a solution and of the solvent $\sigma - \sigma_0$ can be obtained with a flow microcalorimeter (10) to $\pm 7 \times 10^{-5}$ JK^{-1} cm^{-3} on samples of the order of 10 cm³. A commercial version of this instrument (Picker dynamic flow calorimeter, Techneurop Inc.) has a sensitivity improved by a factor of about two.

The apparent molal volumes and heat capacities were calculated from the relations

$$\phi_V = M/d - 1000\,(d - d_0)/dd_0m \tag{3}$$

and

$$\phi_C = Mc_p + 1000\,(c_p - c_p{}^0)/m \tag{4}$$

where M is the solute molecular weight, c_p and d are the heat capacities per unit weight ($c_p = \sigma/d$), and densities of the solutions and $c_p{}^0$ and d_0 are the corresponding properties of the pure solvent or pure mixed solvent. Concentrations of solutes in mixed solvents are often given in aquamolalities (number of moles of solute per 55.51 moles of mixed solvent). However, ϕ_V and ϕ_C data calculated from aquamolalities are exactly the same as those calculated from molalities (m), and only the concentration scales are shifted proportionally. It therefore makes little difference which scale is used, and molalities were used for simplicity.

The actual experimental data on the binary system and the transfer functions from water to U (*1, 2, 6*), TBA (*3*), cyclic ethers and amines (*4*), ACT, DMSO, and ACM (*5*) are available elsewhere. A summary of these data are shown for the four electrolytes LiCl, NaCl, Me₄NBr, and Bu₄NBr in Figures 1–5. The cyclic amines hydrolyze in water, and as a result their ϕ_V and ϕ_C in water are too negative at low concentrations. A hydrolysis correction can be made (*11*), or this effect can be eliminated by measuring the data in about $0.1N$ NaOH rather than in pure H_2O (*4*). The ϕ^0 values of cyclic amines were taken by extrapolation of the data in $0.1N$ NaOH. The excess volumes $\phi_V^{EX} = \phi_V - \phi_V^0$ and heat capacities $\phi_C^{EX} = \phi_C - \phi_C^0$ of the nonelectrolytes in water are shown in Figures 6 and 7. The values of ϕ_V^0, ϕ_C^0, and the initial concentration dependence of ϕ_C and ϕ_V are listed in Table I. It turns out that for the ternary system hydrolysis has little effect on the trends and can usually be neglected (*4*).

The complementary transfer functions $\Delta Y_N^0(W \rightarrow W + NaCl)$ were de-

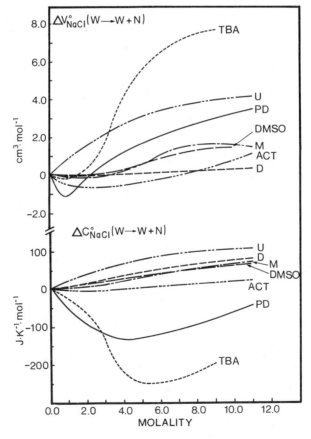

Figure 2. Transfer functions of 0.1m NaCl from water to aqueous mixed solvents at 25°C

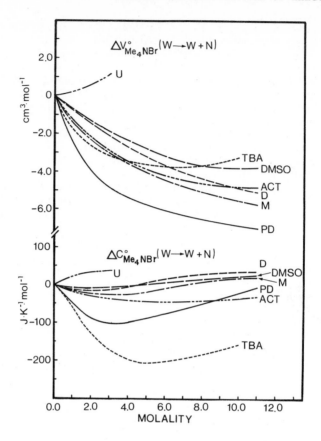

*Figure 3.   Transfer functions of 0.1m Me₄NBr from
water to aqueous mixed solvents at 25°C*

Table I.   Values of $\phi_V{}^\circ$, $\phi_C{}^\circ$, A_V, and B_V of Various Nonelectrolytes
in Water at 25°C

Nonelectrolyte	$\phi_V{}^\circ$ $(cm^3\ mol^{-1})$	A_V $(cm^3\ kg\ mol^{-2})$	$\phi_C{}^{\circ\ a}$ $(JK^{-1}\ mol^{-1})$	$B_C{}^a$ $(JK^{-1}\ kg\ mol^{-2})$
D	80.94	−0.17	220.1	−11.8
THP	91.74	−1.79	385.4	0.5
M (0.1N NaOH)	82.54	−0.48	237.3	−6.6
PZ (0.1N NaOH)	83.31	−1.01	262.1	−1.5
PD (0.1N NaOH)	91.65	−2.42	405.5	17.0
U	44.24	0.12	85.5	5.7
ACM	55.61	−0.12	162.5	−1.4
ACT	66.92	−0.37	240.5	−7.8
DMSO	68.92	−0.26	183.7	−5.5

^{a} Corrected values (29).

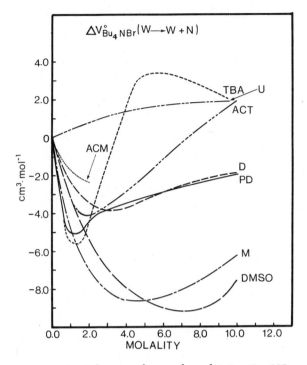

Figure 4. Volumes of transfer of 0.1m Bu_4NBr from water to aqueous mixed solvents at 25°C

termined for all the nonelectrolytes including PZ and THP, and they are compared with the low concentration data of $\Delta Y^0{}_{NaCl}(W \rightarrow W + N)$ in Figures 8 and 9 [The equation for $\Delta C^0{}_{NaCl}(W \rightarrow W + U)$ in Ref. 2 should read $\Delta C^0{}_{NaCl} = 23.2m - 1.85m^2 + 0.05_5m^3$]. In these experiments the nonelectrolyte concentration was kept constant at $0.4m$. As seen from Figures 8 and 9, both sets of data fall essentially on the same line. The small differences observed with some systems probably arise because the transfer functions are not truly standard ($0.1m$ for NaCl and $0.4m$ for N). The experimental slopes are summarized in Table II. At high cosolvent concentration the equivalence between the two complementary transfer functions is lost, as we can see in Figure 10 from the enthalpies, volumes, and heat capacities of the system W–NaCl–TBA (12). Similar observations are also made for the system W–Bu_4NBr–TBA (12). The equivalence of the complementary transfer functions at low concentrations is therefore a general phenomenon for all thermodynamic properties and all E–N–W systems and follows from the cross differentiation rule (13).

Discussion

Solute–Solute Model. The observed transfer functions of electrolytes in the mixed aqueous solvents are not simple, and they show various maxima and minima in the cosolvent concentration dependence—especially with hydrophobic

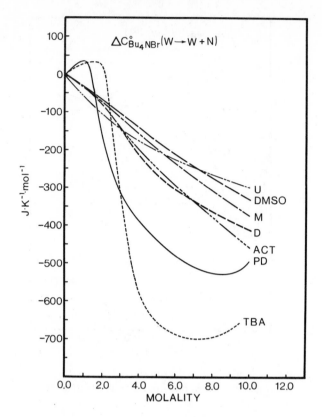

*Figure 5.   Heat capacities of transfer of 0.1m Bu₄NBr
from water to aqueous mixed solvents at 25°C*

Table II.   Transfer Parameters for Various Nonelectrolytes from
Water to Aqueous NaCl Solutions at 25°C

Nonelectrolyte	$\frac{1}{m} \Delta V_N^\circ$ $(W \to W + NaCl)$	$\frac{1}{m} \Delta C_N^\circ$ $(W \to W + NaCl)$
U	0.72	22
ACT	−0.17	−5
DMSO	−0.01	5
D	−0.11	5
THP	−0.06	—
M	−0.15	0
PZ	−0.10	−2
PD	−0.37	−17

electrolytes such as Bu₄NBr.   There has been a strong tendency in the past to interpret mixed solvents in terms of a variable dielectric constant.   For example, with D + W mixtures, a range of dielectric constants can be achieved, and the conductance of various electrolytes can be accounted for satisfactorily from the Fuoss–Onsager equation using a mean dielectric constant *(13)*.   Such an approach

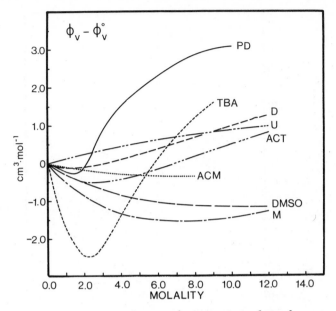

Figure 6. Excess volumes of various nonelectrolytes in water at 25°C; ϕ_V^0 of M and PD were obtained by extrapolation of data above 0.2m to infinite dilution.

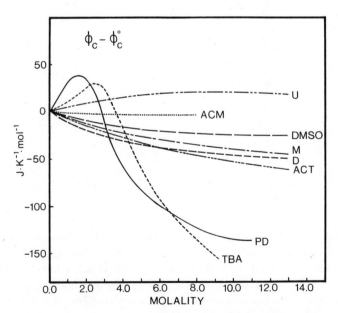

Figure 7. Excess heat capacities of various nonelectrolytes in water at 25°C; ϕ_C^0 of M and PD were obtained by extrapolation of data above 0.2m to infinite dilution.

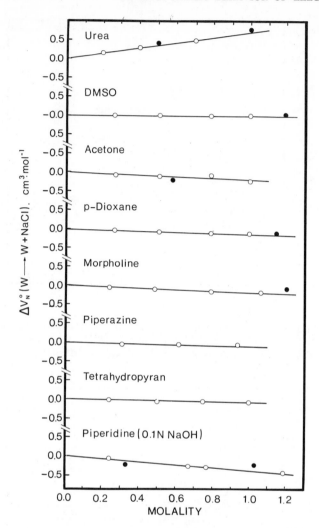

*Figure 8.   Volumes of transfer of the various nonelec-
trolytes at 0.4m from water to aqueous NaCl solutions:
● corresponding $\Delta V^0{}_{NaCl}(W \rightarrow W + N)$*

is not very promising with the present properties.   The dielectric constants of
U + W mixtures are nearly equal to that of pure water, but significant transfer
functions are observed in U + W mixtures.   Also, D has the lowest dielectric
constant of all the present nonelectrolytes, but the transfer functions in this mixed
solvent are among the less abnormal.   Of course with volumes and heat capacities
of transfer, it is not the dielectric constant that is involved but its pressure and
second temperature derivatives, and these properties are not available at present.
Still, it is doubtful that this approach would be very fruitful.

The second approach that has been rather popular with mixed aqueous solvents is to assume that the mixture is more or less structured than that of pure water.   There is much evidence to show that the particular hydrogen-bonded structure of water influences many of the properties of electrolytes in water (*15*). If nonelectrolytes can modify the structure of water (*15*), they can have an indirect effect on the properties of electrolytes.   This explanation has been particularly successful in the case of U + W mixtures (*1, 2*).   Such a simple approach is not as successful with hydrophobic cosolvents.   For example, $\Delta H_E^0(W \to W + TBA)$ are positive for both alkali halides (*16*) and tetraalkylammonium bro-

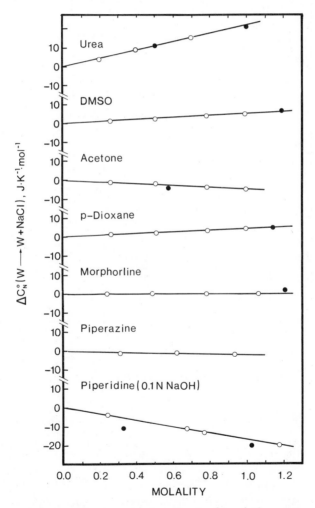

Figure 9.   Heat capacities of transfer of the various nonelectrolytes at 0.4m from water to aqueous NaCl solutions: ● *corresponding* $\Delta C^0_{NaCl}(W \to W + N)$

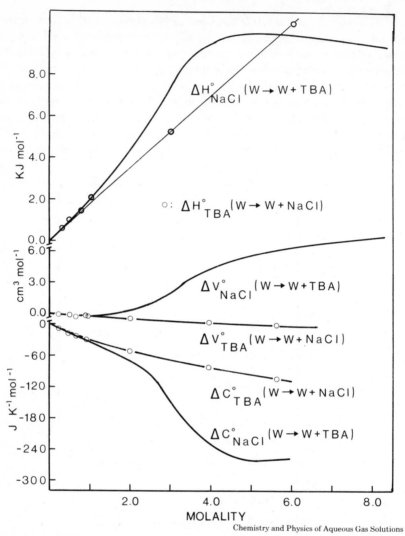

Figure 10. Thermodynamic properties of the system TBA–NaCl–W (12)

mides (*9*), while opposite signs would have been predicted from this approach.

The identity of $\Delta Y_E^0(W \rightarrow W + N)$ with $\Delta Y_N^0(W \rightarrow W + E)$ strongly suggests that the transfer functions are reflecting primarily solute–solute interactions between E and N (*13*). This can be illustrated in the following way (*12*): the thermodynamic function $\overline{Y}_N$ of the nonelectrolyte can be expressed in terms of the concentration of the various components,

$$\overline{Y}_N = \overline{Y}_N^0 + B_{NN}m_N + B_{NE}m_E + B_{NNE}m_N m_E + B_{NEE}m_E^2 + \ldots \quad (5)$$

where the B's are interaction parameters between the various solutes and also include any change in solute–solvent interactions. For simplicity, the electrolyte is not separated into its ionic components even though it is obvious that the interactions of a nonelectrolyte with a cation can be quite different from those with an anion, especially with Bu_4NBr.

If the nonelectrolyte concentration is low enough, all terms involving m_N can be neglected. The function $\overline{Y}_N$ then becomes

$$\overline{Y}_N{}^* = \overline{Y}_N{}^0 + B_{NE}m_E + B_{NEE}m_E{}^2 + \ldots \tag{6}$$

Similarly, the function for the electrolyte is

$$\overline{Y}_E{}^* = \overline{Y}_E{}^0 + B_{EN}m_N + B_{ENN}m_N{}^2 + \ldots \tag{7}$$

The concentration dependence of the transfer functions are thus given by

$$\frac{d}{dm_E}\{\Delta Y_N{}^0(W \rightarrow W + E)\} = d\overline{Y}_N{}^*/dm_E = B_{NE} + 2B_{NEE}m_E + \ldots \tag{8}$$

and

$$\frac{d}{dm_N}\{\Delta Y_E{}^0(W \rightarrow W + N)\} = d\overline{Y}_E{}^*/dm_N = B_{EN} + 2B_{ENN}m_N + \ldots \tag{9}$$

The identical slopes observed in Figures 8–10 suggest that at low concentrations $B_{EN} = B_{NE}$; this means that the interactions of a single ion with the surrounding nonelectrolytes is the same as that of a single nonelectrolyte with the surrounding ions provided only pair interactions are involved. When triplet and higher interactions become important, this identity is lost.

Structural Interactions. All electrostatic and dispersion forces will contribute to the solute–solute interactions, but they alone cannot account for the observed trends, and it is often necessary to make use of the concept of structural hydration interactions. By structural interaction we mean any interaction which involves the three-dimensional structure of the solvent. With hydrophilic ions the structure breaking effect (*17, 18*) is now well understood and generally accepted. With hydrophobic solutes the problem is more complicated. While the high heat capacity and viscosity of hydrophobic nonelectrolytes or electrolytes seem to be observed only with highly structured liquids like water, the nature of the interactions giving rise to these effects is still obscure. Hydrophobic hydration might be similar to a time-averaged clathrate hydrate (*19*); it might be related to the difficulty in forming a cavity in a highly structured solvent (*20, 21*), or it might simply be a surface effect (*22*).

Thermodynamic data never give us any direct information on the molecular nature of the solute–solute or solute–solvent interactions. It is only through a comparison with other systems and through models and theories that the relative importance of the various types of interactions can be established. This comparative approach will therefore be used with the transfer functions.

The simplest cosolvents to understand are the strongly hydrophobic ones. The shapes of ϕ_V^{EX} and ϕ_C^{EX} of TBA and PD in Figures 6 and 7 are characteristic of hydrophobic solutes. By analogy with micelle formation, the large decrease in ϕ_C^{EX} (23) and increase in ϕ_V^{EX} (24) at higher concentrations are probably caused by some association resulting from hydrophobic bonding. The transfer functions $\Delta Y^0_{Bu_4NBr}(W \to W + TBA)$ and $\Delta Y^0_{Bu_4NBr}(W \to W + PD)$ have qualitatively the same shape as ϕ_V^{EX} and ϕ_C^{EX} of TBA and PD. In other words, Bu_4NBr is behaving qualitatively as if it were another TBA or PD molecule. The converse transfer functions $\Delta Y^0_{TBA}(W \to W + Bu_4NBr)$ (12) are also similar in shape to $\Delta Y^0_{Bu_4NBr}(W \to W + TBA)$ as we would expect from this interpretation.

With alkali halide–TBA–W or alkali halide–PD–W systems, the parameters B_{NE} are negative for volumes and heat capacities (see Figures 1–5 and 10). This sign seems to be the one usually observed for the interaction of a hydrophobic with a hydrophilic solute (6). At intermediate cosolvent concentration, $\Delta Y_E^0(W \to W + TBA)$ and $\Delta Y_E^0(W \to W + PD)$ deviate in the direction we would expect for hydrophobic association: the volume increases sharply, and the heat capacity decreases further. Inorganic electrolytes lower the critical micelle concentration of surfactants by salting out the monomers, thus favoring micellization (25); in a similar way, in the co-sphere of a hydrophilic ion, the hydrophobic bonding between the cosolvent molecules may be enhanced.

Significant differences are observed between the ternary and binary systems at high cosolvent concentrations. This is expected since at high concentrations the transfer functions are tending to $\Delta Y_E^0(W \to N)$. If data in the pure cosolvent were available, it would be preferable to speak of excess transfer functions (26, 27) defined by

$$\Delta Y_E^{EX}(W, N \to W + N) = \overline{Y}_E^0(W + N) - X_W\overline{Y}_E^0(W) - X_N\overline{Y}_E^0(N) \quad (10)$$

where X is the respective mole fraction. As a result of the low solubility, ϕ_V^0 and ϕ_C^0 of most electrolytes are unfortunately not known in the pure cosolvents investigated.

Urea, as a cosolvent, is at the other extreme. All the concentration dependences of the binary and ternary systems are quite regular. The excess volume (Figure 6) is positive, which is rarely observed for nonelectrolytes in water. With the exception of the heat capacities of Bu_4NBr, all the parameters B_{EU} are positive for volumes and heat capacities, and the sign of the transfer functions is always opposite what we would expect for the structural hydration contribution to $\overline{V}^0$ and $\overline{C}_p^0$.

Similarly, $\Delta H_U^0(W \to W + E)$ is of opposite sign to the expected structural hydration contribution of E (28). There does not seem to be too much specificity in the interactions involving U which probably acts as a statistical structure breaker; the local U–W interactions are probably not too different from the W–W ones, but the long-range ordering is destroyed. Schrier et al. (28) have interpreted the B_{UE} parameters in terms of a destructure overlap cosphere model. Since urea acts as a structure breaker, such a model will predict the same sign for the transfer functions as the models which assume that the U–W mixtures act as a uniform solvent less structured than pure water.

The other nonelectrolytes (D, ACT, ACM, M, DMSO, and PZ) are intermediate between U and the strongly hydrophobic cosolvent (TBA, PD, and THP). As we can see from Table II and from Figures 1–10, there is not much specificity in $\Delta Y_E^0(W \rightarrow W + N)$. Most of the specificity at higher cosolvent concentration comes from the tendency of $\Delta Y_E^0(W \rightarrow W + N)$ to approach $\Delta Y_E^0(W \rightarrow N)$. If we assume that at high cosolvent concentration $\bar{Y}_E^0(W + N)$ are all tending to their intrinsic value, the present data suggest that all these nonelectrolytes are slightly hydrophobic. This agrees with the interpretation of the properties of the binary systems N–W (*4, 5*). It is not always easy to identify the trends in the functions for the various N; the situation would probably be easier if we had $\Delta Y_E^{EX}(W, N \rightarrow W + N)$ data. Still from the slopes B_{EN}, the deviations from linearity, and the minima and maxima, we can probably conclude that the hydrophobic character of the nonelectrolytes is increasing in the order

$$U < ACM \leq D \leq DMSO < M < PZ < ACT < THP < PD < TBA$$

In general, the transfer functions do not show much specificity for the alkali halides (*3, 12*). The lithium salts are exceptions in that the magnitude of the transfer functions is significantly smaller than those of the corresponding sodium salts, and they show very little specificity for the polar nonelectrolytes. The transfer functions in Figure 1 are tending in a very regular way to the limiting $\Delta Y^0_{LiCl} (W \rightarrow N)$, and the little specificity observed is probably caused by the Cl^- ion. These observations therefore support the suggestion that Li^+ has very little structural hydration (*2*).

The transfer functions of Me_4NBr (Figure 3) also show relatively little specificity if we assume that the high concentration data is tending to $\Delta Y^0_{Me_4NBr}(W \rightarrow N)$. The deviations from this ideal behavior are in the direction we would expect if Me_4NBr is a structure breaker.

Acknowledgment

We thank E. G. Schrier for making his manuscript available to us prior to publication.

Literature Cited

1. Philip, P. R., Desnoyers, J. E., Hade, A., *Can. J. Chem.* (1973) **51**, 187.
2. Desrosiers, N., Perron, G., Mathieson, J. G., Conway, B. E., Desnoyers, J. E., *J. Solution Chem.* (1974) **3**, 789.
3. Avédikian, L., Perron, G., Desnoyers, J. E., *J. Solution Chem.* (1975) **4**, 331.
4. Kiyohara, O., Perron G., Desnoyers, J. E., *Can. J. Chem.* (1975) **53**, 2591.
5. Kiyohara, O., Perron G., Desnoyers, J. E., *Can. J. Chem.* (1975) **53**, 3263.
6. Desrosiers, N., Desnoyers, J. E., *Can. J. Chem.* (1975) **53**, 3206.
7. Picker, P., Tremblay, E., Jolicoeur, C., *J. Solution Chem.* (1974) **3**, 377.

8. Criss, C. M., Cobble, J. W., *J. Am. Chem. Soc.* (1961) **83**, 3223.
9. Mohanty, R. K., Sarma, T. S., Subramanian, S., Ahluwalia, J. C., *Trans. Faraday Soc.* (1971) **67**, 305.
10. Picker, P., Leduc, P.-A., Philip, P. R., Desnoyers, J. E., *J. Chem. Thermodyn.* (1971) **3**, 631.
11. Cabani, S., Conti, G., Lepori, L., *J. Phys. Chem.* (1972) **76**, 1338.
12. Desnoyers, J. E., Perron, G., Morel, J.-P., Avédikian, L., in "Chemistry and Physics of Aqueous Gas Solutions," W. A. Adams, Ed., Electrochemical Society, Princeton, 1975, p 172.
13. Stern, J. H., Lazartic, J., Frost, D., *J. Phys. Chem.* (1968) **72**, 3053.
14. Lind, J. E., Jr, Fuoss, R. M., *J. Phys. Chem.* (1961) **65**, 989.
15. Franks, F., "Water, a Comprehensive Treatise," Vols. 2 and 3, Plenum, New York, 1973.
16. Pointud, Y., Ducros, M., Juillard, J., Avédikian, L., Morel, J.-P., *Thermochim. Acta* (1974) **8**, 423.
17. Frank, H. S., Evans, M. W., *J. Chem. Phys.* (1945) **13**, 507.
18. Lewis, G. N., Randal, M., "Thermodynamics," K. S. Pitzer and L. Brewer, Eds., McGraw-Hill, New York, 1961, p 378.
19. Glew, D. N., Moelwyn-Hughes, E. A., *Discuss. Faraday Soc.* (1953) **15**, 150.
20. Pierotti, R. A., *J. Phys. Chem.* (1963) **67**, 1840.
21. Ibid. (1965) **69**, 281.
22. Aveyard R., Heselden, R., *J. Chem. Soc. Faraday I*, (1975) **71**, 312.
23. Leduc, P.-A., Fortier, J.-L., Desnoyers, J. E., *J. Phys. Chem.* (1974) **78**, 1217.
24. Desnoyers, J. E., Arel, M., *Can. J. Chem.* (1967) **45**, 359.
25. Shinoda, K., Nakagawa, T., Tamemushi, B., Isemura, T., "Colloidal Surfactants," Academic, London, 1963.
26. de Visser, C., Somsen, G., *J. Chem. Thermodyn.* (1973) **5**, 147.
27. de Visser, C., Somsen, G., *J. Solution Chem.* (1974) **3**, 847.
28. Schrier, M. Y., Turner, P. J., Schrier, E. E., *J. Phys. Chem.* (1975) **79**, 1391.
29. Desnoyers, J. E., deVisser, C., Perron, G., Picker, P., *J. Solution Chem.* (1976) **5**, 605.

RECEIVED June 9, 1976. Work supported by the National Research Council of Canada and Environment Canada.

17

Enthalpies of Solution of Several Solutes in Aqueous–Organic Mixed Solvents

C. DE VISSER and G. SOMSEN

Department of Chemistry, Free University of Amsterdam, de Lairessestraat 174, Amsterdam, The Netherlands

Measured enthalpies of solution of n-Bu$_4$NBr *at 5°, 25°, and 55°C in mixtures of DMF and water are compared with those of other tetraalkylammonium bromides, rubidium chloride, and urea, reported earlier. For the latter salts, the enthalpies of solution in the mixtures are almost proportional to the mole fraction of water, but for* n-Bu$_4$NBr *a large endothermic maximum occurs. From this profile it appears to be possible, using a simple hydration model, to calculate both the number of water molecules which surround a hydrophobic alkyl group and the enthalpic effect of the hydrophobic hydration. The latter effect appears to diminish strongly with increasing temperature, whereas the number of water molecules involved to the hydrophobic hydration changes very little with temperature.*

One difference in behavior between the hydrophilic alkali halides and hydrophobic solutes like the larger tetraalkylammonium halides in water is expressed by the enthalpy. The enthalpies of solution of the larger tetraalkylammonium halides in water are more exothermic than those of the corresponding alkali halides but in other solvents, e.g., several amides, propylene carbonate (PC), and dimethylsulfoxide (DMSO), the reverse is true. Generally, this phenomenon is attributed to an enhanced hydrogen bonding in the highly structured solvent water in the vicinity of the tetraalkylammonium ions (hydrophobic hydration) (*1*). This idea is substantiated by the absence of the effect in solvents like *N,N*-dimethylformamide (DMF), PC, and DMSO (*2*), where specific structural effects are not present in the pure solvents.

In order to see how the hydrophobic hydration breaks down when an aprotic cosolvent is added to water and how it starts to contribute when water is added to the aprotic solvent, we measured the enthalpies of solution of hydrophobic and, for comparison, hydrophilic solutes in mixed solvent systems of water and

Table I.   Standard Enthalpies of Solution in kJ mol^{-1} of
Tetra-n-butylammonium Bromide in Mixtures of
N,N-dimethylformamide and Water at 5°, 25°, and 55°C as a
Function of the Mole Fraction of Water, X_{H_2O}

5°C		25°C		55°C	
X_{H_2O}	$\Delta H°$ (sol.)	X_{H_2O}	$\Delta H°$ (sol.)	X_{H_2O}	$\Delta H°$ (sol.)
0	+9.32	0	+12.5	0	+15.90
0.224	+16.31 ± 0.01	0.192	+18.48 ± 0.05	0.124	+19.88 ± 0.04
0.421	+22.10 ± 0.06	0.311	+22.07 ± 0.05	0.321	+25.68 ± 0.04
0.601	+25.86 ± 0.03	0.488	+27.12 ± 0.04	0.519	+31.78 ± 0.05
0.676	+25.69 ± 0.02	0.623	+29.63 ± 0.05	0.681	+35.02 ± 0.05
0.806	+18.45 ± 0.01	0.695	+29.77 ± 0.04	0.778	+34.31 ± 0.05
0.890	+5.26 ± 0.02	0.760	+27.89 ± 0.03	0.874	+29.38 ± 0.02
0.949	−8.19 ± 0.02	0.869	+18.21 ± 0.06	0.930	+23.92 ± 0.03
1	−23.76 ± 0.03	0.945	+5.26 ± 0.02	1	+13.08 ± 0.04
		1	−8.58 ± 0.06		

an aprotic solvent like DMF.   In the present paper, the enthalpies of solution of
tetra-n-butylammonium bromide (n-Bu$_4$NBr) in DMF–water in particular are
considered and the results will be compared with those of other tetraalkylammonium bromides, RbCl, and urea published before (3, 4, 5).   Since it might be
expected that the hydrophobic hydration effect is very sensitive to temperature
changes, the enthalpies of solution of n-Bu$_4$NBr in DMF–water have been
measured at 5°, 25°, and 55°C.

Table II.   Standard Enthalpies of Solution in kJ mol^{-1}
Tetra-n-propylammonium Bromide, of Rubidium Chloride, and
Water as a Function of the

Me$_4$NBr[a]		Et$_4$NBr[a]		n-Pr$_4$NBr[b]	
X_{H_2O}	$\Delta H°$ (sol.)	X_{H_2O}	$\Delta H°$ (sol.)	X_{H_2O}	$\Delta H°$ (sol.)
0	11.3	0	9.4	0	9.9
0.100	16.52 ± 0.12	0.147	12.86 ± 0.06	0.143	13.64 ± 0.02
0.263	20.32 ± 0.03	0.310	15.30 ± 0.04	0.309	17.20 ± 0.05
0.364	21.72 ± 0.03	0.446	16.39 ± 0.05	0.494	20.28 ± 0.01
0.506	23.32 ± 0.01	0.563	15.96 ± 0.01	0.600	20.65 ± 0.01
0.674	22.76 ± 0.03	0.683	13.86 ± 0.03	0.705	18.92 ± 0.02
0.759	22.35 ± 0.04	0.820	9.91 ± 0.05	0.865	9.87 ± 0.03
0.849	21.97 ± 0.01	0.928	7.40 ± 0.04	1	−4.25 ± 0.05
0.949	23.33 ± 0.06	1	6.2		
1	24.57 ± 0.05				

[a] Reference 3.
[b] Reference 6.
[c] Reference 5.
[d] Reference 4.

Experimental and Results

All enthalpy of solution measurements were carried out with an LKB 8700-1 precision calorimetry system. The experimental procedure and tests of the calorimeter have been reported previously (3, 4, 5). The purification of the solvent DMF (Baker Analyzed Reagent) and of all solutes used has been described in the same papers. The solvent mixtures were prepared by weighing and the mole fraction of water in the DMF–water mixtures was corrected for the original water content of the amide as measured by Karl Fischer titration.

The enthalpy of solution measurements of n-Bu$_4$NBr in DMF–water were made in very dilute solutions (0.02–0.001 mole kg^{-1}) so that, in view of the experimental error, any concentration dependence of the enthalpies of solution in these solutions was neglected. Consequently, the enthalpy of solution at infinite dilution, ΔH^0(sol.), was taken to be the average of three or more independent measurements agreeing within 150 J mol^{-1}. Final results of ΔH^0(sol.) of n-Bu$_4$NBr with their mean deviations at 5°, 25°, and 55°C are given in Table I. The results for the other tetraalkylammonium bromides, for RbCl, and for urea in DMF–water at 25°C are summarized in Table II.

Discussion

Previously we found that the enthalpies of solution of n-Bu$_4$NBr in mixtures of DMF and n-methylformamide (NMF) are almost proportional to the solvent composition and may be regarded as close to ideal (6). In this context, ideal behavior means that $\Delta H^{id} = X_A \Delta H_A^0 + X_B \Delta H_B^0$ in which ΔH_A^0 and ΔH_B^0 are the enthalpies of solution in two pure solvents A and B. As a consequence, the

of Tetramethylammonium-, Tetraethylammonium-, and
of Urea at 25°C in Mixtures of N,N-Dimethylformamide and
Mole Fraction of Water, X_{H_2O}

RbCl[c]		urea[d]	
X_{H_2O}	ΔH° (sol.)	X_{H_2O}	ΔH° (sol.)
0	0.89	0	5.85 ± 0.03
0.513	13.22 ± 0.06	0.100	5.44 ± 0.03
0.651	16.10 ± 0.05	0.200	5.48 ± 0.01
0.749	16.34 ± 0.02	0.300	5.75 ± 0.03
0.828	16.13 ± 0.03	0.400	6.32 ± 0.04
0.934	16.04 ± 0.03	0.500	7.31 ± 0.03
1	17.21 ± 0.02	0.601	8.70 ± 0.03
		0.700	10.25 ± 0.01
		0.800	11.89 ± 0.01
		0.928	14.05 ± 0.03
		1	15.28 ± 0.04

excess enthalpy of solution ΔH^E(sol.) is defined as ΔH^E(sol.) $= \Delta H^{exp} - \Delta H^{id}$ $= 0$ for n-Bu$_4$NBr in DMF–NMF.

In DMF–water mixtures, the ΔH^E(sol.) of n-Bu$_4$NBr deviates substantially from zero, but for Me$_4$NBr, urea, and RbCl, ΔH^E(sol.) is small, especially when compared with the values for n-Bu$_4$NBr (*see* Figure 1).

In view of the different behavior of n-Bu$_4$NBr in mixtures of DMF and NMF and of DMF and water, we recently (6) derived an equation for the excess enthalpy of solution in the DMF–water mixture (ΔH^E(M)) by use of a simple hydrophobic hydration model. Summarizing this derivation, we conceived the enthalpies of solution in the DMF–H$_2$O system (ΔH^0(M)) as being the result of two effects: (a) When the hydrophobic hydration of tetraalkylammonium ions is absent, the corresponding enthalpy of solution in pure water ΔH^1(H$_2$O) and in the mixture ΔH^1(M) should be correlated by:

$$\Delta H^1(M) = X_{H_2O}\Delta H^1(H_2O) + (1 - X_{H_2O})\, \Delta H^0(DMF) \tag{1}$$

(b) However, in the real case, we must account for hydrophobic hydration with an enthalpic contribution both in pure water, Hb(H$_2$O) and in the aqueous mixture, Hb(M). Consequently, ΔH^0(M) $= \Delta H^1$(M) $+$ Hb(M). Since also ΔH^0(H$_2$O) $= \Delta H^1$(H$_2$O) $+$ Hb(H$_2$O), the excess enthalpy of solution in the mixture is given by:

$$\Delta H^E(M) = Hb(M) - X\, Hb(H_2O) \tag{2}$$

Values of Hb(M) with increasing mole fraction X of water can be calculated with the aid of the model of Mastroianni, Pikal, and Lindenbaum (7), using the following assumptions:

(a) In aqueous solution a n-Bu$_4$N$^+$ ion is surrounded by a cage of N water molecules and each butyl group is surrounded by a subcage of $N/4$ water molecules. The presence of one or more DMF molecules will prevent the formation of a subcage but will not affect the structures of the subcages around the other butyl groups.

(b) In a mixture of DMF and water, the distribution of solvent molecules surrounding the ion is random so that the probability of a given solvation site being occupied by a water molecule is X_{H_2O}. Hence, this probability is $X^{N/4}$ for $N/4$ sites.

Table III. Values of the Enthalpic Effect of Hydrophobic Hydration Hb(H$_2$O), the Number of Water Molecules per Alkyl Group $N/4$ Correlated with It, and the Mean Deviations of the Calculated from the Experimental Values of ΔH^E of n-Bu$_4$NBr in DMF-Water Mixtures at Different Temperatures

$T(°C)$	$Hb(H_2O)(kJ\ mol^{-1})$	$N/4$	Mean Deviation($kJ\ mol^{-1}$)
5	−65.6	6.0	0.38
25	−52.8	6.4	0.23
55	−34.4	6.7	0.15

(c) The enthalpic effect of hydrophobic hydration is merely the result of subcage formation and each butyl group contributes $\frac{1}{4}$ Hb(H_2O). Consequently, Hb(M) = 4·$X^{N/4}$·($\frac{1}{4}$) Hb(H_2O) = $X^{N/4}$Hb(H_2O) and Equation 2 becomes:

$$\Delta H^E = (X^{N/4} - X) \text{ Hb}(H_2O) \tag{3}$$

Equation 3 can be used in a curve-fitting program (6) to search for the best fit of the experimental data by optimizing both parameters Hb(H_2O) and $N/4$. These calculations were performed on an IBM-1130 computing system.   The curve fitting has been carried out for ΔH^E values of n-Bu$_4$NBr in mixtures of DMF and water at 5°, 25°, and 55°C as seen in Figure 2.   Values of Hb(H_2O) and $N/4$ corresponding to the best fit together with the mean deviation of the calculated ΔH^E values from the experimental ones, are given in Table III.   It should be noted that in the derivation of Equations 2 and 3 clearly it is assumed that any significant deviation of ΔH^E(M) from zero is caused by a change of the hydrophobic hydration of n-Bu$_4$ NBr in the mixed solvents.   In view of the very small values of ΔH^E(M) for non-hydrophobic solutes (Figure 1), this idea has been justified as a first approximation.

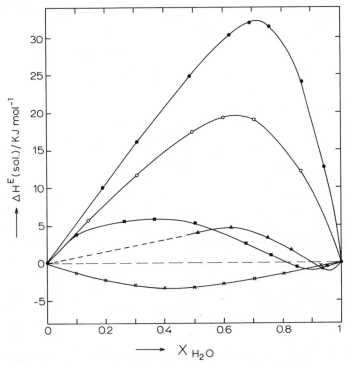

Figure 1.   Excess enthalpies of solution ΔH^E(sol.) of some tetraalkylammonium bromides, rubidium chloride, and urea in mixtures of N,N-dimethylformamide and water as a function of the mole fraction of water, X_{H_2O}. ●, n-Bu$_4$NBr; ○, n-Pr$_4$NBr; ■, Me$_4$NBr; ▲, RbCl; ×, urea.

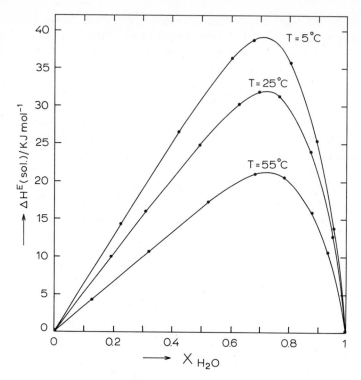

*Figure 2.   Excess enthalpies of solution $\Delta H^E(sol.)$ of n-Bu$_4$NBr
in mixtures of N,N-dimethylformamide and water as a function
of the mole fraction of water at 5°, 25°, and 55°C.*

Some interesting results can be deduced from Table III and Figure 2.

(a) At all three temperatures the fit of the experimental values given in
Figure 2 is very good.   As a consequence, the model given above provides a good
method to describe the experimental results.   This is demonstrated further if we
compare the enthalpies of solution themselves at different temperatures (Figure
3):   below $X_{H_2O} = 0.6$, the three $\Delta H^0(M)$ curves are parallel, indicating indeed
that, in the absence of hydrophobic hydration, the enthalpies of solution in the
DMF–water mixtures are close to ideal, whereas above $X_{H_2O} = 0.6$, deviations
occur which are caused by hydrophobic hydration in water and in water rich
mixtures.

(b) The value of $N/4$ does not change significantly as a function of temperature, whereas the value of Hb(H$_2$O) is almost reduced to 50% going from
5°C to 55°C.   In other words, the number of water molecules involved in the
hydrophobic hydration of n-Bu$_4$NBr is hardly any function of temperature, but
the enthalpic contribution to this hydration diminishes rather strongly with increasing temperature.   The constancy of $N/4$ as a function of temperature (N

is about 25) may support the idea of some clathrate-like structure in solution especially in view of the fact that solid hydrates of n-Bu$_4$NBr also contain clathrate-like structures in which the n-Bu$_4$N$^+$ ion is surrounded by about 32 water molecules (*1*). The decrease of Hb(H$_2$O) with increasing temperature may be caused by decrease of the hydrogen bond energy (*8*). This conclusion about the variation of the strength of the hydrogen bond with temperature supports the assumption necessarily made by Philip and Desnoyers (*9*) to predict the sign of heat capacities of electrolyte solutions from a two state model.

As a consequence of the model employed, values of Hb(H$_2$O) and N ought to be independent of the choice of the cosolvent as long as specific structural effects are absent. Therefore, we applied Equation 3 to the enthalpies of solution of n-Bu$_4$NBr in DMSO–water mixtures (*10*), since DMSO is a dipolar aprotic solvent like DMF. The best fit of the ΔH^E values in this mixture yields: Hb(H$_2$O) = −49.2 kJ mol^{-1} and N/4 = 6.4, in excellent agreement with our values at 25°C given in Table III.

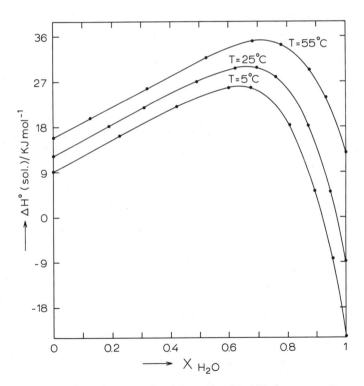

Figure 3.   Enthalpies of solution ΔH^0(sol.) of n-Bu$_4$NBr in mixtures of N,N-dimethylformamide and water as a function of the mole fraction of water at 5°, 25°, and 55°C.

Acknowledgments

We are grateful to W. J. M. Heuvelsland and P. Pel for their valuable assistance and stimulating discussions.

Literature Cited

1. Wen, W. Y., in "Water and Aqueous Solutions," R. A. Horne, Ed., Wiley-Interscience, New York, 1972, p 613.
2. Krishnan, C. V., Friedman, H. L., *J. Phys. Chem.* (1971) **75**, 3606.
3. de Visser, C., Heuvelsland, W. J. M., Somsen, G., *J. Sol. Chem.* (1975) **4**, 311.
4. de Visser, C., Grünbauer, H. J. M., Somsen, G., *Z. Phys. Chem.* (1975) **97**, 69.
5. de Visser, C., van Netten, E., Somsen, G., *Electrochim. Acta* (1976) **21**, 97.
6. de Visser, C., Somsen, G., *J. Phys. Chem.* (1974) **78**, 1719.
7. Mastroianni, M. J., Pikal, M. J., Lindenbaum, S., *J. Phys. Chem.* (1972) **76**, 3050.
8. Eisenberg, D., Kauzmann, W., "The Structure and Properties of Water," Oxford Press, New York, 1969, p 181.
9. Philip, P. R., Desnoyers, J. E., *J. Sol. Chem.* (1972) **1**, 353.
10. Fuchs, R., Hagan, C. P., *J. Phys. Chem.* (1973) **77**, 1797.

RECEIVED July 1, 1975.

Heats of Solution and Dilution of Lithium Perchlorate in Aqueous Acetonitrile

R. P. T. TOMKINS, G. M. GERHARDT, L. M. LICHTENSTEIN, and P. J. TURNER

Department of Chemistry, Rensselaer Polytechnic Institute, Troy, N. Y. 12181

Heats of solution of lithium perchlorate in 20, 40, 60, 80, 90, and 100 wt % acetonitrile–water mixtures at 298.16°K are reported. The heats of dilution were measured for lithium perchlorate in the mixed solvent containing 90 wt % CH_3CN. The heats of transfer (ΔH_{tr}) of lithium perchlorate from water to aqueous acetonitrile were calculated. The results are discussed in terms of the structure of the solvent system and selective solvation properties of the lithium ion.

The heats of solution and dilution of electrolytes in nonaqueous–aqueous solvent mixtures have been limited mostly to alcohol–water systems and a few measurements in dimethylsulfoxide–water and dioxane–water mixtures (1, 2). The structural maximum in aqueous–organic solvents at high water content has been well established by a variety of techniques (3, 4), but few systems have been explored over the whole composition range.

The purpose of this study is to examine the structural features of acetonitrile–water mixtures over the whole composition range using the heats of solution and dilution of lithium perchlorate as a probe. The effect of water on thermodynamic properties such as heats of solution is also of interest.

Acetonitrile has been selected as the solvent in this study since it is a possible candidate for a nonaqueous electrolyte battery (5). From this viewpoint, acetonitrile has several attractive physical properties, as shown in Table I. It has a useful liquid state range and a reasonably low vapor pressure and viscosity at ambient temperature. In addition, many common electrolytes are soluble in acetonitrile. Acetonitrile is a good model solvent for solvation studies, as the molecule is a linear aprotic dipole.

Table I.   Physical Properties of Acetonitrile (1)

Boiling point	81.60° C
Freezing point	−43.84° C
Vapor pressure	88.81 Torr (at 25° C)
Viscosity	0.375 cP (at 15° C)
Dielectric constant	37.5 (at 20° C)

Nonaqueous Electrolytes Handbook

Experimental

Lithium perchlorate was purified as described earlier (6). Acetonitrile was Fisher ACS reagent grade and was used without further purification. Water was distilled twice, the second time in a Corning AG-2 all glass still, and had a specific conductivity at 25°C of 8×10^{-7} ohm^{-1} cm^{-1}. The mixed solvents were prepared by weight as shortly as possible before heat measurements were made.

Heats of solution were measured at 298.16°K using an LKB 8700-1 precision calorimeter. Details of the procedure have been described earlier (6). The principal difference in procedure in this case was that the ampoules were to be filled with solid instead of solution. The ampoules were filled with crushed lithium perchlorate using a one-mm copper wire as a ramrod, and their contents were weighed by difference in a nitrogen-filled drybox. They were placed about four at a time in a small desiccator and transferred to the sealing apparatus and sealed as quickly as possible. Solid sticking to the filling stem was weighed by difference on washing and drying the stem after sealing. Heats of dilution were measured at 298.16°K with an LKB 11700 batch microcalorimeter.

The heats of solution of lithium perchlorate in aqueous acetonitrile were measured at concentrations between 0.01 and 0.1m. The concentration dependence was small compared with the experimental scatter of about 0.1–0.2 kcal mole^{-1}. ΔH_S values are given in Table II. The heats of solution in anhydrous acetonitrile were corrected to infinite dilution using measured heats of dilution (6), and the corrected values were averaged. The heats of dilution were measured for lithium perchlorate in the mixed solvent containing 90% MeCN.

Table II.   Heats of Solution of Lithium Perchlorate in Aqueous Acetonitrile

MeCN (wt %)	$-\Delta H_S$ (kcal mol^{-1})	Mean deviation (kcal mol^{-1})	No. of measurements
0	6.35	0.05	10
20	7.5	0.09	14
40	10.0	0.16	11
60	12.0	0.26	9
80	14.2	0.28	10
90	15.8	0.20	5
100	13.6	0.15	4

Table III. Heats of Dilution of Lithium Perchlorate in 10% Aqueous Acetonitrile

m^a (final)	$m^{1/2}$	$-\Delta H_{Dil}{}^b$ (cal mol^{-1})	ϕ_L (cal mol^{-1})
0.0180	0.134	816	69
0.0448	0.212	776.2	109
0.1124	0.335	712.3	172
0.3049	0.552	543.7	341
1.0172	1.009	0	885

[a] Initial concentration for each dilution is $1.0172m$.
[b] Estimated uncertainty of measurement: ±30 cal mol^{-1}.

The data were analyzed by the method of Harned and Owen (7). There was little difference between a linear and a square root extrapolation at the lowest measured molalities. A square root extrapolation was used. Heats of dilution and ϕ_L values are listed in Table III. Figure 1 shows the dependence of ϕ_L versus $m^{1/2}$.

Discussion

Heats of solution for lithium perchlorate in pure water have been reported previously (8). The heats of transfer to the aqueous mixtures and anhydrous acetonitrile are given in Table IV and Figure 2.

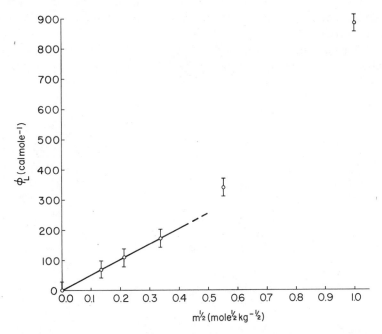

Figure 1. Apparent relative partial molal enthalpies of lithium perchlorate in 9:1 (w/w) acetonitrile–water as a function of $m^{1/2}$

Table IV.   Heats of Transfer of Lithium Perchlorate from Water to
Aqueous Acetonitrile

MeCN (wt %)	Mole Fraction of MeCN X_2	$-\Delta H_t (W \rightarrow S)$ (kcal mol^{-1})
20	0.0989	1.2
40	0.2264	3.7
60	0.3971	5.7
80	0.6372	7.9
90	0.7982	9.5
100	1.000	7.26

The free energies of transfer, ΔG_t, to the aqueous mixtures have not been measured, but from the summation of ionic free energies (9), the free energy of transfer of LiClO$_4$ from water to anhydrous acetonitrile is found to be +7.6 kcal mol^{-1}.   Thus the entropy of transfer (as $T\Delta S$) is −14.9 kcal mol^{-1}.   The fact that the solubility (on a mole fraction basis) of LiClO$_4$ is very nearly the same in acetonitrile as in water is the result of differences in the free energy of the crystalline state in equilibrium with solution (10), i.e., LiClO$_4$·4MeCN or LiClO$_4$·3 H$_2$O and of ion association, and does not reflect the true situation of the ions at high dilution.

The general tendency in aqueous organic solvents (11) is for the free energies of transfer to be monotonic functions of solvent composition.   Often the enthalpy of transfer shows a maximum in the water-rich region, and for salts of small ions there may be a further reversal in the organic-rich region.

The heats of dilution are nearly linear in m$^{1/2}$ below 0.1m, with a slope of about 500 cal kg$^{1/2}$ mole$^{-3/2}$.   This is very close to the theoretical and measured

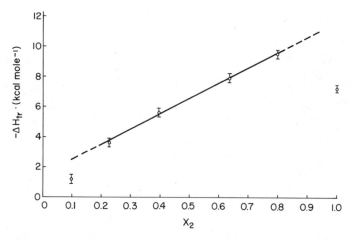

Figure 2.   Heats of transfer $-\Delta H_{tr}$ of lithium perchlorate
from water to aqueous acetonitrile at 298°K

values for 1:1 salts in pure water (*12*). The temperature dependences of the density and dielectric constant of the mixed solvent are not known, so it is impossible to verify that this is the expected limiting slope. If we assume that these missing values are close to those for pure acetonitrile, i.e., $\alpha \sim 0.0014$ K^{-1} and $\partial \ln D / \partial T \sim 0.005$ K^{-1}, then the predicted limiting slope is of the order of 1700 cal kg$^{1/2}$ mol$^{-3/2}$. It is possible that the low value observed is caused by the work required to separate ion pairs on dilution.

Above 0.1m the ϕ_L values of aqueous salts flatten out and may become negative except where small ions are involved. In 90% acetonitrile, the θ_L curve becomes steeper with increasing concentration. This is probably caused by depletion of the free water in the solvent, so that ion solvation by acetonitrile becomes significant. It is also possible that, because ionic electric fields are significant at greater distances in a solvent of lower dielectric constant, the overlapping of ionic co-spheres has a much more drastic effect on the enthalpy than it does in water for this salt.

Apart from a small exothermic region at the water-rich end, ΔH^E for the acetonitrile–water system is strongly positive at 298°K (*13*), indicating that the water structure is significantly broken up by the addition of large amounts of acetonitrile. ΔH^E passes through a maximum at about $X_2 = 0.65$, but ΔH_{tr} apparently does not change direction until later than this.

It seems likely that: (a) the smaller amount of water is still competing successfully with the acetonitrile for solvation sites, (b) this effect is accentuated by the continuing breakup of water structure so that "nonaqueous" water is a far more effective solvent than "aqueous" water, and (c) ordering by the ions is much more significant in a weakly associated aprotic solvent. While this general type of behavior has been found before for small ions in mixed solvents, these results indicate that the reversal of the effect in acetonitrile–water occurs at much lower water content than was observed where both components are highly structured, e.g., for HCl and HClO$_4$ in aqueous ethylene glycol (*14, 15*), for KBr in aqueous formic acid (*16*), and for HCl in aqueous ethanol (*17*). In contrast, HBr and HI in aqueous methanol (*18, 19*) do seem to behave in a similar manner in this region, although HCl does not (*20*). In each of these (HBr and HI) cases, it is possible that the movement of ΔH_{tr} is controlled by a large and sudden increase in ΔG_{tr}. This may apply to aqueous acetonitrile also.

The path of ΔH_{tr} above $X_2 = 0.8$ is especially interesting. In order to gain a reliable interpretation in this region, further work is planned to investigate the effects of small quantities of water on the thermodynamic properties of electrolytes.

Acknowledgment

We wish to thank R. H. Wood of the University of Delaware for the use of the LKB 11700 batch microcalorimeter in his laboratory.

Literature Cited

1. Janz, G. J., Tomkins, R. P. T., "Nonaqueous Electrolytes Handbook," Vol. I, New York, Academic, 1972; Vol. II, New York, Academic, 1973.
2. Covington, A. K., Dickinson, T., Eds. "Physical Chemistry of Organic Solvent Systems," London, Plenum, 1973.
3. Feakins, D., in "Physico-Chemical Processes in Mixed Aqueous Solvents," F. Franks, Ed., New York, American Elsevier, 1967.
4. Franks, F., Ives, D. J. G., *Q. Rev. Chem. Soc.* (1966) **20**, 1.
5. Jasinski, R., "High Energy Batteries," New York, Plenum, 1967.
6. Tomkins, R. P. T., Turner, P. J., *J. Chem. Eng. Data* (1976) **21**, 153.
7. Harned, H. S., Owen, B. B., "The Physical Chemistry of Electrolytic Solutions," New York, Reinhold, 1958, p 334.
8. Parker, V. B., "Thermal Properties of Aqueous Uni-Univalent Electrolytes," NSRDS-NBS 2, Washington, D.C., 1965.
9. Abraham, M. H., *J. Chem. Soc., Faraday I* (1973) 1375.
10. Tomkins, R. P. T., Turner, P. J., *J. Chem. Eng. Data* (1975) **20**, 50.
11. Bennetto, H. P., Willmott, A. R., *Q. Rev. Chem. Soc.* (1971) **25**, 501.
12. Young, T. F., Seligmann, P., *J. Am. Chem. Soc.* (1938) **60**, 2379.
13. Morcom, K. W., Smith, R. W., *J. Chem. Thermo.* (1969) **1**, 503.
14. Stern, J. H., Nobilione, J. M., *J. Phys. Chem.* (1968) **72**, 3937.
15. Stern, J. H., Nobilione, J. M., *J. Phys. Chem.* (1969) **73**, 928.
16. Ivanova, E. F., Fesenko, V. N., *Zh. Fiz. Khim* (1969) **43**, 1006.
17. Butler, J. A. V., Robertson, C. M., *Proc. R. Soc.* (1929) **125 A**, 694.
18. Schwabe, K., Ferse, E., *Ber. Bunsenges. Phys. Chem.* (1966) **70**, 849.
19. McIntyre, J. M., Ph.D. Thesis, University of Arkansas, 1968.
20. Slansky, C. M., *J. Am. Chem. Soc.* (1940) **62**, 2430.

RECEIVED September 4, 1975.

Standard Gibbs Free Energy of Transfer of n-Bu$_4$NBr from Water to Water–Organic Solvent Mixtures as Deduced from Precise Vapor Pressure Measurements at 298.15°K

CLAUDE TREINER and PIERRE TZIAS

Laboratoire d'Electrochimie, E.R.A. 310 Université Pierre et Marie Curie, Bât. F., 4 Place Jussieu, 75230 Paris Cedex 05 France

The standard free energy of transfer ΔG^0_t of tetrabutylammonium bromide (n-Bu$_4$NBr) from water to water–acetone and from water to water–tetrahydrofuran mixtures has been deduced from precise vapor pressure measurements of dilute solutions. The ΔG^0_t function is negative from water to the water-rich region in both chemical systems and becomes positive for high organic solvent compositions. With ΔH^0_t values obtained from calorimetric measurements, the standard molar entropy of transfer ΔS^0_t has been calculated in the case of the water–acetone mixtures. The standard free energy of transfer is so-called entropy controlled in the water-rich region and enthalpy controlled in the organic-rich region, ΔH^0_t and ΔS^0_t being positive in the whole solvent concentration range. These findings are rationalized assuming that ΔG^0_t for n-Bu$_4$NBr may be considered the sum of two effects of opposite sign: a nonspecific cavity effect caused by the cation and a specific effect caused by the bromide ion.

It has been shown recently (*1*) that the transfer of urea from water to water-rich water–tetrahydrofuran (THF) mixtures is an entropy-controlled phenomenon ($T|\Delta S^0_t| > |\Delta H^0_t|$), but from water to water–THF mixtures of mole fraction $Z_2 > 0.20$ it is an enthalpy-controlled phenomenon ($|\Delta H^0_t| > T|\Delta S^0_t|$); moreover, the transfer is energetically favorable in the former case and unfavorable in the latter. A minimum in the standard free energy transfer function with solvent composition (or a maximum in solubility) in the case of aqueous binary mixtures has been observed previously, e.g., for molecules like some amino acids and polynucleotides, β-lactoglobulin, and uric acid (2, 3, 4, 5). We found the same kind of behavior in the case of simple 1-1 electrolytes as suggested by the work of De Ligny and co-workers on the solubilities of some tetraalkylammonium salts in water–organic solvent mixtures (6). The thermodynamic properties of these electrolytes have been studied intensively in water and, to some extent, in binary aqueous mixtures. These studies have been reviewed recently (7, 8). In the latter case, most of the available information has been deduced from enthalpy and heat capacity measurements (9–15). We present here the first results obtained on the standard molar free energy of transfer ΔG^0_t of tetrabutylammonium bromide (n-Bu$_4$NBr) from water to water–THF and to water–acetone mixtures as deduced from precise vapor pressure measurements in dilute solutions at 25°C. The water–THF mixtures were chosen for comparison purposes because of our previous work on thiourea and urea in the same solvent system, and the water–acetone mixtures were chosen because standard enthalpy measurements were available for n-Bu$_4$NBr in these solvents (12) so that the standard entropy function could be calculated.

We shall now outline briefly some of the characteristics of the experimental method used.

Theory

Grunwald and Bacarella (16) have shown that the rate of change of the standard chemical potential $\overline{G}_0$ of a solute with water mole fraction Z_1 in a binary solvent mixture can be expressed by the relation:

$$\frac{1000}{M_{12}} \left(\frac{\partial \ln \dfrac{\alpha 1}{\alpha 2}}{\partial m} \right)_{Z_1} = \frac{1}{RT} \frac{\partial \overline{G}_0}{\partial Z_1} + 2 \left(\frac{\partial \ln \gamma_\pm}{\partial Z_1} \right)_m - 2mr \left(\frac{\partial \ln \gamma_\pm}{\partial m} \right)_{Z_1} \quad (1)$$

$\alpha_1 = a_1/a_1{}^0$ and $\alpha_2 = a_2/a_2{}^0$ are the ratio of the activities of each solvent of the binary mixture before and after addition of the solute of molality m at constant Z_1; M_{12} is the molecular weight of the binary solvent; $M_{12} = M_1 Z_1 + M_2 Z_2$ and $r = M_1 - M_2/M_{12}$; $\gamma_\pm$ is the solute activity coefficient.

The activities are defined by the following set of relations, assuming that the vapor behaves as a perfect gas:

$$a_1 = \frac{P(1-y)}{P_1{}^0}, \, a_1{}^0 = \frac{P^0(1-y_0)}{P_1{}^0}, \, a_2 = \frac{Py}{P_2{}^0}, \, a_2{}^0 = \frac{P^0 y_0}{P_2{}^0} \quad (2)$$

where P_0 and y_0 are respectively the total pressure and mole fraction vapor phase composition of the solvent mixture before addition of the solute, P and y the same quantities after addition of the solute. The reference state for the solute is infinitely dilute solution in the binary solvent mixture of Z_1 composition ($\gamma_\pm \to 1$ when $m \to 0$), $\partial \overline{G}_0/\partial Z_1$ is expressed in the mole fraction scale. Although the assumption made in relation 2 does not introduce any appreciable error in the final calculations (*1*), we have used the second virial coefficients of the gas mixtures in the case of water–THF mixtures as these data were available (*see* Ref. *16*).

If one assumes that in a dilute solution the activity coefficient of a 1–1 electrolyte is given by the Debye–Hückel law,

$$\ln \gamma_{\pm} , - \frac{\psi q}{1 + \psi a}$$

then the determination of $\partial \overline{G}_0 / \partial Z_1$ rests upon the measurement of the total pressure P and the vapor phase composition y. We have recently proposed the following procedure (*18*).

The activities a_1 and a_2 are related through the Gibbs–Duhem equation:

$$\frac{1000}{M_{12}} (Z_1 \, d\ln a_1 + Z_2 \, d\ln a_2) = - 2 \, dm - 2 \, d\ln \gamma_{\pm} \tag{3}$$

Assuming again that $\gamma_{\pm}$ follows the Debye–Hückel law, the total pressure P is measured as a function of the solute concentration, then the vapor phase y, the only unknown in Equation 4, can be calculated, and hence the activities a_1 and a_2 can also be calculated, provided the activities a_1^0 and a_2^0 of each solvent prior to the addition of the solute are known; $\partial \overline{G}^0 / \partial Z_1$ can be obtained next from Equation 1. Finally, integration of $\partial \overline{G}^0 / \partial Z_1$ with respect to Z_1 leads to the standard molar free energy of transfer ΔG^0_t between $Z_1 = 1$ (if water is chosen as the reference solvent) and any value of Z_1.

Experimental

The apparatus and the experimental procedure have been described in detail (*1, 18*). The solvent mixture is degassed in a Pyrex cell by freezing with liquid nitrogen, pumping, and melting. A cup-dispensing device permits the addition of several cups containing the salt samples. The vapor pressure is measured with a precision Texas Instruments gauge of a sensitivity of 0.005 Torr. The temperature of the oil-bath in which the cell is immersed in $T = 298.15 \pm 0.003°$K. n-Bu$_4$NBr (O.S.I. puriss.) was dried several days under vacuum at 50°C; the limiting conductance of the electrolyte in water at 298.15°K was chosen as the purity criterion: $\Lambda_0 = 97.46 \pm 0.02$ as compared to $\Lambda_0 = 97.46 \pm 0.05$ (*19*), $\Lambda_0 = 97.23$ (*20*) and $\Lambda_0 = 97.50 \pm 0.01$ (*21*). THF (Merck pro analysi) and acetone (Carlo Erba purr.) were dried and distilled without further purification. The water content was of the order of 0.01%.

Results

Tables I and II present the variation of total pressure P and vapor phase composition y with salt molality obtained for each Z_1 composition for the two binary systems studied. For each addition of salt to the solvent a new value of

Table I. Characteristic Parameters for Vapor–Liquid Equilibrium: $n\text{-Bu}_4\text{NBr}$ in Water–Acetone Mixtures at $298.15°\text{K}$

Wt % acetone	Z_1	m (mol kg^{-1})	P (Torr)	y	$\dfrac{\partial \bar{G}°}{\partial Z_1}$ (kcal mol^{-1})	$-\Delta G°_t$ (cal mol^{-1})
4.910	0.98425	0	48.91	0.52167		
		0.01166	48.84	0.52111	5.90	
		0.02401	48.75	0.52049	5.89	
		0.03424	48.69	0.51998	5.87	
		0.04707	48.60	0.51936	5.88	
					Av. 5.89 ± 0.02	85
7.941	0.97394	0	62.75	0.61330		
		0.00614	62.68	0.61299	6.53	
		0.01299	62.62	0.61265	6.28	
		0.01819	62.56	0.61234	6.37	
		0.02373	62.51	0.61209	6.37	
					Av. 6.39 ± 0.15	150
11.964	0.95955	0	79.05	0.69837		
		0.00571	78.98	0.69815	5.41	
		0.01322	78.85	0.69773	6.73	
		0.02005	78.76	0.69741	6.52	
		0.02653	78.68	0.69719	6.02	
					Av. 6.17 ± 0.81	240
19.80	0.92887	0	103.33	0.80595		
		0.00500	108.22	0.80578	5.89	
		0.01886	107.93	0.80533	5.67	
		0.02529	107.79	0.80510	5.73	
					Av. 5.76 ± 0.06	430
29.971	0.88281	0	141.99	0.84773		
		0.00952	141.72	0.84747	5.37	
		0.01772	141.50	0.84726	5.04	
		0.02632	141.30	0.84706	4.74	
		0.03549	141.01	0.84677	5.04	
					Av. 5.05 ± 0.28	680
39.986	0.82873	0	160.90	0.86880		
		0.01116	160.60	0.86858	3.84	
		0.01889	160.36	0.86839	4.10	
		0.02838	160.12	0.86821	3.85	
		0.03841	159.84	0.86799	3.84	
					Av. 3.91 ± 0.16	915
50.061	0.76281	0	172.35	0.88090		
		0.00551	172.22	0.88082	2.74	
		0.01145	172.09	0.88075	2.37	
		0.01704	171.96	0.88066	2.39	
		0.02401	171.79	0.88055	2.42	
					Av. 2.48 ± 0.18	1110

Table I (*Continued*)

60.084	0.68171	0	183.10	0.89007		
		0.00634	183.01	0.89005	0.38	
		0.01103	182.94	0.89003	0.32	
		0.01863	182.83	0.88999	0.27	
		0.02411	182.75	0.88996	0.22	
				Av.	0.30 ± 0.07	1230
80.226	0.44277	0	199.90	0.90745		
		0.00613	199.91	0.90756	−3.67	
		0.01075	199.92	0.90763	−3.72	
		0.01646	199.93	0.90773	−3.84	
		0.02223	199.95	0.90787	−3.90	
				Av.	−3.78 ± 0.24	835

Table II.   Characteristic Parameters for Vapor–Liquid Equilibrium: n-Bu₄NBr in Water–Tetrahydrofuran Mixtures at 298.15°K

Wt % THF	Z_1	m (mol kg⁻¹)	P (Torr)	y	$\dfrac{\partial \bar{G}_0}{\partial Z_1}$ (kcal mol⁻¹)	$-\Delta G^0{}_t$ (cal mol⁻¹)
5.418	0.98590	0	60.33	0.5886		
		0.00468	60.28	0.58883	7.04	
		0.00992	60.22	0.58800	7.47	
		0.02005	60.12	0.58747	7.15	
				Av.	7.22 ± 0.18	95
7.611	0.97942	0	74.38	0.6553		
		0.00950	74.23	0.65475	8.25	
		0.02007	74.06	0.65409	8.27	
		0.03908	73.78	0.65305	7.80	
				Av.	8.11 ± 0.27	145
8.25	0.97703	0	78.87	0.6772		
		0.01203	78.66	0.67650	8.34	
		0.01876	78.55	0.67614	8.04	
		0.02556	78.42	0.67571	8.33	
				Av.	8.24 ± 0.17	165
15.01	0.95776	0	108.06	0.7725		
		0.00461	107.94	0.77230	7.69	
		0.01071	107.79	0.77205	7.24	
		0.01685	107.63	0.77177	7.26	
		0.02265	107.47	0.77149	7.39	
				Av.	7.40 ± 0.17	320

Table II (*Continued*)

19.981	0.94132	0	125.18	0.8039		
		0.00555	125.05	0.80374	4.75	
		0.01130	124.81	0.80342	7.35	
		0.01622	124.64	0.80320	7.43	
		0.02104	124.58	0.80315	6.00	
				Av.	6.38 ± 1.58	435
30.04	0.90318	0	144.41	0.8280		
		0.00706	144.21	0.82781	4.81	
		0.01399	144.01	0.82763	4.83	
		0.02058	143.82	0.82745	5.01	
				Av.	4.88 ± 0.12	650
40.02	0.85772	0	155.61	0.8386		
		0.01636	155.22	0.8383	2.59	
		0.02431	155.07	0.83820	2.95	
		0.03184	154.84	0.83802	2.54	
				Av.	2.69 ± 0.24	820
50.01	0.80014	0	157.39	0.8425		
		0.00723	157.31	0.84248	−0.04	
		0.01445	157.22	0.84243	−0.02	
		0.02103	157.13	0.84240	−0.13	
		0.02706	157.08	0.84238	−0.33	
				Av.	0.13 ± 0.19	895
60.239	0.72557	0	160.60	0.8450		
		0.00501	160.61	0.84508	−3.22	
		0.01390	160.63	0.84519	−3.39	
		0.01962	160.65	0.84528	−3.62	
				Av.	−3.41 ± 0.14	770

$\partial \overline{G}_0 / \partial Z_1$ can be calculated and should be independent of concentration if our procedure is correct. The results presented in Tables I and II show that this is indeed the case. Finally $\Delta G^0{}_t$ is obtained by analytical integration. The variation of $\partial \overline{G}_0 / \partial Z_1$ with Z_1 is fitted to two second-degree polynomial functions, one for the region of the maximum (Figure 1) and one for the rest of the curve:

$$\frac{\partial \overline{G}_0}{\partial Z_1} = a + bZ_1 + cZ_1{}^2 \tag{4}$$

Each of the functions is then integrated with $\Delta G^0{}_t = 0$ for $Z_1 = 1$.

Table III presents the values of the constants used in the calculations. The y_0 data have been obtained from the variation of P_0 with Z_1 by numerical integration of the Gibbs–Duhem equation using the Runge–Kutta method (*22, 23*). The comparison of the P_0 values of Table I with those obtained by some previous workers (*24, 25*) shows that our results are at most higher by 0.5–1 Torr.

Finally one should take into account in Equation 1 the degree of dissociation α of the ions into ion pairs, assuming that the chemical potential of the free ions is given by the following relations:

$$\overline{G} = \overline{G}_0 + 2RT \ln m\alpha + 2RT \ln \gamma'_{\pm} \qquad (5)$$

where

$$\ln \gamma' = -\frac{\psi q \alpha^{1/2}}{1 + \psi q \alpha^{1/2}} \quad \text{and} \quad q = \frac{e^2}{2DkT} \qquad (6)$$

However this would necessitate the determination of the derivative of α with respect to Z_1. This quantity cannot be measured with enough accuracy, especially when the association constant K_A is small (this is the case in the high dielectric constant values of the solvents studied: $D \geqslant 30$). We have made two conductance runs, in a 59.02% water–THF mixture and in a 81.90% water–AC

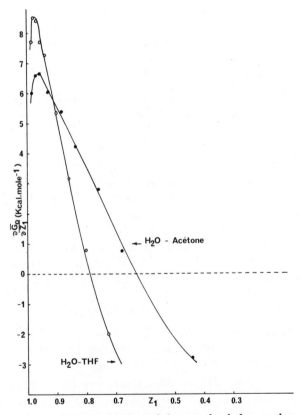

Figure 1. Rate of change of the standard chemical potential of n-Bu₄NBr *with solvent composition vs. water mole fraction at 298.15°K*

Table III. Physical Constants of the Aqueous Binary Mixtures at 298.15°K (1)

Wt %	P_0	y_0	D	$\dfrac{\partial \log D}{\partial Z_1}$
		water–acetone		
0	23.75	0	78.54	—
4.91	44.91	0.52167	75.9	2.27
7.94	62.75	0.61330	74.1	2.25
11.96	79.05	0.69837	71.8	2.22
19.80	108.33	0.80595	67.0	2.15
29.97	141.99	0.84773	60.8	2.07
39.99	160.90	0.86880	54.6	1.96
50.06	172.35	0.88090	48.2	1.82
60.08	183.10	0.89007	41.8	1.67
80.23	199.90	0.90745	29.6	1.20
		water–THF		
0	23.75	0	78.54	—
5.45	59.33	0.5886	75.2	3.42
7.61	74.38	0.6553	73.6	3.41
8.25	78.87	0.6772	73.1	3.39
15.01	108.06	0.7725	68.6	3.39
19.98	125.18	0.8039	64.9	3.34
30.04	144.41	0.8280	57.2	3.35
40.00	155.61	0.8386	48.4	3.32
50.01	157.39	0.8425	40.3	3.28
60.24	156.60	0.8450	31.9	3.23

mixture, which correspond to D values equal respectively to 28.9 and 32.0. The values of K_A are 22 ± 1 and 29 ± 1—the apparatus, experimental procedure, and conductance equation used have been described elsewhere (26, 19). The error introduced by assuming that K_A is equal to zero can be evaluated by comparison with the careful analysis of the solubility of some 1–1 electrolytes in similar aqueous binary systems made by De Ligny et al. (6). It can be considered to be of the order of 30 cal mol^{-1} at most for the highest values of K_A (and the lowest value of solvent dielectric constant).

The accuracy of the ΔG^0_t values can then be estimated as:

(a) the extrapolation of $\partial \overline{G}^0 / \partial Z_1$ to $Z_1 = 1$ outside the polynomial fitting involves an error difficult to determine with precision but may be estimated to be of the order of ± 10 cal mol^{-1} in the present case;

(b) the error introduced by the integration procedure is negligible;

(c) adding the errors on each $\partial \overline{G}^0 / \partial Z_1$ value and the $K_A = 0$ assumption, leads to an estimated absolute error on ΔG^0_t of the order of 20 cal mol^{-1} for the points nearest to pure water to about 100 cal mol^{-1} for the point corresponding to the richest organic solvent mixture for each solvent system.

Discussion

The only data with which our ΔG^0_t values can be compared are those of De Ligny and co-workers who used the solubility method.   Unfortunately there is only one value which can be compared with our experimental results: the standard free energy of transfer of $n\text{-Bu}_4\text{NBr}$ from water to a 20 wt % water–acetone mixture: $\Delta G^0_t = -0.8$ kcal mol^{-1} (*6*, *15*).   This value is too negative compared with our $\Delta G^0_t = -430$ cal mol^{-1}; the results concerning the tetrapropyl ammonium salts as deduced from the solubility method are still more negative in the water-rich region than our ΔG^0_t values for $n\text{-Bu}_4\text{NBr}$.   Difficulties with the solubility method for the determination of free energies of transfer in the case of electrolytes with large ions have been pointed out before (*28*, *29*).

The general features of the experimental results are shown in Figures 1–4. As a first approximation, the variation of total pressure P with solute molality may be represented by a linear relationship: $P = P_0 + km$.   Figure 2 compares

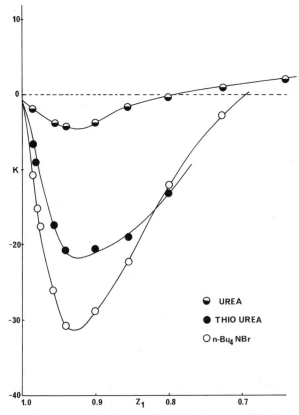

Figure 2. Rate of change of total pressure with solute concentration vs. water mole fraction in water–THF mixtures at 298.15°K

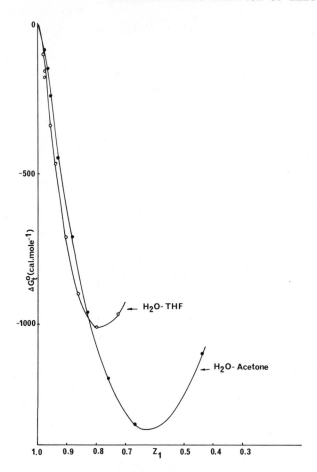

Figure 3. Variation of the standard molar free energy of transfer of n-Bu₄NBr from water to aqueous binary mixtures at 298.15°K

the variation of the slope k with solvent composition for n-Bu$_4$NBr, urea, and thiourea in water–THF mixtures at 298.15°K. In the water-rich region the total pressure P decreases $(k < 0)$ when the solute is added to the solvent mixtures. In the case of NaCl (*18*), an increase of the pressure is observed, a common feature for alkali halides in aqueous binary mixtures (*27, 30*), and the minima appear for the same Z_1 value for the three different solutes.

This figure was drawn from purely experimental results; Figure 1 compares the variation of $\partial \overline{G}^0/\partial Z_1$ for n-Bu$_4$NBr in water–THF and water–acetone mixtures. This graph gives us much more information. In mixtures very rich in water $\partial \overline{G}^0/\partial Z_1$ goes through a maximum; this corresponds to an inflection point in the integrated function $\Delta G^0_t = f(Z_1)$. Then $\partial \overline{G}^0/\partial Z_1$ changes sign, which means that ΔG^0_t goes through a minimum (Figure 3); urea and thiourea present the same characteristics in water–THF mixtures.   The maximum in the $\partial \overline{G}^0 / \partial Z_1$ functions occurs at the same Z_1 value for the electrolyte and the two nonelectrolytes.

We shall discuss now the variation of the three main thermodynamic functions with solvent composition for the case of n-Bu$_4$NBr–water–acetone system and shall extend this discussion to the n-Bu$_4$NBr–water–THF system. Figure 4 and Table IV present the results obtained.   The figure was constructed as follows: first the standard enthalpy of transfer ΔH^0_t, obtained by Ahluwalia and co-workers (*12*) from pure water to $Z_2 = 0.30$, was used in order to get the standard entropy of transfer function from the relation:

$$T \Delta S^0_t = \Delta H^0_t - \Delta G^0_t$$

The results obtained are found in Table IV.   Unfortunately there are no data

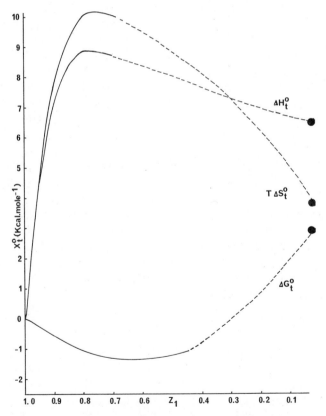

Figure 4. Thermodynamic transfer functions for n-Bu$_4$NBr from water to water–acetone mixtures at 298.15°K

Table IV.   Standard Free Energy, Enthalpy, and Entropy of Transfer
of n-Bu$_4$NBr from Water to Water–Acetone Mixtures at 298.15°K

Z_1	$-\Delta G^0_t$ (cal mol^{-1})	ΔH^0_t (cal mol^{-1})	ΔS^0_t (cal mol^{-1} deg^{-1})
0.98	104	1817	6.44
0.96	232	3698	13.18
0.9426	346	5005	17.95
0.90	591	7080	25.73
0.85	829	8300	30.62
0.80	1014	8898	33.25
0.70	1235	8701	33.33
0.0	−3100	6400	11.1

available for ΔH^0_t from water to the acetone-rich mixtures nor for the standard
free energy and entropy of transfer from water to pure acetone.   However, fairly
reasonable values can be obtained from the following considerations.   Abraham
(29) has shown that assuming $\Delta G^0_t = 0$ for the transfer of Me$_4$N$^+$ ion from water
to all solvents, ΔG^0_t for transfer of Et$_4$N$^+$ from water to N-methylformamide,
methanol, ethanol, and acetonitrile is equal to -1.43 ± 0.03 kcal mol^{-1} and to
-1.52 kcal mol^{-1} from water to acetone.   This seems to be the only "experimental" value for the transfer of a tetraalkylammonium ion from water to pure
acetone.   The same constancy holds for the transfer of the Pr$_4$N$^+$ ion from water
to the solvents listed above, $\Delta G^0_t = -3.6 \pm 0.2$ kcal mol^{-1}.   We have thus assumed that the same constancy holds for Bu$_4$N$^+$: $\Delta G^0_t = -5.9 \pm 0.4$ kcal mol^{-1}
is the average of values obtained for the transfer from water to methanol and to
acetonitrile (molality scale).   Taking the value of 10.7 kcal mol^{-1} for the bromide
ion given by the same author we get in the mole fraction scale ΔG^0_t (Bu$_4$NBr)
$= 3.1$ kcal mol^{-1}.   From the compilation of Abraham ΔS^0_t of n-Bu$_4$NBr is $+11$
cal deg^{-1} mol^{-1}, hence $\Delta H^0_t = 6.4$ kcal mol^{-1}.

The variation of ΔH^0_t vs. Z_1 shown in Figure 4 is similar to the corresponding
function for the same electrolyte in water–dimethylsulfoxide (14) or various
water–amide mixtures (13); so even large errors (of the order of one or two kcal
mol^{-1}) on the estimated values of ΔG^0_t and $T\Delta S^0_t$ for the transfer from water
to pure acetone should not change the shape of the curves and the main qualitative
results which can be deduced from them.

(a) The transfer of n-Bu$_4$NBr is favorable from water to the water-rich
region and unfavorable from water to the organic-rich region.

(b) The transfer is entropy-controlled in the water-rich region ($T|\Delta S^0_t| >
|\Delta H^0_t|$) and enthalpy-controlled ($T|\Delta S^0_t| < |\Delta H^0_t|$) in the acetone-rich re-
gion.

(c) ΔH^0_t and ΔS^0_t are positive over the whole solvent concentration range.
This behavior is quite different from that of most common 1–1 electrolytes in
aqueous binary mixtures.   For example (31–35), the transfer of alkali halides
and halogenated acids from water to these solvent mixtures is often first endo-

thermic, but the transfer becomes exothermic in organic-rich mixtures. In this respect urea possesses some of the characteristics of the thermodynamic properties of electrolytes (36). Clearly the two ions Br^- and Bu_4N^+ contribute very differently to the transfer process.

If we adopt some reasonable extrathermodynamic assumption (37) in order to get the "individual free energy of transfer" of the ions, we find that $\Delta G^0_t(-)$ of the bromide ion is positive from water to acetone and that $\Delta G^0_t(+)$ of the $n\text{-}Bu_4N^+$ is negative over the whole solvent concentration range. In terms of preferential solvation models we would say that Br^- is solvated by the water molecules and $n\text{-}Bu_4N^+$ by the acetone molecules.

The same situation is found for $n\text{-}Bu_4NBr$ in water–THF mixtures. Single ion $\partial\overline{G}^0/\partial Z_1$ values have been obtained for several electrolytes in a 40.00 wt % water–THF mixture using a number of different extrathermodynamic assumptions (38); they all led to the same single ion $\partial\overline{G}_0/\partial Z_1$ values within less than 20%, a result which gives some confidence in the extrathermodynamic approach. The following table compares $\partial\overline{G}_0/\partial Z_1$ for $n\text{-}Bu_4N^+$ and Br^- for a water–THF and a water–acetone mixture at the same composition and for the same dielectric constant of the medium. The single ion values for Br^- in water–acetone were calculated from Bax et al. (37), and for $n\text{-}Bu_4N^+$ using our own data.

Comparison of $\partial\overline{G}^0/\partial Z_1$ (in kcal mol^{-1}) mole fraction scale:

	water–THF	*water–acetone*	*water–acetone*
	($Z_1 = 0.8572$, $D = 48.4$)	($Z_1 = 0.8572$, $D = 57.7$)	($Z_1 = 48.4$, $D = 0.7628$)
Br^-	−17.2	−12.6	−12.7
$n\text{-}Bu_4N^+$	19.8	17.4	15.2
$n\text{-}Bu_4NBr$	2.6	4.8	2.5

Note that these single ion values were obtained from entirely different extrathermodynamic assumptions: elaborate extrapolation procedure in the case of the water–acetone mixtures, and tetraphenylboron assumption for the water–THF mixture. $n\text{-}Bu_4N^+$ and Br^- showed similar behavior in the two binary systems studied; this might be the consequence of the similarity between the thermodynamic properties of the two aqueous binaries; e.g., both are typically aqueous systems with $|\Delta H^E| < T|\Delta S^E|$.

Finally the minimum observed for the ΔG^0_t function of $n\text{-}Bu_4NBr$ in both chemical systems may be interpreted as follows: the variation of $\Delta G^0_{t(+)}$ and $\Delta G^0_{t(-)}$ with solvent composition are generally not linear, so the minimum can be looked upon as the consequence of the opposite behavior of anion and cation towards the solvent molecules. However this interpretation does not take into account the fact that at low organic solvent mole fraction one has: $\Delta H^0_t < T\Delta S^0_t$. This observation means that a structural (nonspecific) effect predominates in the water-rich region. This effect might be related to the cavity effect as can be evaluated from Pierrotti's scaled particle theory because the large size of the

n-Bu_4N^+ ion and its hydrophobic character precludes any strong ion-dipole interactions. In the case of the bromide ion the cavity effect should obviously be negligible and the specific effect predominates ($\Delta H^0_t > T\Delta S^0_t$). The same explanation might hold for the observed maximum of solubility of surfactants in hydroorganic solvent mixtures (39).

We suggest therefore that for a minimum to occur in the standard free energy of transfer function of a solute in aqueous binaries we need a structural (nonspecific) effect (related here to the hydrophobic character of the cation) in the water-rich region and superposed to it, classical solute–solvent interactions with predominant water–solute interactions.

Our interpretation resembles somewhat that proposed by Roseman and Jencks (2) in order to describe the behavior of nonelectrolytes like uric acid in aqueous binary mixtures, though these authors seem to refuse to consider the respective magnitude of entropy and enthalpy as indications of structural effects, as we have done.

It is pertinent at this stage of the discussion to recall various studies (40, 41) on the effect of urea on water structure. It has been shown, for example, that the behavior of urea in water can be explained by an essentially structural model. This conclusion could well be related to our own findings on the behavior of urea in water–THF mixtures, which is qualitatively similar to that of n-Bu_4NBr. The predominant structural effect $T\Delta S^0_t > \Delta H^0_t$ is here in both cases in the water-rich region, and the preferential solvation of urea by water becomes the predominant effect at higher THF composition, as does the Br^- ion for n-Bu_4NBr.

Finally, the qualitatively similar behavior of uric acid in water–t-BuOH mixtures, and urea and thiourea in water–THF mixtures might be due to the similarity between the three organic solutes, but it is clear from a quantitative comparison of the standard free energy data that the balance between nonspecific and specific effects is very sensitive to changes in the solute molecule and could very easily be shifted to predominant specific or nonspecific effects.

Acknowledgment

We are grateful to F. Franks for reading the manuscript and making helpful comments.

Literature Cited

1. Treiner, C., Tzias, P., *J. Solution Chem.* (1975) **4**(6), 471.
2. Roseman, M., Jencks, W. P., *J. Am. Chem. Soc.*, (1975) **97**, 631.
3. Herskovits, T. T., Harrington, J. P., *Biochemistry* (1972) **11**, 4800.
4. Winnek, P. S., Schmidt, C. L. A., *J. Gen. Physical* (1936) **19**, 773.
5. Inone, H., Timasheff, S. N., *J. Am. Chem. Soc.* (1968) **90**, 1890.
6. De Ligny, C. L., Bax, D., Alfenaar, M., Elferink, M. G. L., *Recl. Trav. Chim. Pays-Bas.*, (1969) **86**, 1183.
7. Sarma, T. S., Ahluwalia, J. C., *Chem. Soc. Rev.* (1973) **2**, 203.
8. Wen-Yang Wen, in "Water and Aqueous Solutions," Horne R. A., *Ed.*, Wiley, New York, 1972.

9. Bhatnagar, N., Campbell, A. N., *Can. J. Chem.* (1974) **52**, 203.
10. Kim, L. M., Mishenko, K. P., Poltoratskii, G. M., *Russ. J. Phys. Chem.* (1973) **47**, 1058.
11. Fuchs, R., Bear, J. L., Rodewald, R. F., *J. Am. Chem. Soc.* (1969) **91**, 579.
12. Mohanty, R. K., Sarma, T. S., Subramanian, S., Ahluwalia, J. C., *Trans. Faraday Soc.* (1970) **66**, 1073.
13. De Visser, C., Somsen, G., *J. Solution Chem.* (1974) **3**, 847.
14. Mastroiani, M. J., Pikal, M. J., Lindenbaum, S., *J. Phys. Chem.* (1972) **76**, 3050.
15. Mohanty, R. K., Alhuwalia, J. C., *J. Solution Chem.* (1972) **1**, 531.
16. Treiner, C., Bocquet, J. F., Chemla, M., *J. Chim. Phys.* (1973) **70**, 72.
17. Grunwald, E., Bacarella, A. L., *J. Am. Chem. Soc.* (1958) **80**, 3840.
18. Treiner, C., *J. Chim. Phys.* (1973) **70**, 1183.
19. Martel, R. W., Kraus, C. A., *Proc. Natl. Acad. Sci., U.S.* (1955) **41**, 9.
20. Treiner, C., Justice, J. C., *J. Chim. Phys.* (1967) **64**, 1516.
21. Evans, D. F., Kay, R. L., *J. Phys. Chem.* (1966) **70**, 366.
22. Jambon, C., Clechet, P., *Bull. Soc. Chim. Fr.* (1971) **4**, 1213.
23. Van Ness, H. C., "Classical Thermodynamics of Nonelectrolyte Solutions," Pergamon, London, 1964.
24. Rhim, J. N., Park, S. S., Lee, H. O., Kiche, J. *J. Korean Inst. Chem. Eng.* (1974) **12** (3), 179.
25. Taylor, A. E., *J. Phys. Chem.* (1900) **4**, 367.
26. Justice, J. C., Bury, R., Treiner, C., *J. Chim. Phys.* (1968) **65**, 1708.
27. Treiner, C., Bocquet, J. F., Chemla, M., *J. Chim. Phys.* (1973) **70**, 472.
28. Salomon, M., *J. Electroanal. Soc.* (1971) **118**, 1609.
29. Abraham, M. H., *J. Chem. Soc. Faraday Trans. 1* (1973) **69**, 1375.
30. Johnson, A. I., Furter, W. F., *Can. J. Chem. Eng.* (1960) **6**, 78.
31. Feakins, D., Voice, P. J., *J. Chem. Soc. Faraday Trans. 1* (1972) **68**, 1390.
32. Pointud, Y., Juillard, J., Morel, J. P., Avedikian, L., *Electrochim. Acta* (1974) **17**, 1921.
33. Roy, R. N., Bothwell, A. L. M., *J. Chem. Eng. Data* (1971) **16**, 747.
34. Morel, J. P., Morin, J., *J. Chim. Phys.* (1970) **67**, 2018.
35. Bates, R. G., *in* "Hydrogen-Bonded Solvent Systems," A. K. Covington and P. Jones, *Eds.*, Taylor and Francis Ltd., 1968, and references herein.
36. Rupley, J. A., *J. Phys. Chem.* (1964) **68**, 2002.
37. Bax, D., De Ligny, C. L., Remijnse, A. G., *Rec. Trav. Chim. Pays-Bas* (1972) **91**, 1225.
38. Treiner, C., Finas, P., *J. Chim. Phys.* (1974) **71**(1), 67.
39. Shiramaya, K., Matuura, R., *Bull. Chem. Soc. Jpn.* (1965) **38**(3), 373.
40. Frank, H. S., Franks, F., *J. Chem. Phys.* (1968) **69**, 1704.
41. Finer, E. G., Franks, F., Tait, M. J., *J. Am. Chem. Soc.* (1972) **94**, 4424.

RECEIVED June 16, 1975.

20

Equations for the Work Function, Activity Coefficient, and Osmotic Coefficient for Particles Having Ion–Dipole Characteristics

EDWARD S. AMIS

University of Arkansas, Fayetteville, Ark. 72701

GEORGE JAFFE[1]

Louisiana State University, Baton Rouge, La. 70803

We explain theoretically why the ammonium salts, quaternary ammonium iodides, and thallous nitrate, give J values ($j = 1 - $ osmotic coefficient) which rise above those of the limiting case of the Debye-Hückel theory. The purpose was accomplished by deriving an equation for the work function of an ion–dipole and applying the equation to the derivation of equations for the activity coefficient and for the osmotic coefficient of an ion–dipole. These equations could be applied to simple ions by eliminating terms involving dipole moments, and to dipolar molecules by eliminating terms involving only ionic charges. Successful applications of the equations for activity and osmotic coefficients were made to pertinent data on ions, dipoles, and ion–dipoles.

Debye and Hückel (*1*) have derived an expression for the work function of an ion in an ion atmosphere in solution. They and others (*1, 3, 4*) have applied this function to various phenomena in liquid media. The authors (*2*) have previously deduced, in a similar way, the field around a dipole and have combined it with Onsager's (*5*) theory of polar liquids to obtain an equation that explains the electrostatic effects on the rates of reaction between ions and dipolar molecules (*2*). The equation has been applied (*2, 6, 7, 8*) to the rates of several ion–dipolar molecular reactions.

---

[1] Deceased.

In this paper the authors propose to make a general derivation for the work function of a particle having the characteristics of an ion–dipole. Further, it is planned to apply this result to obtain equations for the activity coefficient and osmotic coefficient for an ion–dipole particle. The theory will be checked by applying it to published data on the activity and osmotic coefficients for ions, dipole, and ion–dipoles.

Theoretical

The following calculations are based on the Debye–Hückel (*1*) expression for the field of an ion and on the corresponding expression for the field of a dipole as formulated by the authors (*2*). However, it will be necessary to modify the field of Debye and Hückel for the interior of the ion to make it conform to the model used by the authors.

We shall assume that there are s different species of ions. Let $\epsilon_j = \epsilon z_j$ be the charge on an ion of species j, and let n_j be their number per cm^3. Then the parameter κ of the Debye–Hückel theory is given by

$$\kappa^2 = (4\pi\epsilon^2/DkT) \sum_{j=1}^{s} n_j z_j^2 \tag{1}$$

If the ions are idealized as spheres of radius a_j, the field of Debye and Hückel is given by

$$\psi_0 = Ae^{-\kappa r}/r, \qquad r \geq a \tag{2}$$

$$\psi_i = \epsilon z/r + B, \qquad r \leq a \tag{3}$$

Here A and B are constants which are to be determined. (In Equations 2 and 3 we have omitted the index j and will continue to do so as long as no ambiguity arises.)

In the original treatment of Debye and Hückel these constants were determined under the assumption that the ion had a point charge at $r = 0$ and that the interior of the ion had the same dielectric constant D as the solvent. In the Onsager (*5*) theory of dipolar liquids it is assumed that the molecule is represented by a spherical cavity in the liquid with a singularity at its center. The characteristics of the molecule are its electric moment in vacuum μ_0 and its polarizability α. This is to be related to an internal refractive index n by

$$\alpha = [(n^2 - 1)/(n^2 + 2)]a^3$$

Since our treatment of the ionic atmosphere around a dipolar molecule makes use of the Onsager model, it becomes necessary to adopt a similar model for the ion. Consequently we are going to assume that the ion is also represented by a spherical cavity in the surrounding dielectric with a point charge at its center. Then the constants by the ordinary boundary conditions become

$$A = (\epsilon z/D)e^{\kappa a}/(1 + \kappa a) \tag{5}$$

and

$$B = -(\epsilon z/a)\{1 - 1/D + \kappa a/[D(1 + \kappa a)]\} \tag{6}$$

In the same way the field around a dipole is represented by (2)

$$\psi'_0 = A'(e^{-\kappa r}/r^2)(1 + \kappa r) \cos \theta, \qquad r \geq a \tag{7}$$

$$\psi'_i = (m/r^2 + B'r) \cos \theta, \qquad r \leq a \tag{8}$$

where the angle θ is counted from the direction of the dipole. The constants involved are given by

$$A' = 3 \, me^{\kappa a}/[D(2 + 2\kappa a + \kappa^2 a^2) + (1 + \kappa a)] \tag{9}$$

$$B' = (m/a^3)[D(2 + 2\kappa a + \kappa^2 a^2) - 2(1 + \kappa a)]/$$
$$[D(2 + 2\kappa a + \kappa^2 a^2) + (1 + \kappa a)] \tag{10}$$

and

$$m = \mu_0(n^2 + 2)[D(2 + 2\kappa a + \kappa^2 a^2) + (1 + \kappa a)]/$$
$$3[D(2 + 2\kappa a + \kappa^2 a^2) + n^2(1 + \kappa a)] \tag{11}$$

For what follows, it will be convenient to assume that every particle of the s species carries both a charge ϵz and a dipole. Then its field will be given by

$$\Psi_0 = \psi_0 + \psi'_0, \qquad r \geq 0 \tag{12}$$

$$\Psi_i = \psi_i + \psi'_i, \qquad r \leq 0 \tag{13}$$

The simpler case of an ion can be obtained from the general formula by setting $\mu_0 = 0$, and correspondingly, the case of a pure dipole by setting $z = 0$.

We aim to derive the expression for the work function and for the activity coefficient for an ion–dipole of the characteristic kind. In doing so we follow the procedure of Debye (9, see Ref. 10), and we calculate the work which is necessary to charge the entire system starting from a given zero state.

To avoid infinite values of the energy (in which some of the finer details of the calculation would be lost), we now eliminate the singularities for the immediate neighborhood of $r = 0$. Therefore we assume that the ionic strength consists of little charged spheres of radius ρ and that the dipoles are formed by pairs of such charges separated by a distance $2d$. The change of the electric moment connected with polarization will be expressed as a change of d. Thus

$$\mu_0 = 2\epsilon d_0 \tag{14}$$

represents the permanent moment, and

$$m = 2\epsilon d \tag{15}$$

the modified moment (Equation 11) produced by the polarization of the medium and by the ionic atmosphere.

Furthermore we shall have completely defined circumstances if we assume the following gradation of lengths:

$$\rho \ll d \ll a. \qquad (16)$$

Under these restrictions our Equations 2, 3, 7, and 8 do not become affected except in the immediate neighborhood of the charges. Of course the assumption $d \ll a$ will not correspond to actual conditions under which d, though smaller than a, is comparable in size.   This deficiency can be avoided by treating d as a finite quantity from the start.   Then the solution for ψ becomes an infinite series of terms corresponding to multipole singularities (at $r = 0$).   Our Equations 2, 3, 7, and 8 represent the first two terms of this series, and it can be shown by straight-forward calculations that the error involved in the omission of the higher terms is not significant.

Now we proceed to the work necessary to charge the entire system.   Let every charge ϵ have obtained the fractional value $\lambda\epsilon$ where as usual $0 \leq \lambda \leq 1$. Then the potential for the interior will be given (from Equations 2, 3, 5, 6, 7, 8, 9, 10, and 11) by

$$\Psi_i(\lambda) = \frac{\lambda \epsilon z}{r} - \frac{\lambda \epsilon z}{a} + \frac{\lambda \epsilon z}{rD} - \frac{\lambda \epsilon z}{aD}\frac{\lambda \kappa a}{1 + \lambda \kappa a} + \frac{\lambda \epsilon}{r_1} - \frac{\lambda \epsilon}{r_2} + \lambda B'(\lambda) r \cos\theta \qquad (17)$$

Here r_1 and r_2 represent the distances from the two charges (at $r = d$, $\theta = 0$, π) into which the dipole singularity has been resolved.   Furthermore it has been taken into account that κ is proportional to λ.   Now we find the work for increasing the charge on the ionic part by $\epsilon z d\lambda$ and increasing the charge on the dipole by $\pm \epsilon d\lambda$ as

$$dA' = \frac{(2 + z^2)\lambda d\lambda \epsilon^2}{\rho} - \frac{z^2\epsilon^2\lambda d\lambda}{a}\left(1 - \frac{1}{D}\right)$$
$$- \frac{z^2\epsilon^2\kappa\lambda^2 d\lambda}{D(1 + \lambda\kappa a)} - \frac{\epsilon^2\lambda d\lambda}{d} + 2\epsilon B'(\lambda)d\lambda \qquad (18)$$

It is not difficult to perform the integration in a complete way.   However, we shall be satisfied to state the results up to terms in κ^2 only.   Thus we find

$$A' = \frac{(2 + z^2)\epsilon^2}{2\rho} - \frac{\epsilon^2 z^2}{2a}\left(1 - \frac{1}{D}\right) - \frac{\epsilon^3}{\mu_0}\alpha - \frac{\mu^2_0}{a^3}\beta - \frac{\kappa\epsilon^2 z^2}{3D}$$
$$+ \kappa^2 a^2\left\{\frac{\epsilon^2 z^2}{4aD} + \frac{\epsilon^3}{\mu_0}\gamma - \frac{\mu^3_0}{a^3}\delta\right\} \qquad (19)$$

Here the coefficients α, β, γ, and δ are functions of D and n^2 defined by

$$\alpha = 3(2D + n^2)/[(n^2 + 2)(2D + 1)], \qquad (20)$$

$$\beta = (D - 1)(2D + 1)(n^2 + 2)^2/[9(2D + n^2)^2], \qquad (21)$$

$$\gamma = 3(n^2 - 1)D/[2(n^2 + 2)(2D + 1)^2], \tag{22}$$

and

$$\delta = (n^2 + 2)^2 D[D(4n^2 + 2) - n^2 + 4]/[36(2D + n^2)^3] \tag{23}$$

As zero state of our system we chose that state in which all constituents are dispersed to infinite dilution within a dielectric of dielectric constant D_0 (which value may or may not be chosen as $D_0 = 1$). Then we obtain the contribution to the work function by imagining the system charged (in the way calculated above) at the given concentrations in a system of dielectric constant D and subsequently discharged at infinite dilution in a medium of dielectric constant D_0. Adding over all species of particles we obtain an expression for the work function which can be written in the form

$$A = A' - A'_0 = \sum_{i=1}^{s} N_i w_i \tag{24}$$

where w_i is defined by

$$w_i = \frac{\epsilon^2_i}{2a_i}\left[\frac{1}{D} - \frac{1}{D_0}\right] - \frac{\epsilon^3_i}{\mu_{0i}}[\alpha_i(D) - \alpha_i(D_0)] - \frac{\mu^2_{0i}}{a^3_i}[\beta_i(D) - \beta_i(D_0)]$$

$$-\kappa a_i \frac{\epsilon^2_i}{3a_i D} + \kappa^2 a^2_i\left[\frac{\epsilon^2_i}{4a_i D} + \frac{\epsilon^3_i}{\mu_{0i}}\gamma_i(D) - \frac{\mu^2_{0i}}{a^3_i}\delta_i(D)\right] \tag{25}$$

The subscripts i which have been added refer to the different species of ion–dipoles, and the symbols $\alpha_i(D), \ldots$ indicate that the respective coefficients have to be calculated with the value of D and n^2_i. From the way our work function A has been derived, it is evident that it contains the contributions which are caused by the presence of the solutes and by the change in dielectric constant of the solvent. The contributions which result from the first term in Equation 19 and which represent the work which is required to build up the ion–dipole in a standard environment (e.g., a vacuum) have disappeared from Equation 24 (being identical in A' and A'_0). This self-energy of the particles is without interest for the present investigation and depends, of course, in a decisive way on the underlying model.

It will be convenient to follow Debye's procedure further and to introduce activity potentials h_i defined by Equation 26:

$$\log h_i = w_i/kT; \tag{26}$$

then the activity coefficient f_i becomes

$$\log f_i = \log h_i + \sum_{j=1}^{s} \frac{\partial \log h_j}{\partial \log N_i} \tag{27}$$

We will refer the index 0 to the solvent. It should be pointed out that $\log h_0 = 0$ but that κ depends on N_0 because of the relations

$$n_i = N_i/V \tag{28}$$

and

$$V = N_0 v_0 + \sum_{i=1}^{s} N_i v_i \tag{29}$$

Here v_0 and v_i represent the changes in the total volume caused by adding, respectively, a molecule of solvent or a particle of the ith sort to the solution (10).

Proceeding now in the usual manner we finally solve for the activity coefficients of the solvent

$$\log f_0 = \frac{\kappa v_0}{6kDT} \sum_{j=1}^{s} n_j \epsilon^2_j - \frac{\kappa^2 v_0}{kT} \sum_{j=1}^{s} n_j$$

$$\times \left[\frac{\epsilon^2_j a_j}{4D} + \frac{\epsilon^3_j a^2_j}{\mu_{0j}} \gamma_j(D) - \frac{\mu^2_{0j}}{a_j} \delta_j(D) \right] \tag{30}$$

The activity coefficients for the solutes can be simplified by neglecting terms of the order v_i/V. We will split the activity coefficient into the part for infinite dilution and the part depending on concentration. This results in

$$\log f_i = \log f_{i0} + \log f_{i\kappa} \tag{31}$$

where

$$\log f_{i0} = \frac{\epsilon^2}{2kTa_i} \left[\frac{1}{D} - \frac{1}{D_0} \right] - \frac{\epsilon^3}{kT\mu_{0i}} [\alpha_i(D) - \alpha_i(D_0)]$$

$$- \frac{\mu^2_{0i}}{kTa^3_i} [\beta_i(D) - \beta_i(D_0)] \tag{32}$$

and

$$\log f_{i\kappa} = -\frac{\kappa \epsilon^2_i}{2DkT} + \frac{\kappa^2}{kT} \left[\frac{\epsilon^2_i a_i}{4D} + \frac{\epsilon^3_i a^2_i}{\mu_{0i}} \gamma_i(D) - \frac{\mu^2_{0i}}{a_i} \delta_i(D) \right]$$

$$+ \left[\frac{\kappa^2 \epsilon^2_i}{kT \sum\limits_{k=1}^{s} N_k \epsilon^2_k} \right] \sum_{j=1}^{s} N_j \left[\frac{\epsilon^2_j a_j}{4D} + \frac{\epsilon^3_j a^2_j}{\mu_{0j}} \gamma_j(D) - \frac{\mu^2_{0j}}{a_j} \delta_j(D) \right] \tag{33}$$

Equation 32 expresses the influence of the dielectric constant of the medium in the case of infinite dilution of solutions. The first term is attributed to the ionic charge and is of the Born form. Born (11) obtained a corresponding term in A (*see* Equation 24) in his derivation of the heat of hydration, and Scatchard (12) introduced it in the theory of activities. The second and third terms represent the influence of the dipolar part. Their form is essentially affected by the use of the Onsager model (5).

The part of the activity coefficients depending on κ (Equation 33) can be simplified further if suitable average values are introduced. If the restriction $d \ll a$ is sufficiently well fulfilled, the term in γ_i is small compared with the term in δ_i. The latter therefore represents the main influence caused by the presence of dipoles since the terms in ϵ_i are from ionic charges. They are identical with terms of corresponding order in the Debye–Hückel theory.

We shall now derive the osmotic coefficient j, making use of the well-known relation

$$j = 1 - g = \frac{1}{P}\frac{\partial A}{\partial V} \tag{34}$$

where

$$P = \sum_{i=1}^{s} n_i kT \tag{35}$$

represents the osmotic pressure of the ideal dilute solution (13).

Making use of the relation

$$\frac{\partial \kappa}{\partial V} = -\frac{\kappa}{2V} \tag{36}$$

which follows from Equation 1, we find from Equations 24, 34, and 35

$$j = 1 - g = (1/\sum_{i=1}^{s} n_i)$$

$$\times \left\{ \frac{\kappa^3}{24\pi} - \frac{\kappa^2}{kT} \sum_{i=1}^{s} n_i \left[\frac{\epsilon^2_i a_i}{4D} + \frac{\epsilon^3_i a^2_i}{\mu_{0i}} \gamma_i(D) - \frac{\mu^2_{0i}}{a_i} \delta_i(D) \right] \right\} \tag{37}$$

Again, this expression might be simplified by the introduction of mean values of a, $1/a$, etc.

Applications

Activity Coefficient of a Dipolar Molecule. Our first application will be to the activity coefficients of a dipolar molecule as a function of the dielectric constant of the solvent. Harned and Ross (14) have determined the activity coefficient of methyl acetate in dioxane–water mixtures of various compositions at 25°C. Equation 32 can be applied to this data, and since the particles have no ionic charges, the first term can be omitted. For the difference between the activity coefficients of methyl acetate in dioxane–water mixtures and those in water we have from Equation 32

$$\log_{10} f_{i(D-H_2O)} - \log_{10} f_{i(H_2O)} = -\frac{\epsilon^3_i}{2.303 kT\mu_{0i}} [\alpha_i(D)_{D-H_2O}$$

$$- \alpha_i(D)_{H_2O}] - \frac{\mu^2_{0i}}{2.303 kTa^3_i} [\beta_i(D)_{D-H_2O} - \beta_i(D)_{H_2O}] \tag{38}$$

where α and β are defined by Equations 20 and 21, respectively.

Choosing the reasonable value of 1.50 for the square of the internal refractive index of methyl acetate and 1.85×10^{-18} (15, 16, 17) for the moment of the compound, we find the agreement between the calculated and observed values of the differences of molal activity coefficients for methyl acetate recorded in Table I. This moment 1.85×10^{-18} is the average of values given for ethyl acetate at 25°C by Smythe et al. (15), Miller and Sack (16), and Krehma and Williams (17). We felt justified in using this reasonable value of the moment for methyl acetate in solution since adding to the length of the hydrocarbon chain does not increase materially the moment of a molecule unless there is simultaneously an increase in distance between polar groups in the molecule. For example, in the gas phase at 25°C, the μ values of methyl and ethyl acetates are 1.72 $\times 10^{-18}$ and 1.78×10^{-18}, respectively, within an estimated accuracy of $\pm 5\%$ (26). The radius a_i has been taken as 0.732 Å. The values of the constants used in these calculations were those given by Rysselberghe (18) except for the values of the dielectric constants of the mixed solvents. These were taken from Åkerlöf (19). All future calculations will be based on like values of the constants involved. The value of the index of refraction ($n = 1.35935$; $n^2 = 1.848$) of liquid methyl acetate recorded in the literature could have been used, but the agreement between calculated and observed values of ($\log_{10} f_{i(D-H_2O)} - \log_{10} f_{i(H_2O)}$) would have extended only to a dielectric constant of 34.25 using μ_{0i} as 1.85×10^{-18} and a_i as 0.732 Å; the agreement with $n^2 = 1.50$ extends fairly acceptably to a dielectric constant of 25.95. It is probable that n, μ_{0i}, and a_i all have values that vary with the solvent because of the difference in polarization, solvation, etc. in different media (5, 7, 20), but since there are so many variables, it is much easier to accept some reasonable value or values for one or two of these and then determine whether the theoretical calculations result in acceptable magnitudes for the other quantities involved. The fact that calculated and observed data differ at low dielectric constants corresponds to observations made in other cases of electrostatic phenomena (5, 7, 18, 21).

Table I. Comparison of Calculated and Observed Values of $\log f_{i(D-H_2O)} - \log f_{i(H_2O)}$ for Methyl Acetate in Dioxane–Water Mixtures

D	Calculated $\log f_{i(D-H_2O)} - \log f_{i(H_2O)}$	Observed $\log f_{i(D-H_2O)} - \log f_{i(H_2O)}$
78.3	0.0000	0.0000
69.68	0.0246	0.0269
60.79	0.0500	0.0585
51.90	0.0944	0.0991
42.96	0.1381	0.1549
34.25	0.2280	0.2321
25.95	0.3470	0.3307
17.69	0.6031	0.4301
10.71	1.2432	0.5272
5.605	2.9953	0.5884

Table II. Calculated and Observed Values of the Activity
Coefficients of Potassium, Sodium, Lithium, and Hydrogen Chlorides[a]

D	Wt. % MeOH	KCl		NaCl		LiCl		HCl	
		$\log f_\infty$ (Calc.)	$\log f_\infty$ (Obs.)	$\log f_\infty$ (Calc.)	$\log f_\infty$ (Obs.)	$\log f_\infty$ (Calc.)	$\log f_\infty$ (Obs.)	$\log f_\infty$ (Calc.)	$\log f_\infty$ (Obs.)
78.3	0	0.000	0.000	0.000	0.000	0.000	0.000	0.000	0.000
74.05	10	0.144	0.171	0.135	0.160	0.095	0.118	0.039	—
69.16	20	0.323	0.345	0.303	0.324	0.220	0.239	0.088	0.095
64.28	30	0.528	0.528	0.495	0.492	0.360	0.361	0.144	0.150
59.59	40	0.757	0.733	0.710	0.667	0.516	0.487	0.206	0.208
54.90	50	1.025	0.931	0.962	0.853	0.700	0.620	0.276	0.278
50.09	60	1.352	1.204	1.268	1.059	0.922	0.769	0.368	0.367
45.00	70	1.775	1.504	1.663	1.282	1.210	0.936	0.484	0.476
40.14	80	2.278	1.866	2.135	1.557	1.553	1.124	0.622	0.640
35.70	90	2.857	2.307	2.679	1.870	1.948	1.369	0.780	0.956

[a] At infinite dilution in MeOH–H_2O mixtures corresponding to a water value assumed to be unity.

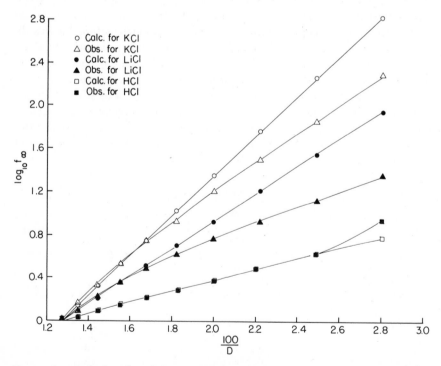

Figure 1.   Calculated and observed values of $\log_{10} f_\infty$ for KCL, LiCl, and HCl vs. $100/D$ for the solvent

Activity Coefficients of Uni-Univalent Chlorides.   The second application is to the activity coefficients of electrolytes as a function of the dielectric constant of the media.   Returning to Equation 32, for electrolytes at infinite dilution only the Born term remains, that is,

$$\log_{10} f_{i0} = \frac{\epsilon_i^2}{4.606kTa_i} \left[\frac{1}{D} - \frac{1}{D_0} \right] \tag{39}$$

since there are no dipoles involved.

Åkerlöf (22) has determined the activity coefficients of potassium, sodium, lithium, and hydrogen chlorides at infinite dilution in MeOH–H$_2$O mixtures in relation to the corresponding water values (assumed to be unity) in each case. In this calculation there is only one parameter involved.   The agreement between calculated and observed values of $\log_{10} f_\infty$ was obtained using a_0 as 0.652 Å, 0.695 Å, 0.955 Å, and 2.39 Å, respectively, for the potassium, sodium, lithium, and hydrogen chloride.   The data for KCl, NaCl, LiCl, and HCl are presented in Table II, and data for three of them are plotted in Figure 1.   In the calculations, Åkerlöf's values for the dielectric constants for the mixed solvent were used.

Since electrostatic data generally conform to theory better at higher dielectric constants of the solvent, the a_i values were chosen to make agreement between calculated and observed values of the activity coefficients better in this region. Åkerlöf's value for the activity coefficient of HCl in 90 wt % methanol appears to be in error since, unlike the other activity coefficient, it is higher than the calculated value, and the observed plot curves upward instead of downward in this solvent region.   However, the agreement between calculated and observed data for HCl is unusually good up to this concentration of methanol in the solvent.   In all cases fair agreement over the whole range of dielectric constants could have been obtained by proper selection of the parameter a_i.   The values of a_i listed here for the chlorides considered are in opposite order from those found by Born (11) for the positive ions involved.   Table III gives the comparison.

Born's ionic radii pertain to the unhydrated ions.   Our values for a_i for the alkali metal chlorides are in the same direction as the degree of hydration (23) of the metal ions of these chlorides as is shown by columns two and three of Table III.   Since each of these positive ions is accompanied by a chloride ion bearing

Table III.   The Values of a_i Used in These Calculations Compared with the Radii Found by Born for the Corresponding Positive Ions and Also for Their Atoms

Substance	Degree of Hydration of Positive Ions	Our a_i Values, for the Chlorides	Born's Atomic Radii	Born's Ionic Radii
KCl	5.4	0.652	2.62	2.00
NaCl	8.4	0.695	2.10	1.59
LiCl	14.0	0.955	1.72	1.49
HCl	1.0	2.39	1.68	0.625

four water of hydration molecules, our values of a_i could be related to the degree of hydration with respect to the alkali metal ions. It does not seem unreasonable to assume that the electrostatic forces causing hydration will be larger for smaller unhydrated ions. The values of a_i used are apparently small except for HCl. These small values of a_i could indicate the importance of higher-order terms omitted in the derivation. On the whole however, especially at higher dielectric constants, agreement between calculation and observation are satisfactory considering the approximations made in the derivation. In the case of HCl the agreement, except for the 90 wt % MeOH, is remarkable.

Osmotic Coefficients of Mixtures of Ions with Ion–Dipoles. In the two preceding applications of the theory formulated in this paper we have made use of the limiting cases where either the ion character or the dipole character of the particles being considered was absent. In the third application we will apply the equation for the osmotic coefficient to substances which yield both particles possessing ion–dipole characteristics and particles which are only ionic in character. The ammonium salts have been observed (24) to give j values (j = 1 − osmotic coefficient) which in dilute regions rise above the values of the limiting case of the Debye–Hückel theory. Also the quaternary ammonium iodides and thallous nitrate give values of j which are above the limiting law values and which increase with the increasing concentration of salt (25). There has been no explanation of these phenomena. Scatchard and Prentiss (24) state frankly that they can give no reason for their results. Ebert and Lang (25) suggest that the effect is caused by the quaternary ammonium ions possessing moments.

Table IV.   Comparison between Calculated and Observed j Values
for Quaternary Ammonium Iodides

Salt	M	$\sqrt{\mu}$	$j = 1-g$ (Calc.)	$j = 1-g$ (Obs.)	$j = 1-g$ D.H.L.L.	a_i (Å)	μ_{oi} (D.U.)
N(CH$_3$)$_4$I	0.0442	0.2102	0.097	0.104	0.078	0.9	9
	0.0502	0.2241	0.105	0.104	0.083		
	0.0878	0.2963	0.148	0.129	0.110		
	0.0896	0.2993	0.150	0.129	0.111		
	0.0919	0.3032	0.153	0.146	0.113		
N(C$_2$H$_5$)$_4$I	0.1242	0.3524	0.147	0.158	0.131	1.0	7
	0.3130	0.5595	0.249	0.245	0.208		
	0.4529	0.6730	0.310	0.300	0.250		
	0.5443	0.7378	0.346	0.338	0.275		
N(C$_3$H$_7$)$_4$I	0.0479	0.2189	0.101	0.113	0.081	1.1	9
	0.0564	0.2375	0.111	0.103	0.088		
	0.0691	0.2629	0.125	0.135	0.098		
	0.1055	0.3248	0.160	0.161	0.121		
	0.1217	0.3489	0.178	0.182	0.130		
	0.1973	0.4442	0.244	0.236	0.165		
	0.2863	0.5351	0.313	0.287	0.199		
	0.3856	0.6210	0.385	0.326	0.231		

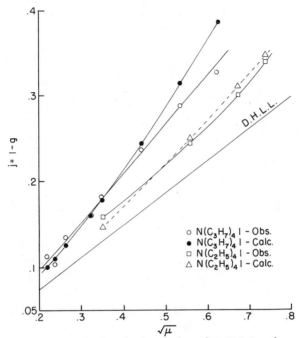

*Figure 2.   Calculated, observed, and D.H.L.I. values
of $j = 1 - g$ for $N(C_3H_7)_4I$ and $N(C_2H_5)_4N(C_2H_5)_4I$
vs. $\sqrt{\mu}$*

If it can be assumed that ammonium and substituted ammonium ions are polar as well as ionic, then the data can be explained by Equation 37. Using an internal refractive index of the substituted ammonium ions equal to that of liquid ammonia, $1.325^{16.5}$ (*26*) we have calculated and observed results which agree as indicated in Table IV. The moments in vacuum and the a_i values used in these calculations are specified in the tables. The results for tetraethyl- and tetra-propylammonium iodide are given in Figure 2. The large value of the moment in vacuum need not be disturbing since a simple calculation will show that if the internal refractive index of the ions concerned be chosen, say, as great as 2, then the moment can be decreased (e.g., from $\mu_{0i} = 10$ D.U. to $\mu_{0i} < 5$ D.U.) with only a slight difference in agreement between calculated and observed results. A reasonable value of refractive index was chosen as that of liquid ammonia to facilitate calculations as explained in "Applications." The fact that quaternary ammonium chlorides and bromides do not exhibit these enhanced j values indicates that the ionic and not the dipolar effects are predominant for these salts throughout the whole concentration range studied. The data for thallous nitrate can be accounted for using similar values of a_i, n, and μ_{0i}.

The ammonium chloride and ammonium sulfate data in Table V are presented as a comparison of observed and calculated values of j using for the cal-

Table V. Calculated and Observed j Values for Ammonium Salts in the Region of Dilute Solutions

Substance	M	$\sqrt{\mu}$	$j = 1 - g$ (Calc.)	$j = 1 - g$ (Obs.)	$j = 1 - g$ D.H.L.L.
NH_4Cl	0.001	0.0316	0.0131	0.0140	0.0118
	0.002	0.0447	0.0193	0.0203	0.0166
	0.005	0.0707	0.0331	0.0321	0.0263
	0.01	0.1000	0.0506	0.0411	0.0372
$(NH_4)_2SO_4$	0.001	0.0548	0.0456	0.0471	0.0408
	0.002	0.0775	0.0674	0.0689	0.0576
	0.005	0.1225	0.1154	0.0972	0.0912
	0.01	0.1732	0.1774	0.1228	0.1289

culations a_i as 0.50 Å, μ_{0i} as 10.0 D.U. and $n = 1.325$. Again an increase of n to a value of 2 would reduce the value of μ_{0i} to less than 5 D.U.   In the case of these salts it is evident that the dipole influences are predominant only in dilute solutions, while in concentrated solutions the ionic characteristics prevail.   Similar results would be obtained for other ammonium salts; e.g., about the same values of a_i, μ_{0i}, and n would apply for ammonium bromide and ammonium iodide.

Again we wish to emphasize that it is the modified moments $(\mu_{0i}/\gamma(D)$ and $(\sqrt{\delta(D)}\,\mu_{0i})$ arising from the use of Onsager's theory of the moments in liquids that are important.   With reasonable values of internal refractive index, the moments in vacuum can be reduced.   Furthermore, our theory does account for values of j greater than those predicted by the Debye–Hückel limiting law if the ions can be assumed to have dipole characteristics.

Literature Cited

1. Debye, P., Hückel, E., *Phys. Z.* (1923) **24**, 185.
2. Amis, E. S., Jaffé, G., *J. Chem. Phys.* (1942) **10**, 598.
3. Gronwall, T. H., LaMer, V. K., Sandved, K., *Phys. Z.* (1928) **29**, 558.
4. LaMer, V. K., Gronwall, T. H., Grieff, L. J., *J. Phys. Chem.* (1931) **35**, 2245.
5. Onsager, L., *J. Am. Chem. Soc.* (1936) **58**, 1486.
6. Amis, E. S., "Solvent Effects on Reaction Rates and Mechanisms," Academic, New York, 1967, pp. 43–46.
7. Quinlan, J. E., Amis, E. S., *J. Am. Chem. Soc.* (1955) **71**, 4187.
8. Amis, E. S., Hinton, J. F., "Solvent Effects on Chemical Phenomena," Academic, New York, 1973, pp. 247–259.
9. Debye, P., *Phys. Z.* (1924) **25**, 97.
10. Falkenhagen, Hans "Electrolytes," R. P. Bell, Trans., Oxford University, Oxford, 1933.
11. Born, M., *J. Phys.* (1920) **1**, 45.
12. Scatchard, G., *Chem. Rev.* (1932) **10**, 229.
13. Falkenhagen, Hans, "Electrolytes," Translated by R. P. Bell, Oxford University, Oxford, 1933, pp. 115, 253.
14. Harned, H. S., Ross, A. M., Jr., *J. Am. Chem. Soc.* (1941) **63**, 1933.
15. Smythe, C. D., Dornte, R. W., Wilson, E. B., *J. Am. Chem. Soc.* (1931) **53**, 4242.
16. Miller, H., Sack, H., *Phys. Z.* (1930) **31**, 815.
17. Krehma, J. J., Williams, J. W., *J. Am. Chem. Soc.* (1927) **49**, 2408.
18. Rysselberghe, P. V., *J. Am. Chem. Soc.* (1943) **65**, 1249.

19. Åkerlöf, G., *J. Am. Chem. Soc.* (1932) **54,** 4125.
20. Amis, E. S., LaMer, V. K., *J. Am. Chem. Soc.* (1939) **61,** 905.
21. Amis, E. S., *J. Am. Chem. Soc.* (1941) **63,** 1606.
22. Åkerlöf, G., *J. Am. Chem. Soc.* (1930) **52,** 2353.
23. MacInnes, D. A., "The Principles of Electrochemistry," Reinhold, New York, 1939, p. 93.   New ed.: Dover Publications, New York, 1961.
24. Scatchard, G., Prentiss, S. S., *J. Am. Chem. Soc.* (*1932*) *54, 2696*.
25. Ebert, L., Lang, J., *Z. Phys. Chem. Abt. A. Haber Bd.* (1928) **139,** 584.
26. "Handbook of Chemistry and Physics," 54th Ed., Chemical Rubber Co., Cleveland, 1973–1974.

RECEIVED June 9, 1975.

21

Viscosity of Dilute Solutions of Salts in Mixed Solvents

ROBERT A. STAIRS

Department of Chemistry, Trent University, Peterborough, Ontario, Canada K9J 7B8

The viscosity of solutions of KI in water–methanol (W–M) and of $LiClO_4$ in W–M and in water–acetone (W–A) at 25°C are interpreted (through the Jones–Dole B coefficient) as showing preferential solvation of both solutes by water over methanol but little discrimination by $LiClO_4$ between water and acetone. Values of B ranged between −0.16 and +0.68 for KI in W–M, 0 and 0.80 for $LiClO_4$ in W–M, both showing minima at about 0.15 mole fraction of methanol and essentially linear from the aqueous value 0.044 to 1.14 in pure acetone for $LiClO_4$ in W–A. Upper limits to primary solvation numbers calculated from these ranges and the molar volumes of the solvents were: $LiClO_4$ in W–A, 7; in W–M, 16; KI in W–M, 15. Precision was insufficient for significant A coefficients to be obtained.

For a suspension of solid spheres in a medium of viscosity η_0, Einstein (1) has derived the following expression for the relative vicosity of the suspension over that of the medium:

$$\eta/\eta_0 = 1 + 2.5\phi$$

where ϕ is the volume fraction of the suspended spheres. In a salt solution there will be effects of interionic forces and effects of the ionic field on the local viscosity of the medium, but also an effect like that of Einstein's suspended spheres caused simply by the volume of the solvated ions. Here let us consider a solute M in a solvent of mixed composition, A and B, in which both A and B can interact with M to form solvates or complexes MA_n, MB_n and mixed complexes $MA_{n-q}B_q$. If the molar concentration of M in the solution is C moles/l., and all of it is complexed,

$$C = [MA_n] + [MA_{n-1}B] + \ldots + [MB_n].$$

The volume fraction $\phi = C\overline{V}$ where $\overline{V}$ is the mean effective molar volume of the complexed forms of M (in liters):

$$\overline{V} = \frac{\sum\limits_{q=0}^{n} [MA_{n-q}B_q]V_{MA_{n-q}B_q}}{\Sigma[MA_{n-q}B_q]}$$

Let the concentration $[MA_{n-q}B_q] = C_q$ and assume the partial molar volume of $MA_{n-q}B_q$ to be $V_0 + (n-q)V_A + qV_B$ (where V_0 need not necessarily equal V_M); then

$$\overline{V} = V_0 + nV_A + \frac{\Sigma qC_q}{\Sigma C_q}(V_B - V_A)$$

$$= V_0 + nV_A + \overline{q}(V_B - V_A)$$

One may now write for the stepwise substitution of B for A, i.e., for the reaction

$$MA_{n-q+1}B_{q-1} + B \rightleftarrows MA_{n-q}B_q + A$$

the equilibrium constant (in terms of C_q and the mole fractions of A and B in the solvent):

$$K_q = \frac{C_q x_A}{C_{q-1} x_B}.$$

It is tempting to assume that all the K_q are equal, but even if the MA bonds are unaffected by the presence of B in the complex and vice versa, this should not be so, for an argument through the probability of replacing one of the q B's in a complex by A, etc., leads to the relation:

$$K_q = \frac{(n-q+1)}{q}K,$$

where K is the value the constant would have were there only one coordination site. [See Ref. 2. Note that my K is their $K^{1/n}$.] Virtue in this case brings the reward of simplicity, for on expressing the sums in the expression for $\overline{q}$ in terms of constants of this kind and the quantity

$$y = K\frac{x_B}{x_A}$$

we obtain

$$\overline{q} = \frac{ny}{1+y}$$

Values of $\overline{q}/n$ for various K are plotted in Figure 1.

Viscosity data for fairly dilute salt solutions are usually discussed in terms of the Jones–Dole equation (3)

$$\eta/\eta_0 = 1 + A\sqrt{C} + BC$$

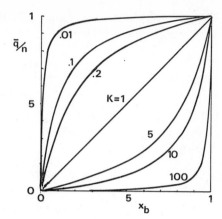

Figure 1.　Mean fraction $\bar{q}/n$ of solvation sites occupied by B molecules in solvates of the type $MA_{n-q}B_q$ as a function of solvent composition for different values of the discrimination constant K. (K is defined in the text, and measures the tendency of A to be favored over B at a single site.)

in which η is the viscosity of the solution, η_0 the viscosity of the pure solvent, C the molar concentration, and A and B constants. Both were originally empirical, but the significance of A has been elucidated by Falkenhagen and co-workers (4) and shown to arise from long-range interionic interactions. The B coefficient is considered to arise from ion–solvent interactions in three ways. (a) An Einstein contribution B_E arises from the simple bulk of the solvated ions. (b) The solvent is affected in the neighborhood of the ions by electrostriction. In water this tends, by breaking down the open hydrogen-bonded clusters of molecules, to reduce the viscosity, and hence it is a negative term. (c) Certain ions, by the way they fit into the water structure, may enhance it either directly by tetrahedral coordination (Li^+, NH_4^+) or in a curious roundabout way by interacting with water so weakly that the neighboring molecules form a local structure of a different sort about the ion (hydrophobic interaction, characteristic of larger carboxylate anions).

　　　Now, for a solute with molar volume $\overline{V}$ l. mol^{-1} and molar concentration C we can write for the volume fraction $\phi = \overline{V}C$. The Einstein equation becomes

$$\eta/\eta_0 = 1 + 2.5\,\overline{V}C$$
$$= 1 + B_E C$$

which is the Jones–Dole equation with $A = 0$ and $B = B_E$. Hence

$$B_E = 2.5(V_0 + nV_A + \bar{q}(V_B - V_A))$$

Since $\bar{q}$ ranges from 0 to n, the possible variation of B from this cause is $2.5n(V_A - V_B)$. To estimate the magnitude of the effect, let us take $n = 4$, and $V_A - V_B = 0.040$ l. This gives a variation of 0.40 in B. Typical values of B for simple electrolytes in aqueous solution run from about -0.1 to $+0.2$, so this expected variation should be easily detectable. It is clear that for a solute such as LiCl in mixtures of water with a less polar liquid, n is by no means constant over the whole range of solvent compositions, and it is effectively larger in pure water because of the structure-making effect of Li^+. Perhaps over a limited range of solvent composition at the water-poor end, however, it might be roughly constant.

Finlay (5) states that heats of hydration of typical crystalline salts tend to cluster in the vicinity of three kcal per mole of water. If we estimate the energy of substitution of a less polar molecule for water as half that figure, i.e., 1.5 kcal, we obtain an estimate for K of about ten. From the corresponding curve in Figure 1 it may be seen that 70% of the effect should happen in the range of solvent compositions from 0.8 to 1.0 mole fraction of the less polar constituent. Unfortunately, recent workers (6, 7, 8) have avoided just this region. We have now made such measurements on solutions of KI and $LiClO_4$ in water–methanol, and of $LiClO_4$ in water–acetone, extending the range of solvent compositions into the organic-rich region.

Experimental

Rather than rigorously drying the acetone and methanol solvents prior to adding water, we used reagent grades (Fisher) dried roughly with anhydrous calcium sulfate and distilled with protection from moist air. The distillates were stored under dry nitrogen and transferred by nitrogen pressure. The stock acetone so prepared contained 1.2% moisture (density at 25.0°C, 0.7877), and the methanol contained <0.1% moisture (density at 25.0°C, 0.7868 g/ml.). Solvent mixtures were made from these and water by direct weighing, and their concentrations were calculated with appropriate corrections. Salts were not fused but oven-dried at 110°C. KI was then assumed to be dry, but $LiClO_4$ was analyzed for moisture by loss on ignition at 250°C (12.95 ± 0.16% H_2O). The appropriate correction was made for the concentrations of this salt.

Viscosity measurements were made with two Cannon–Ubbelohde viscometers, and timing was by an optical device actuating an electronic timer (Wescan Instruments, Inc.). An air thermostat was used. The viscometers were calibrated with redistilled air-saturated water over the range 10°–50°C. The kinetic-energy correction was used in the form:

$$\nu = \eta/\rho = K_1 t + K_2/t$$

with one viscometer. With the other the correction K_2/t was negligible.

Densities required for the calculation of viscosity were obtained in three ways. The densities of the salt solutions were measured by Archimedean displacement (9) of a borosilicate glass bob weighed in air, water, and the solvents and solutions. The results were expressed as linear functions of the molar concentration of salt, and the slopes obtained are recorded in Table II. Densities of the solvent mixtures were taken from published tables (10, 11). Densities of the stock acetone and methanol were measured by a conventional pycnometer to greater precision than the salt solutions, and they were compared with the literature values for analysis and with the values obtained by displacement as a check. The two methods agreed within 0.0005 g/ml.

Assuming this figure to represent the precision of the density values and combining it with the timing errors found, which ranged between 0.04 in 50 sec and 0.1 in 150 sec, we estimated the relative precision in viscosity as 0.1%. The resulting precision of the Jones–Dole A coefficient was insufficient to permit much significance to be attached to the calculated values, but the B ceofficients were estimated to be precise to ± 0.015 for values ranging from -0.2 to $+1.0$. We estimated the accuracy of the absolute viscosities to be no better than $\pm 1\%$, owing mainly to difficulty in reproducing the settings of the air thermostat. Internal consistency within a run was much better than that, as noted.

Fitting the Jones–Dole equation to the data was not done in the usual manner in the linearized form $(\eta/\eta_0 - 1)/C^{1/2} = A + BC^{1/2}$, since this casts a heavy burden on the required precision of η_0 and also inflates the errors at the lower-concentration end. Instead, we used the form:

$$\eta = \eta_0 + aC^{1/2} + bC$$

in which η_0, $a(= A\eta_0)$ and $b(= B\eta_0)$ are all treated as adjustable parameters in a least-squares procedure. The figure for η_0 recorded in Table II is this parameter, not the measured value of η at $C = 0$.

Results

Table I contains the viscosities obtained for the solvent mixtures and the salt solutions. Table II summarizes the results for the solutions and contains the viscosity of each solvent mixture without added salt, the constants A and B of the Jones–Dole equation, the value of the density–concentration coefficient $d\rho/dC$, and the density of the solvent mixture.

The viscosities of the solvent mixtures are in general agreement with literature values (12, 13) and will not be discussed further. The A coefficients, as noted above, are not expected to be highly significant, but the occasional negative value among them prompts the following comment. While the theory (4) does not allow for negative values of this constant, and there is therefore good reason to dismiss them as being in error, it is not clear to me that this must necessarily be so in a ternary solution.

The values of B obtained (together with some from other authors, as noted) are plotted in Figure 2. Each of the methanol–water curves shows a minimum in the water-rich region which is attributed by Werblan et al. (6) to enhancement of the water structure with increasing amounts of organic solvent up to a point. The absence of the minimum where acetone replaces methanol seems to support this interpretation. Attributing the whole of the change in B, between the lowest point (the value at the water end or in the minimum) and the value at the water-poor extreme, to the change in the effective volume of the solvated solute, i.e., taking it as a measure of $2.5\,n\,(V_A - V_B)$, one can obtain a solvation number n which can be compared with solvation numbers derived in other ways. This

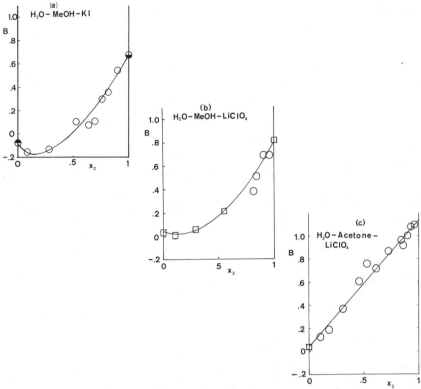

Figure 2.   Viscosity B coefficients for salt solutions in mixed water–organic solvents, as a function of the mole fraction of the organic component in the solvent.   O, this work; ◐, Ref. 18; ◑ Ref. 16; □ Ref. 6.

is done in Table III in which these solvation numbers are listed together with values of primary hydration numbers calculated from ionic values listed by Bockris and Reddy (*14*).   The numbers are roughly double those from other measurements, which is not unexpected as the whole change in B surely does not arise from this cause.   The differences among ions seem reasonable: $Li^+ > Na^+$ and $Cl^- > ClO_4^-$.

The values of the constant K which indicate the tendency of the solutes to discriminate in favor of water over the organic component are in principle obtainable from the form of the curves in the organic-rich region.   Here, rough comparison of the curves in Figure 1 with the curves in the water–methanol systems suggests that, for KI, K lies somewhat between 2 and 5, and that for $LiClO_4$, it is slightly larger, perhaps between 3 and 6. K for $LiClO_4$ in water–acetone appears to be close to unity, suggesting that there is no net discrimination between water and acetone by this salt.   Covington (*15*) has pointed out that this probably represents cancellation of opposing effects, for one ion may favor water and the other acetone.

Appendix

Table I. Viscosities of Solutions, cP

$H_2O-MeOH-KI$

x_2:							
0.0813		0.285		0.525		0.639	
c	η	c	η	c	η	c	η
0	1.255	0	1.624	0	1.351	0	1.159
0.0137	1.251	0.0088	1.622	0.0055	1.358	0.0132	1.162
0.0268	1.250	0.0190	1.619	0.0125	1.351	0.0196	1.163
0.0313	1.250	0.0315	1.616	0.0236	1.353	0.0281	1.168
0.0493	1.247	0.0397	1.615	0.0278	1.352	0.0451	1.165
0.0559	1.249	0.0460	1.620	0.0369	1.355	0.0502	1.168
0.0652	1.244	0.0547	1.612	0.0424	1.355	0.0592	1.172
0.0787	1.245	0.0599	1.612	0.0507	1.354	0.0693	1.170
0.0804	1.241	0.0655	1.612	0.0619	1.358	0.0730	1.172
0.0908	1.235	0.0758	1.608	0.0749	1.358	0.0793	1.173
0.1033	1.241					0.0846	1.173
0.1047	1.235						

$H_2O-MeOH-KI$

x_2:							
0.689		0.756		0.816		0.899	
c	η	c	η	c	η	c	η
0	1.079	0	0.889	0	0.881	0	0.755
0.0023	1.082	0.0046	0.891	0.0052	0.885	0.0078	0.758
0.0081	1.083	0.0101	0.892	0.0126	0.886	0.0145	0.765
0.0158	1.084	0.0193	0.896	0.0208	0.886	0.0182	0.764
0.0213	1.085	0.0223	0.899	0.0259	0.889	0.0223	0.764
0.0281	1.088	0.0270	0.903	0.0333	0.892	0.0307	0.767
0.0352	1.088	0.0341	0.905	0.0373	0.893	0.0345	0.767
0.0402	1.091	0.0432	0.896	0.0426	0.898	0.0398	0.773
0.0463	1.091	0.0498	0.909	0.0494	0.898	0.0439	0.776
0.0516	1.086	0.0573	0.908	0.0555	0.901	0.0508	0.780
		0.0689	0.912	0.0671	0.901	0.0555	0.775
		0.0794	0.916	0.0784	0.906	0.0610	0.781
		0.0853	0.918	0.0905	0.907		

$H_2O-MeOH-LiClO_4$

x_2:							
0.816		0.847		0.907		0.964	
c	η	c	η	c	η	c	η
0	0.803	0	0.761	0	0.681	0	0.596
0.0100	0.812	0.0081	0.767	0.0116	0.689	0.0076	0.601
0.0205	0.818	0.0190	0.774	0.0191	0.690	0.0175	0.607
0.0254	0.819	0.0272	0.777	0.0267	0.693	0.0328	0.614

Table I. (*continued*)

H₂O−MeOH−LiClO₄

x_2:	0.816		0.847		0.907		0.964	
c	η	c	η	c	η	c	η
0.0355	0.824	0.0412	0.783	0.0366	0.700	0.0497	0.622
0.0456	0.827	0.0592	0.792	0.0435	0.703	0.0668	0.629
0.0582	0.835	0.0774	0.800	0.0631	0.712	0.0770	0.633
0.0682	0.840	0.0930	0.808	0.0731	0.716	0.0972	0.642
0.0783	0.845	0.1015	0.810	0.0875	0.722		
0.0895	0.849						
0.0966	0.852						

H₂O−Acetone−LiClO₄

x_2:	0.1025		0.185		0.310		0.465	
c	η	c	η	c	η	c	η
0	1.377	0	1.351	0	1.067	0	0.737
0.0040	1.376	0.0081	1.354	0.0087	1.072	0.0082	0.741
0.0108	1.379	0.0173	1.358	0.0254	1.082	0.0233	0.749
0.0195	1.378	0.0354	1.364	0.0370	1.088	0.0283	0.753
0.0314	1.380	0.0540	1.371	0.0535	1.095	0.0380	0.755
0.0433	1.381	0.0703	1.374	0.0752	1.105	0.0522	0.764
0.0796	1.383	0.0794	1.377	0.1083	1.119	0.0556	0.769
—	—	0.0980	1.384	—	—	0.0744	0.775
						0.0857	0.779
						0.0909	0.782

H₂O−Acetone−LiClO₄

x_2:	0.535		0.616		0.729		0.849	
c	η	c	η	c	η	c	η
0	0.626	0	0.524	0	0.424	0	0.352
0.0051	0.630	0.0041	0.527	0.0118	0.429	0.0089	0.356
0.0143	0.635	0.0127	0.530	0.0247	0.435	0.0193	0.360
0.0283	0.642	0.0220	0.534	0.0343	0.438	0.0308	0.364
0.0371	0.647	0.0337	0.539	0.0472	0.443	0.0418	0.369
0.0454	0.651	0.0423	0.542	0.0580	0.448	0.0544	0.373
0.0608	0.659	0.0503	0.545	0.0658	0.451	0.0666	0.376
0.0743	0.666	0.0622	0.550	0.0779	0.455	0.0757	0.380
0.0888	0.674	0.0823	0.558	0.0818	0.456	0.0854	0.384
		0.0967	0.564	0.0921	0.461		
				0.0975	0.463		

Table I. (*continued*)

$H_2O-Acetone-LiClO_4$

0.864		0.902		0.930		0.962	
c	η	c	η	c	η	c	η
0	0.345	0	0.332	0	0.320	0	0.306
0.0110	0.351	0.0161	0.337	0.0061	0.323	0.0154	0.312
0.0184	0.353	0.0303	0.342	0.0186	0.328	0.0262	0.316
0.0273	0.357	0.0403	0.346	0.0319	0.332	0.0363	0.318
0.0394	0.361	0.0520	0.350	0.0415	0.337	0.0587	0.328
0.0509	0.365	0.0699	0.357	0.0552	0.341	0.0733	0.332
0.0649	0.370	0.0824	0.362	0.0711	0.346	0.0767	0.332
0.0723	0.372	0.0897	0.365	0.0791	0.349	0.0857	0.336
0.0769	0.374					0.0929	0.338

Table II. Solvent Viscosity, Jones–Dole Coefficients, and Dependence of Density on Salt Concentration at the Various Solvent Compositions

System	$x_2{}^a$	$\eta, {}_0 cP^a$	A	B	$\dfrac{d\rho}{dC}$	ρ_0
$H_2O-MeOH-KI$	0^b	0.890	0.005	−0.0755	0.122	0.9971
	0.0813	1.255	0.017	−0.162	0.121	0.9760
	0.285	1.624	0.008	−0.130	0.120	0.9295
	0.525	1.351	−0.006	0.104	0.116	0.8765
	0.639	1.159	0.024	0.078	0.114	0.8540
	0.689	1.076	0.027	0.111	0.115	0.8460
	0.756	0.889	0.028	0.300	0.118	0.8305
	0.816	0.881	0.008	0.353	0.122	0.8192
	0.889	0.755	0.010	0.543	0.130	0.8035
	1.000^b	0.544	0.016	0.675	0.150	0.7866
$H_2O-MeOH-LiClO_4$	0.816	0.804	0.060	0.385	0.091	0.8192
	0.847	0.760	0.041	0.520	0.091	0.8146
	0.907	0.681	0.006	0.695	0.091	0.8032
	0.964	0.596	0.036	0.695	0.091	0.7929
$H_2O-Acetone-$ $LiClO_4$	0^c	0.890	0.013	0.044	0.055	0.9971
	0.1025	1.3773	−0.012	0.128	0.068	0.9590
	0.185	1.3511	0.016	0.192	0.080	0.9309
	0.310	1.0668	0.028	0.369	0.092	0.8949
	0.465	0.7371	0.022	0.604	0.103	0.8596
	0.535	0.626	0.034	0.758	0.109	0.8453
	0.616	0.5242	0.020	0.717	0.113	0.8322
	0.729	0.424	0.027	0.870	0.117	0.8150
	0.849	0.352	0.034	0.959	0.115	0.7999

10. *Int. Crit. Tables*, Vol. III, McGraw-Hill, N. Y. and London, 1928, pp 112, 115.
11. Timmermans, J., "The Physico-Chemical Constants of Binary Systems," Vol. 4, Interscience, New York, 1960, pp 40–42, 156–161.
12. *Int. Crit. tables*, Vol. V, McGraw-Hill, N.Y., 1928, pp 17, 19, 22.
13. Timmermans, J., "The Physico-Chemical Constants of Binary Systems," Vol. 4, Interscience, New York, 1960, pp 43–44, 162–164.
14. Bockris, J. O'M., Reddy, A. K. N., "Modern Electrochemistry," Plenum, N. Y., 1970, p 131.
15. Covington, A. K., private communication, 1975.
16. Jones, G., Fornwalt, H. J., *J. Am. Chem. Soc.* (1935) **57**, 2041–2045.
17. Kaminsky, M., *Discuss. Faraday Soc.* (1957) **24**, 171–179.
18. Laurence, V. D., Wolfenden, J. H., *J. Chem. Soc.* (1934) 1144–1147.

RECEIVED August 1, 1975. Work supported by Trent University and the National Research Council of Canada.

344

proton abstrac
terpretation o
Neverthe
to reactions i
proton. For
(8, 9, 11, 12)

$$J_- = -$$

y_i activity c
ence in hyd
and J_ can
to indicato

Acidity Functions in
Bases in Mixed Solve

T. J. M. POUW, W. J. BOVER, and P. ZUMAN

Department of Chemistry, Clarkson College of

used for e
Prev
from equ
studied a
compou
with hy
consecu
compli
In th
establis
Equati
scale J
Roche

The addition of hydroxide ions to s
(ArCHO + OH⁻ ⇌ ArCH(OH)O⁻
acidity scales in water-ethanol and
containing sodium hydroxide as a base.
mixtures are linearly correlated with
constants. The independence of react
vent composition confirms that substitu
suitable J_ indicators for hydroxide soluti
and water–DMSO mixtures. Dependenc
dium hydroxide concentration is only sligh
nol up to 90% and at a constant sodium h
tion shows only small increase between 90
J_ increases more with increasing DMSO
the effect is much smaller than that of DM
based on proton abstraction from aniline.

the r
0, an
of th
betw
chlo
of s
ram
aci
the
su
fis
fu

Ｔhe electrochemical oxidation of aromatic aldehyde
in strongly alkaline media. Acidity functions for str
solutions of alkali metal and quaternary ammonium hydr
to dissociation of proton (H_), are well established (2, 3).
and diphenylamines (4, 5) and indoles (6) were used as aci
establishment of such scales, but whether an acidity scale b
indicator can be rigorously applied to acid–base equilibria in
different acidic groups for reactions in strongly alkaline me
tionable. For substituted anilines, behavior both parallel (7
(8) to the H_ scale based on indole derivatives has been report
solubility of anilines in aqueous solutions of alkali metal hydroxi
of the aniline derivative with more than one hydroxide ion, irre
tution reactions (9), and the possibility of hydroxide ion additi

343

Table II. *(continued)*

System	$x_2{}^a$	$\eta,\ _0cP^a$	A	B	$\dfrac{d\rho}{dC}$	ρ_0
H$_2$O–Acetone–	0.864	0.345	0.047	0.919	0.115	0.7982
LiClO$_4$	0.902	0.332	−0.005	1.000	0.115	0.7940
	0.930	0.320	0.019	1.084	0.113	0.7910
	0.962	0.306	0.019	1.095	0.109	0.7877

[a] Mole fraction of the organic component in the mixed solvent and the corresponding viscosity in the absence of added salt.
[b] From Refs. *12* and *17*.
[c] From Ref. *6*.

Table III. Primary Solvation Numbers. Comparison of Results from the Variation of B (visc.) with the Most Probable Values by Other Methods

Solvent	H$_2$O–acetone (this method)	H$_2$O–methanol (this method)	H$_2$O[b] (various methods)
Salt			
LiClO$_4$	7.14	15.9	6
NaClO$_4$	—	10.5[a]	5
NaCl	—	15.6[a]	6
KI	—	15.4	4

[a] From Ref. *6*.
[b] From Ref. *14*, assigning 1 to ClO$_4{}^-$.

Acknowledgments

The author thanks Jacob K.-T. Lai for assistance with the measurements.

Literature Cited

1. Einstein, A., *Ann. Phys. Leipzig* (1911) **34**, 591.
2. Covington, A. K., Newman, K. E., ADV. CHEM. SER. (1976) **155**, 153–196.
3. Jones, G., Dole, M., *J. Am. Chem. Soc.* (1929) **51**, 2950–2964.
4. Falkenhagen, H., Dole, M., *Phys. Z.* (1929) **30**, 611 (simple version); Falkenhagen, H., Vernon, E. L., *Phys. Z.* (1932) **33**, 140 (full theory).
5. Finlay, G. R., Phillipson, A., *Can. Chem. Conf., 58th*, Toronto, Canada, May 28, 1975.
6. Werblan, L., Rotowska, A., Minc, S., *Electrochim. Acta* (1971) **16**, 41–49.
7. Lee, M. D., Lee, J. J., Lee, I., *J. Korean Inst. Chem. Eng.* (1973) **11**, 164–173.
8. Feakins, D., Freemantle, D. J., Lawrence, K., *J. Chem. Soc. Faraday Trans. 1* (1974) **70**, 795–806.
9. Archimedes, Περιοχουμενων I prop. 7. Recent editions: (1) T. L. Heath, Ed., "The Works of Archimedes," Dover, N. Y., 1950 (reprint of Cambridge edition of 1897/1912); (2) "Archimède," texte établi et trad. par Ch. Mugler, vol iii, Paris (Les Belles Lettres) 1971.

10. *Int. Crit. Tables,* Vol. III, McGraw-Hill, N. Y. and London, 1928, pp 112, 115.
11. Timmermans, J., "The Physico-Chemical Constants of Binary Systems," Vol. 4, Interscience, New York, 1960, pp 40–42, 156–161.
12. *Int. Crit. tables,* Vol. V, McGraw-Hill, N.Y., 1928, pp 17, 19, 22.
13. Timmermans, J., "The Physico-Chemical Constants of Binary Systems," Vol. 4, Interscience, New York, 1960, pp 43–44, 162–164.
14. Bockris, J. O'M., Reddy, A. K. N., "Modern Electrochemistry," Plenum, N. Y., 1970, p 131.
15. Covington, A. K., private communication, 1975.
16. Jones, G., Fornwalt, H. J., *J. Am. Chem. Soc.* (1935) **57**, 2041–2045.
17. Kaminsky, M., *Discuss. Faraday Soc.* (1957) **24**, 171–179.
18. Laurence, V. D., Wolfenden, J. H., *J. Chem. Soc.* (1934) 1144–1147.

RECEIVED August 1, 1975.   Work supported by Trent University and the National Research Council of Canada.

Acidity Functions in Solutions of Strong Bases in Mixed Solvents

T. J. M. POUW, W. J. BOVER, and P. ZUMAN

Department of Chemistry, Clarkson College of Technology, Potsdam, N. Y. 13676

The addition of hydroxide ions to substituted benzaldehydes (ArCHO + OH⁻ ⇌ ArCH(OH)O⁻) is used to establish J₋ acidity scales in water-ethanol and water-DMSO mixtures containing sodium hydroxide as a base. The pK-values in such mixtures are linearly correlated with Hammett substituent constants. The independence of reaction constant ρ of solvent composition confirms that substituted benzaldehydes are suitable J₋ indicators for hydroxide solutions in water-ethanol and water-DMSO mixtures. Dependence of J₋ values on sodium hydroxide concentration is only slightly affected by ethanol up to 90% and at a constant sodium hydroxide concentration shows only small increase between 90 and 98% ethanol. J₋ increases more with increasing DMSO concentration, but the effect is much smaller than that of DMSO on H₋ values based on proton abstraction from aniline.

The electrochemical oxidation of aromatic aldehydes (*1*) must be studied in strongly alkaline media. Acidity functions for strongly alkaline aqueous solutions of alkali metal and quaternary ammonium hydroxides, corresponding to dissociation of proton (H_-), are well established (*2, 3*). Substituted anilines and diphenylamines (*4, 5*) and indoles (*6*) were used as acid–base indicators for establishment of such scales, but whether an acidity scale based on one type of indicator can be rigorously applied to acid–base equilibria involving structurally different acidic groups for reactions in strongly alkaline media remains questionable. For substituted anilines, behavior both parallel (*7*) and nonparallel (*8*) to the H_- scale based on indole derivatives has been reported. The limited solubility of anilines in aqueous solutions of alkali metal hydroxides, the reactions of the aniline derivative with more than one hydroxide ion, irreversible substitution reactions (*9*), and the possibility of hydroxide ion addition rather than proton abstraction—in particular for nitro derivatives (10)—complicate the interpretation of experimental data.

Nevertheless, validity of acidity function H_- cannot be expected to extend to reactions involving addition of hydroxide ions rather than dissociation of a proton. For addition of hydroxide ions, the acidity function J_- has been defined (8, 9, 11, 12) as in Equation 1, where a_i are activities, c_i concentrations,

$$J_- = -\log(a_H + y_{SHOH^-}/y_{SH}a_w) = pK_2 + pK_w + \log C_{SHOH^-}/C_{SH} \quad (1)$$

y_i activity coefficients, and the subscript w corresponds to water. If the difference in hydration of SH and $SHOH^-$ is neglected, the relationship between H_- and J_- can be expressed as in Equation 2, where the primed values correspond to indicators

$$J_- = H_- + \log a_w + \log (y_{SH}y'_{S^-}/y'_{SH}y_{SHOH}) \quad (2)$$

used for establishing the H_- scale.

Previous attempts (8, 10, 12, 13, 14) to develop the J_- acidity function scale from equilibrium measurements have been unsuccessful. Most frequently studied addition reactions of hydroxide ions are those involving nitroaromatic compounds (formation of Meisenheimer complexes). Measurements of equilibria with hydroxide ions involving nitro compounds were found complicated by consecutive reactions (8, 9, 10), by uptake of a second hydroxide ion (13), or by complicated changes in absorption spectra (13, 14).

In the absence of equilibrium measurements, attempts have been made to establish the J_- scale based on kinetic date. The attempt (15) to approximate Equation 2 by $J_- = H_- + \log a_w$ was only partly successful. Recently, a kinetic scale $J_-(k)$ based on the Zucker–Hammett hypothesis (16) has been proposed by Rochester (17). In Equation 3, which relates kinetic and acidity data, k is

$$J_-(k) = \log k - \log (k_0 K_w) = -\log (a_H + y^{\neq}/a_w y_{SH}) \quad (3)$$

the rate constant at a given sodium hydroxide concentration, k_0 is at [NaOH] = 0, and $y^{\neq}$ is the activity coefficient of the transition state in the rate limiting step of the particular reaction used to measure $J_-(k)$. The aromatic S_N2 reactions between hydroxide ions and 2,4-dinitroanisole, 2,4-dinitrophentole, and 1-chloro-2,4-dinitrobenzene gave the same acidity function $J_-(k)$ for concentrations of sodium hydroxide up to $5M$. A comparison with the rate constants of chloramine and carbon disulfide S_N2 hydrolyses indicates the need for a separate J_- acidity function. An analogous function denoted as H_{R^-} has been derived for the addition of methoxide and ethoxide ions to α-cyanostilbenes in dimethyl sulfoxide–methanol (and ethanol) mixtures (18). Nevertheless, retroaldolization fission made it impossible to use this reaction for determination of the acidity function in aqueous media.

In the course of polarographic and spectrophotometric investigations of isomeric phthalaldehydes, it has been observed that substituted benzaldehydes undergo addition of hydroxide ions in a reversible reaction as shown in Reaction 4.

$$ArCHO + OH^- \overset{K_2}{\rightleftharpoons} ArCH(OH)O^- \qquad (4)$$

Equilibrium constants of Reaction 4 have been measured (*19*) for a series of substituted benzaldehydes by comparing the differences in the spectra of the benzaldehyde (ArCHO) and the geminal diol anion (ArCH(OH)O⁻). With increasing hydroxide ion concentration, the weak carbonyl band at 300 nm decreases and the medium intensity band at 250–280 nm (corresponding to the electronic transition involving the aromatic ring conjugated with the carbonyl group) decreases. Usually absorbance in the latter region was measured to determine the ratio $C_{ArCH(OH)O^-}/C_{ArCHO}$. From measured values of this ratio in aqueous solutions containing varying concentrations of sodium hydroxide, the equilibrium constants K_2 were determined by the standard overlap procedure (*20*) which assumes similar activity coefficient ratios.

From determined values of constants K_2 for individual indicators and the expression of log $C_{ArCH(OH)O^-}/C_{ArCHO}$ at a given C_{OH^-}, values of acidity constants J₋ were calculated by Equation 1, with an accuracy of ± 0.03J₋ unit at J₋ <16.3 and of about ±0.1J₋ unit at higher sodium hydroxide concentrations. The values of J₋ obtained have shown a good agreement with values J₋(*k*) obtained by kinetic measurements (*15, 17*) for solutions containng 4*M* sodium hydroxide or less (for which values J₋(*k*) were available). The relationship between J₋ and H₋ functions is similar to that of J₀′ (or J₀‴) and H₀. In more dilute solutions of sodium hydroxide both functions, J₋ and H₋ follow a similar trend of the dependence on base concentration. Differences between J₋ and H₋ in solutions containing 10*M* or more concentrated sodium hydroxide can be attributed to change in water activity (a_w in Equation 2), but meaningful discussion must be postponed until information will be available about the variation of activity coefficients of individual species with sodium hydroxide concentration.

In organic solvents the acidity functions H₋ corresponding to hydrogen dissociation from neutral indicator acids were reported for solutions of alkali metal alkoxides in various alcohols (*2*), using nitroanilines (*21*), aminobenzenecarboxylic acids (*22*), or indols (*23*) as indicators. For addition reactions of methoxide and ethoxide ions to neutral indicator acids, acidity functions J₋ (also denoted as H_R) based on use of nitrobenzenes (*21*) and α-cyanostilbenes (*18*) as indicators in methanol and dimethylsulfoxide–methanol and –ethanol mixtures were reported. Recently (*24*) the acidity function J₋ (denoted as J_M) was derived for methoxide ion solutions in methanol using substituted benzaldehydes as indicators. These scales involve arbitrary choice of water as the solvent for determination of the dissociation constant of the anchoring acid.

For mixtures of organic solvents with water, the available information (*2*) is derived only from reactions involving dissociation of hydrogen ion, leading to acidity function H₋. Measurements for solutions containing a constant concentration of a base and a varying ratio of water and the organic solvent were carried out using sodium alkoxides as bases in mixtures of water and alcohols (25). Tetramethylammonium hydroxide was also used as the base in mixtures of water with pyridine (4, 7), tetramethylenesulfone (4, 7), dimethylsulfoxide (4, 7, 26), and dimethylformamide (27). When 0.005M sodium ethoxide was used, a relatively modest increase of the value of H₋ (from 11.74 to 13.35) was observed (25) when the ethanol concentration was increased from 0 to 100 mole %. In solutions containing 0.011M tetramethylammonium hydroxide and increasing dimethylsulfoxide (DMSO) concentration, the increase in the value of H₋ was found to be much more dramatic, from 12.0 in water to 26 in 99.5 mole % DMSO. The increase in the value of H₋ with DMSO concentration was smaller at concentrations below 85 mole %, but very steep at higher DMSO concentrations (26).

Substituted benzaldehydes have proved useful as acid–base indicators for reactions involving the addition of hydroxide ions n strongly alkaline aqueous media (19). It seemed logical to extend their use to solutions of sodium hydroxide in water–ethanol and water–DMSO mixtures. In ethanol–water, it was of interest whether the competition between addition of hydroxide and ethoxide ions will be reflected in the dependence of the J₋ function on ethanol concentration. In water–DMSO mixtures, it was important to investigate whether the radical change at higher DMSO concentrations, observed for H₋ values and attributed ·to changes in solvation of the hydroxide ion, will be observed for the addition reaction as well.

The J₋ function was determined at constant sodium hydroxide concentration (0.01M), and varying ethanol or DMSO content and measurements were carried out to define the acidity function J₋ in solutions containing fixed percentages of the organic solvent component and varying concentrations of sodium hydroxide. Such scales provide the possibility of preparing solutions of known J₋ in mixtures containing a given concentration of the organic component and thus seem to be of practical importance (e.g., for electroanalytical measurements). They have rarely been reported, even for the H₋ function.

Furthermore, equilibrium constants K_2 for the formation of the adduct corresponding to Reaction 4 were determined by the overlap procedure in solutions containing fixed concentrations of the organic component, and the effect of solvent composition on the Hammett reaction constant ρ was followed.

Experimental

Most of the benzaldehydes employed were obtained from Aldrich Chemical Co. (Milwaukee). Purity was checked chromatographically and by measurement of boiling or melting points. The few benzaldehydes whose purity proved to be unsatisfactory were recrystallized from ether or ethanol.

Two sets of 0.01M stock solutions of the benzaldehydes were prepared, one in absolute ethanol and one in DMSO, to be used for the experiments in water–ethanol and water–DMSO mixtures, respectively. Both DMSO (Baker Chemical Co.) and ethanol were used without purification.

Sodium hydroxide stock solutions were prepared of three different concentrations, viz., 0.1, 1.0 and 10M. The 0.1 and 1.0M solutions were obtained by diluting Baker reagent grade Dilut-it standardized solutions. The 10M solution was prepared by dilution of 50% Baker Analyzed sodium hydroxide (18.86M). Carbonate free water was used for all dilutions and the solutions were protected from contact with air.

UV spectra were recorded with a Unicam-SP-800-A (Pye Unicam, Cambridge, England) recording spectrophotometer, using matched quartz cells (10 mm optical path).

All solutions used for the measurement of spectra were prepared by mixing adequate volumes of the hydroxide and benzaldehyde stock solutions together with an appropriate amount of water and ethanol or DMSO. All these solutions were made up to a total volume of 10 ml.

In the majority of cases, the absorbance at 250–280 nm was measured in solutions containing 1×10^{-4} M of the benzaldehyde studied. In solutions containing higher concentrations of DMSO, these benzenoid absorption bands were overlapped by a cut-off caused by solvent absorption. In those cases, the absorbance corresponding to $n - \pi^*$ transition of the carbonyl group was measured. Because of the lower extinction coefficient of this absorption band, measurements were then carried out in $5 \times 10^{-4}M$ benzaldehyde solutions.

The ionization ratios $C_{ArCH(OH)O^-}/C_{ArCHO}$ needed were calculated from experimentally accessible absorbancies, using Equation 5

$$C_{ArCH(OH)O^-}/C_{ArCHO} = (A_0 - A)/(A - A_R) \qquad (5)$$

where A_0 is the absorbance of the benzaldehyde solution at such a hydroxide ion concentration or in a buffer of such a pH that no addition of OH$^-$ to benzaldehyde occurs, and A_R is the residual absorbance in a solution where all of the benzaldehyde is present as the anion ArCH(OH)O$^-$. A is the absorbance at the given OH$^-$ concentration. Unless otherwise stated, the values of A, A_0 and A_R were measured at the wave length of maximum benzenoid absorption (250–280 nm). The value of A_R was usually obtained by an extrapolation procedure.

Values of $C_{ArCH(OH)O^-}/C_{ArCHO}$ for each benzaldehyde derivative were measured at 10–15 different sodium hydroxide concentrations in solutions containing fixed ethanol or DMSO concentrations ranging from 1 to 90 vol %. Since spectra obtained in the presence of 1% ethanol were indistinguishable from spectra recorded in purely aqueous solutions, it was possible to use absorbancies obtained in 1% ethanolic solutions for the calculation of pK_2(H$_2$O) values. Ionization ratios were also determined in benzaldehyde solutions containing a constant concentration of sodium hydroxide (0.01M) and an ethanol or DMSO content which was varied between 1 and 98 vol %.

The addition of ethanol appears to have an appreciable influence on the absorptivity of substituted benzaldehydes. Generally the molar absorptivity decreases by about 40% when the ethanol content of the solution is increased from 1 to 90 vol %. Moreover, there is a slight shift of both the benzenoid and the carbonyl band to shorter wavelengths. These changes must be taken into consideration when absorbancies in solutions containing varying concentrations of ethanol are compared. No such effects were observed in the study of solutions containing varying amounts of DMSO.

Unless otherwise stated, spectra were time independent over the 3–5 min needed for recording the spectra.

Results

pK_2-Values. Attention is focused first on the values of equilibrium constant K_2 of substituted benzaldehydes in individual mixed solvents with reference to a standard state in those particular solvents. For this purpose, the ratios $C_{ArCH(OH)O^-}/C_{ArCHO}$ were determined in each solvent mixture as a function of hydroxide concentration. For benzaldehydes with electronegative substituents, the value of the equilibrium constant K_2 defined by Equation 6

$$K_2 = (C_{ArCH(OH)O^-}/C_{ArCHO}C_{OH^-})(f_{ArCH(OH)O^-}/f_{ArCHO}f_{OH^-}) \qquad (6)$$

can be obtained by extrapolation of the plot of $[\log (C_{ArCH(OH)O^-}/C_{ArCHO}) - \log C_{NaOH}]$ against concentration of sodium hydroxide to $C_{NaOH} = 0$ (i.e., $\mu = 0$).

For benzaldehydes with higher pK_2 values, the overlap procedure (20) can be used. Values obtained by both procedures in the individual solvent mixtures are summarized in Tables I and II.

In every solvent system studied, pK_2 values above a certain limit (dependent on the solvent system) could not be measured because of either limited solubility of sodium hydroxide or changes of the spectra with time (indicating competitive processes at high organic solvent concentrations).

Variations in the differences, Δ, between $pK_2(H_2O)$ and pK_2 (mixed solvent) for each individual solvent composition are relatively small (Tables I, II), indicating that Reaction 4 for different substituted benzaldehydes is influenced almost equally by the change in solvent composition. This fact, together with the existing evidence (19) that for aqueous hydroxide solutions substituted benzaldehydes form a suitable set of J_- indicators, proves that substituted benzaldehydes can be used also for the establishment of J_- scales in water–ethanol and water–DMSO mixtures.

J_- for Hydroxide Solutions in Aqueous Ethanol. From the $pK_2(H_2O)$ values and values of $\log C_{ArCH(OH)O^-}/C_{ArCHO}$ at a given C_{OH^-} in a given solvent mixture, it is possible to calculate J_- values for the solvent mixture under consideration using Equation 1 where pK_w is the autoprotolytic constant of water and $pK_2(H_2O)$ is inserted for pK_2. This definition expresses J_- values with reference to a standard state in pure water, and therefore basicities of sodium hydroxide solutions in mixed solvents can be compared to basicities of sodium hydroxide solutions in water by J_- values.

Table I. pK_2 Values for Substituted Benzaldehydes in Water–Ethanol Mixtures

			10% EtOH		50% EtOH		90% EtOH	
Benzaldehyde	$\sigma_x{}^a$	1% EtOH	pK_2	Δ^b	pK_2	Δ^b	pK_2	Δ^b
p-NO$_2{}^c$	+0.78	−1.05	−1.26	0.21	−1.28	0.23	−1.84	0.79
3,5 diCl^c	+0.74	−0.91	−0.97	0.06	—	—	—	—
m-NO$_2{}^c$	+0.71	−0.81	−1.11	0.30	−1.06	0.25	−1.11	0.30
m-CN^c	+0.68	−0.64	−0.79	0.15	−0.85	0.21	−0.95	0.31
p-CN^c	+0.66	−0.84	−1.01	0.17	—	—	−1.25	0.41
3,4 diCl	+0.60	−0.19	—	—	—	—	—	—
p-CF$_3$	+0.55	−0.32	−0.46	0.14	−0.68	0.36	−0.88	0.56
m-CF$_3$	+0.41	−0.07	—	—	—	—	—	—
m-Cl	+0.37	+0.12	−0.26	0.38	−0.05	0.17	−0.46	0.58
m-F	+0.34	+0.22	−0.03	0.25	0.00	0.22	−0.41	0.63
p-Cl	+0.23	+0.54	+0.27	0.27	+0.27	0.27	−0.11	0.65
p-COOH	+0.13	+0.38	—	—	—	—	—	—
m-OCH$_3$	+0.11	+0.76	+0.55	0.21	—	—	—	—
p-F	+0.06	+0.09	—	—	—	—	+0.43	0.56
Bzh	0.00	+1.05	—	—	—	—	+0.48	0.57
m-CH$_3$	−0.07	+1.18	—	—	—	—	—	—
p-CH$_3$	−0.17	+1.48	—	—	—	—	—	—
p-OCH$_3$	−0.27	+2.04	—	—	—	—	—	—
p-OH	−0.52	—	—	—	—	—	—	—
m-OH	−0.71	+2.12	—	—	—	—	—	—

^{a} Hammett substituent constant.
$^b \Delta = pK_2(H_2O) - pK_2$ (solvent mixture).
$^c pK_2$ values of these compounds were obtained by extrapolation of a[log $(C_{\text{ArCH(OH)O}^-}/C_{\text{ArCHO}}) - \log C_{\text{NaOH}}]$ *vs.* C_{NaOH} plot to $C_{\text{NaOH}} = 0$.

Table II. pK_2 Values for Substituted Benzaldehydes in Water–DMSO Mixtures

Benz-	10% DMSO		50% DMSO		80% DMSO		90% DMSO	
aldehyde	pK_2	Δ^a	pK_2	Δ^a	pK_2	Δ^a	pK_2	Δ^a
p-NO$_2{}^b$	−1.14	0.09	−1.29	0.24	−2.03	0.98	−2.35	1.30
m-NO$_2{}^b$	−0.94	0.13	—	—	—	—	—	—
m-CN^b	−0.90	0.26	—	—	—	—	—	—
p-CN^b	−0.92	0.08	−1.10	0.26	−1.67	0.83	−2.18	1.34
p-CF$_3$	−0.43	0.11	−0.67	0.35	−1.27^c	0.95	−1.64^c	1.32
m-Cl	0.00	0.12	−0.21	0.33	−1.21^c	(1.33)		
m-F	—	—	−0.09	0.32				
p-Cl	+0.49	0.06	+0.28	0.26				
m-OCH$_3$	+0.64	0.12						
Bzh	+0.86	0.19						
m-CH$_3$	1.20	(−0.02)						

$^a \Delta = pK_2(H_2O) - pK_2$ (mixed solvent).
^{b} For these compounds, pK_2 was found by extrapolation of the [log $(C_{\text{ArCH(OH)O}^-}/C_{\text{ArCHO}}) - \log C_{\text{NaOH}}]$ plot.
^{c} Determined in $5 \times 10^{-4}M$ benzaldehyde solutions.

Table III. J_- of Solutions of NaOH in H_2O/EtOH

$$J_- \equiv 14 + pK_2 \,(as\ determined\ in\ water) + log\, \frac{C_{ROH^-}}{C_R}$$

			J_-	
C_{NaOH}	1% EtOH	10% EtOH	50% EtOH	90% EtOH
0.01	11.91	—	—	—
0.05	12.74	—	—	—
0.1	13.07	13.20	13.57	13.41
0.2	13.45	13.66	13.73	13.92
0.3	13.68	13.80	13.86	14.14
0.4	13.89	13.94	14.00	14.27
0.5	13.97	14.04	14.14	14.41
0.6	14.04	14.12	14.22	14.53
0.7	14.10	14.20	14.28	14.60
0.8	14.18	14.22	14.34	14.67
0.9	14.24	14.30	14.40	14.73
1.0	14.28	14.34	14.48	14.77
1.5	14.52	14.56	14.70	$C_{NaOH} = 1.32$ 14.95
2.0	14.69	14.72	14.80	
2.5	14.85	14.92	14.96	$C_{NaOH} = 1.89$ 15.15
3.0	14.95	15.06	15.16	
3.5	15.08	15.22		
4.0	15.20	15.38		
4.5	15.32	15.50		
5.0	15.47	15.62		
6.0	15.69			
7.0	15.69			
8.0	16.16			
9.0	16.34			
10.0	16.54			
11.0	17.20			
12.0	17.45			

Calculated values (Table III) of J_- in ethanol–water mixtures show a dependence on sodium hydroxide concentration (Figure 1) resembling that in water.

To investigate the influence of ethanol concentrations higher than 90 vol % on the values of the J_- function, two series of measurements were carried out. The sodium hydroxide concentration was kept constant at 0.01 and 0.1M, respectively, and the concentration of ethanol changed, using p-nitrobenzaldehyde (for the 0.01M NaOH solutions) and m-trifluoromethyl- and m-chlorobenzaldehyde (for the 0.1M NaOH solutions) as indicators. After correction for medium effects caused by ethanol, the slight decrease in benzenoid absorption observed resulted in a small increase in J_- with increasing ethanol concentration (Figure 2), paralleling the trend calculated for 1M sodium hydroxide solutions from Figure 1.

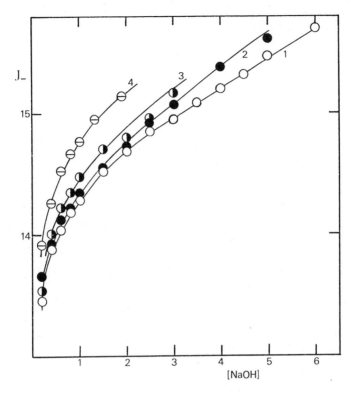

Figure 1. Dependence of the J_ acidity function on sodium hydroxide concentration in water–ethanol mixtures of different composition. Curve 1 (O): 1 vol % EtOH; curve 2 (●): 10 vol % EtOH; curve 3 (◑): 50 vol % EtOH; curve 4 (⊖): 90 vol % EtOH.

J_ for Hydroxide Solutions in Aqueous DMSO. J_ values for solutions containing fixed amounts of DMSO and varying sodium hydroxide concentrations were determined (Table IV) using Equation 1. These show a similar trend for all DMSO concentrations investigated (Figure 3).

The effect of DMSO contents above 90 vol % was studied in mixtures where the sodium hydroxide concentration was kept constant at $0.01M$ and the DMSO content varied. p-Nitro, p-cyano, p-trifluoromethyl, m-chloro, m-fluoro, p-chloro, and m-anisaldehyde were used as indicators. At DMSO concentrations higher than 90 vol %, some of the spectra (particularly those of m-Cl, m-F, p-Cl, and m-OCH$_3$ benzaldehyde) became time dependent and extrapolation of absorbance measurements to zero time became necessary. Hence, the calculated J_ values displayed a larger average deviation (Table V) at these higher DMSO concentrations. Below 80 vol % DMSO, the average deviation was hardly ever higher than 0.05. The dependence of J_ on DMSO concentration was compared with that of H_ (Figure 4).

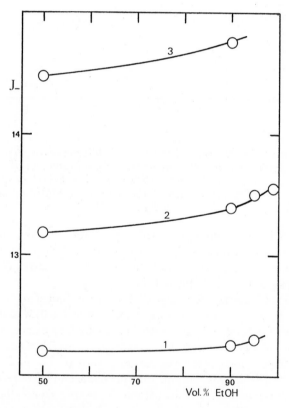

*Figure 2. Influence of ethanol on the J_ value of
0.01M (curve 1) and 0.1M (curve 2) sodium hydroxide.   The two points of curve 3 are taken from
Figure 1 for 1M sodium hydroxide.*

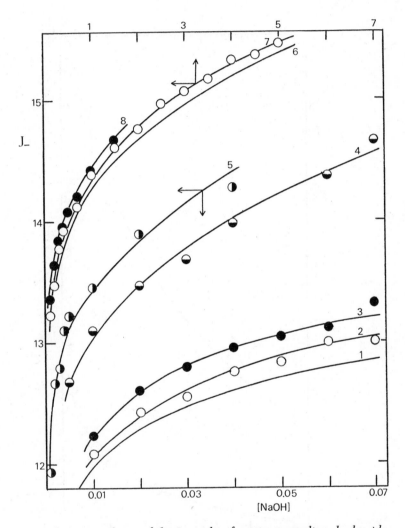

Figure 3. Dependence of the J₋ acidity function on sodium hydroxide concentration in water–DMSO mixtures of different composition. Curves 1 and 6: aqueous solutions. Curves 2 and 7 (○): 10 vol % DMSO. Curves 3 and 8 (●): 50 vol % DMSO. Curve 4 (◐): 80 vol % DMSO. Curve 5 (◑): 90 vol % DMSO. The [NaOH] scale on top of the figure refers to curves 6, 7, and 8, the one on the bottom to curves 1, 2, 3, 4, and 5.

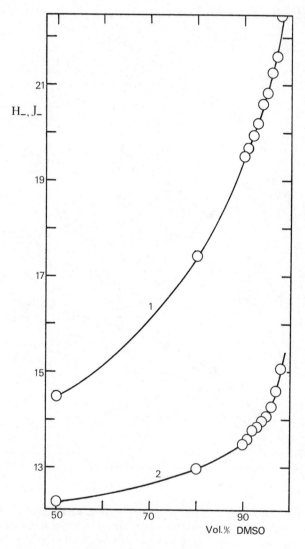

Figure 4. Acidity functions H_- (curve 1) and J_- (curve 2) for water–DMSO mixtures containing 0.01M base (tetramethylammonium hydroxide in case of H_- and sodium hydroxide in case of J_-).

Table IV. J_ of Solutions of NaOH in Water–DMSO Mixtures

C_{NaOH}	$J_-(10\%$ DMSO)	$J_-(50\%$ DMSO)	C_{NaOH}	$J_-(80\%$ DMSO)	$J_-(90\%$ DMSO)
0.01	12.08	12.23	0.001	—	11.93
0.02	12.43	12.60	0.002	—	12.66
0.03	12.54	12.79	0.003	—	12.79
0.04	12.75	12.94	0.004	—	13.10
0.05	12.82	13.04	0.005	12.67	13.22
0.06	12.99	13.11	0.006	—	13.20
0.07	12.98	13.31	0.007	—	13.24
0.1	13.22	13.35	0.008	—	13.24
0.2	13.47	13.63	0.01	13.09	13.45
0.3	13.78	13.84	0.02	13.46	13.89
0.4	13.92	13.95	0.03	13.68	—
0.5	13.99	14.07	0.04	13.98	14.27
0.6	14.01	14.19	0.06	14.37	—
0.7	14.12	14.21	0.07	14.65	—
0.8	14.27	14.25			
0.9	14.24	14.35			
1.0	14.38	14.40			
1.5	14.61	14.67			
2.0	14.76				
2.5	14.97				
3.0	15.07				
3.5	15.17				
4.0	15.33				
4.5	15.37				
5.0	15.46				

Table V. J_ in Water–DMSO Mixtures Containing 0.01M NaOH

% DMSO (V/V)	J_-	n^a	Av. Dev.	Indicators Used
50	12.26	2	0.04	p-NO$_2$, p-CN
80	12.97	2	0.09	p-NO$_2$, p-CN
90	13.48	6	0.13	p-NO$_2$, p-CN, p-CF$_3$[b], m-Cl[b], m-F[b], p-Cl
91	13.58	3	0.11	m-Cl[b], m-F[b], p-Cl
92	13.77	4	0.16	p-Cf$_3$[b], m-Cl[b], m-F[b], p-Cl
93	13.84	3	0.12	m-Cl[b], m-F[b], p-Cl
94	13.98	4	0.09	m-Cl[b], m-F[b], p-Cl, p-OCH$_3$[b]
95	14.08	4	0.14	p-CN, m-Cl[b], m-F[b], p-Cl
96	14.28	4	0.11	m-Cl[b], m-F[b], p-Cl, m-OCH$_3$[b]
97	14.62	4	0.18	m-Cl[b], m-F[b], p-Cl, m-OCH$_3$[b]
98	15.07	3	0.06	m-F[b], p-Cl, m-OCH$_3$[b]

[a] Number of measurements.
[b] Measurements with these compounds were carried out in $5 \times 10^{-4}M$ benzaldehyde solution.

Discussion

Structural Effects and Solvent. The effect of solvent on the equilibrium of Reaction 4 can be first discussed in terms of effects on the susceptibility to substituent effects. The values of pK_2, characterizing this equilibrium, are a satisfactorily linear function of the Hammett constants σ_x as shown by the values of the correlation coefficient r (Table VI). The values of reaction constant ρ are practically independent of the ethanol concentration (Table VI), as was already indicated by the almost constant value of the difference (Δ) between $pK_2(H_2O)$ and pK_2 (mixed solvent) for a given composition of the mixed solvent (Table I). The same situation is indicated for DMSO mixtures (Table II) by the small variations in Δ for any given solvent composition. In this case, the number of accessible pK_2 values was too small to allow a meaningful determination of reaction constants ρ. The structural dependence for various water–ethanol mixtures is thus represented by a set of parallel lines. The shifts between these lines are given by the differences between the $pK_2{}^H$ values (pK_2 of Reaction 4 for the unsubstituted benzaldehyde) in the different solvent mixtures.

Table VI   Influence of Ethanol Percentage on the Free Energy
Relationship

$$pK_2 = -\rho\, \sigma_x + pK_2{}^H$$

% EtOH (V/V)	ρ[a]	Std. Dev. (ρ)	$pK_2{}^H$[b]	Std. Dev. ($pK_2{}^H$)	n[c]	r[d]
1	2.65	0.09	1.08	0.05	18	−0.987
10	2.58	0.17	0.82	0.10	10	−0.984
50	2.84	0.18	0.97	0.11	7	−0.991
90	2.58	0.20	0.52	0.11	10	−0.973

[a] Reaction constant.
[b] pK_2 for the unsubstituted benzaldehyde.
[c] Number of measurements.
[d] Linear correlation coefficient.

Provided that the influence of the water–ethanol composition on the reaction involving addition of hydroxide ions to benzaldehydes can be characterized by any parameter Y_i (the application of Y_- used for benzoic acid dissociations in ethanol–water mixtures (28) might be doubtful), application of the relation $\rho_{Y_i} - \rho_{Y_0} = C(Y_i - Y_0)$ would indicate that the value of C (0.638) for benzoic acids (28) and −0.573 for anilines (29) is close to zero for the benzaldehyde reaction (4).

The Validity of the J₋ Function. For aqueous sodium hydroxide solutions the validity of the J_- function, describing the basicity of a solution by its ability to add hydroxide ions to a carbonyl group to form an anion of the geminal diol, has been proved earlier (19). For all substituted benzaldehydes studied in water–ethanol or water–dimethylsulfoxide (DMSO), the value of log $(C_{ArCH(OH)O^-}/C_{ArCHO})$ determined from absorbance measurements was found to be a linear function of the J_- function with a slope varying between 0.95 and 1.05 in the region of J_- values where measurements were possible. The Cannizzaro reaction or other consecutive processes did not affect the measurements at 25°C except in solutions containing the highest concentrations of DMSO. Careful measurements with derivatives bearing electronegative substituents did not indicate any evidence of formation of dianions of the geminal diol $(ArCH(O^-)_2)$. Hence, benzaldehydes are simpler indicators than cyanostilbenes (18), where competition of carbanion formation, even if not predominant, cannot be completely excluded. The behavior of the benzaldehydes is also simpler than that of nitroaromatic compounds where measurements of equilibria leading to the formation of Meisenheimer complexes were complicated by consecutive reactions (9, 10, 15), by uptake of a second hydroxide ion (13), or by complicated changes in the absorption spectra (13, 14).

Comparison of Aqueous and Water–Ethanol Solutions. The effect of the presence of ethanol in aqueous solutions of sodium hydroxide is usually small. This is shown by the similar shape of the dependence of J_- on sodium hydroxide concentration (Figure 1) and by the small differences in J_- values obtained at the different constant ethanol concentrations up to 90 vol % (Table III). Even when the concentration of sodium hydroxide was kept constant (e.g., 0.1M), the difference between J_- values in 90 vol % ethanol and 98 vol % ethanol was only 0.16 J_- units (Figure 2). In this range of ethanol concentrations, it is necessary to consider the competitive influence of ethoxide ions, the addition of which would result in a decrease of the C_6H_5CO- absorbance indistinguishable from the decrease caused by hydroxide ion addition. In 90 vol % ethanol, the ratio of hydroxide and ethoxide concentrations is about 1:1, while in 98 vol % ethanol, it is possible to extrapolate (30) that about 90% of the base will be present as the ethoxide ion.

The relatively modest increase in the value of the J_- function upon increasing the ethanol content of a sodium hydroxide solution from 90 to 98 vol % indicates that either the nucleophilic reactivity of the ethoxide ion under these conditions does not differ substantially from that of the hydroxide ion (while solvation of the hydroxide ion and the geminal diol anion is similar to solvation of the ethoxide ion and the hemiacetal anion) or that compensation of effects takes place.

The procedure by which Yagil and Anbar calculated theoretical H_- values (31) can be applied to the J_- function as well, leading to Equation 7, where C_{H_2O}

$$J_- = 14 + \log C_{OH^-} - n \log C_{H_2O} \qquad (7)$$

is the free water concentration and n can be considered (31) either as the hydration number of the hydroxide ion or as the difference in hydration numbers between $(ArCHO + OH^-)$ and $ArCH(OH)O^-$. Calculations of J_- with $n = 3$ and $n = 4$ were made for aqueous sodium hydroxide solutions, using for C_{H_2O} the expression $C_{H_2O} = d - 0.001(18n + 40) C_{OH^-}$ in which d is the density of the solution (Table VII). Comparison of the calculated values with the experimental J_- values for aqueous sodium hydroxide solutions seems to indicate a change in n with the concentration of sodium hydroxide. In solutions which are

Table VII. Theoretical and Experimental Values of J_- for
Aqueous NaOH Solutions

C_{NaOH}	$J_-(exp)$	$J_-(theor.)_{n=3}$[a]	$J_-(theor.)_{n=4}$[a]
1	14.28	14.08	14.12
2	14.69	14.46	14.58
3	14.95	14.73	14.90
4	15.20	14.95	15.20
5	15.47	15.16	15.51
6	15.69	15.35	15.82
7	15.96	15.56	16.17
8	16.16	15.78	16.58
9	16.34	16.01	17.09
10	16.54	16.27	17.71
11	17.20	16.50	18.61
12	17.45		

[a] $J_-(theor.) = 14 + \log C_{OH^-} - n \log C_{H_2O}$ where $C_{H_2O} = d - 0.001 (18n + 40) C_{OH^-}$ (d is density of the solution).

$2M$ sodium hydroxide or less, the best agreement between calculated and experimental J_- values is obtained for n larger than 4, between $3M$ and $5M$ for $n = 4$, and between $6M$ and $8M$ for $3 < n < 4$. Although the results for aqueous sodium hydroxide solutions with concentrations $9-11M$ also suggest a value of n between 3 and 4, it is more likely that in this concentration range a further dehydration takes place causing an increase in the activity of the OH^- ion. Under those circumstances, it is no longer appropriate to calculate J_- values by Equation 7, which takes into account only the mass action effect of the decrease in free water concentration.

An attempt to apply Equation 7 to the calculation of J_- values in ethanol–water mixtures containing base, using the expression $C_{H_2O} = d - 0.0011 C_{OH} - (18n + 40) - 0.0079x$ (x is the vol % of ethanol present), failed especially for the higher ethanol percentages where it led to J_- values considerably higher than actually found. This again indicates that the decrease in free water concentration which takes place on adding increasing amounts of ethanol is obscured since ethanol in many respects displays a behavior similar to the water it replaces.

Comparison of Aqueous and Water–DMSO Solutions. The increase in J_- with increasing DMSO concentration cannot be ascribed to a change in one single physical or chemical property. It is necessary to consider the change in dielectric constant, the effect of hydrogen bonding between DMSO and water (particularly at high DMSO concentrations), the change in water activity, dspersion interactions, and the effect of DMSO on the structure of water and on the hydration of the hydroxide ion. These aspects of the increase in the basicity of solutions containing a constant amount of base with increasing DMSO content have been adequately discussed by Dolman and Stewart (26).

The reason for the considerably smaller effect of DMSO concentration on the J_- acidity function describing hydroxide ion addition to benzaldehydes than the effect on the H_- function obtained from measurements involving proton abstraction from anilines and diphenylamines requires further discussion.

Some idea about the origin of this difference might be obtained by comparing Equation 8 for H_- (32)

$$H_- = 14 + \log C_{OH^-} - (n + 1) \log a_{H_2O} + \log \frac{f_{HA}f_{OH^-}}{f_{A^-}} \qquad (8)$$

with a similar expression, Equation 9, which can be derived for the J_- function:

$$J_- = 14 + \log C_{OH^-} - n \log a_{H_2O} + \log \frac{f_{ArCHO}f_{OH^-}}{f_{ArCH(OH)O^-}} \qquad (9)$$

From Equations 8 and 9, the difference between H_- and J_- is found to be

$$H_- - J_- = -\log a_{H_2O} + \log \frac{f_{HA}}{f_{A^-}} - \log \frac{f_{ArCHO}}{f_{ArCH(OH)O^-}}$$

If it is assumed, in a first approximation, that the activity coefficient ratios f_{HA}/f_{A^-} and $f_{ArCHO}/f_{ArCH(OH)O^-}$ do not differ very much, the difference between H_- and J_- for a given solution would be expressed by $H_- - J_- = -\log a_{H_2O}$. However a calculation of the difference by the latter formula, using the only available information (33) about the activity of water in water–DMSO mixtures, leads to values much smaller than those found in practice. This would indicate that the ratio f_{HA}/f_{A^-}, for the H_- indicator acids, is considerably larger than the ratio of $f_{ArCHO}/f_{ArCH(OH)O^-}$ for the benzaldehydes used as J_- indicators. Probably this difference in activity coefficient ratios is caused by a larger extent of charge delocalization in the A^- anion compared to the geminal diol anion, which causes an extra stabilization of A^- (manifested by a decrease in f_{A^-}) by dispersion interaction with DMSO.

The acidity function J_- (denoted as J_M) for addition of methoxide ions to benzaldehydes (24) increases with base concentration much less steeply than acidity function obtained for additions of methoxide ion to polynitrobenzenes (21) or α-cyanostilbenes (18). No single acidity scale can be applied to strongly basic media, and a given acidity scale can be used only in connection with reaction of the same type.

Acknowledgment

One of us (T.P.) is indebted to the "Netherlands American Commission for Educational Exchange" for financial support by a Fulbright–Hayes grant.

Literature Cited

1. Bover, W. J., Zuman, P., *J. Electrochem. Soc.* (1975) **122**, 368.
2. Bowden, K., *Chem. Reviews* (1966) **66**, 119; *see also* "Solute–Solvent Interactions," J. F. Coetzee and C. D. Ritchie, Eds., M. Dekker, New York, 1969.

3. Rochester, C. H., *Q. Rev. Chem. Soc.* (1966) **20**, 511.
4. Stewart, R., O'Donnell, J. P., *Can. J. Chem.* (1964) **42**, 1694.
5. Stewart, R., Dolman, D., *Can. J. Chem.* (1967) **45**, 925.
6. Yagil, G., *J. Phys. Chem.* (1967) **71**, 1034.
7. Stewart, R., O'Donnell, J. P., *Can. J. Chem.* (1964) **42**, 1681.
8. Rochester, C. H., *Trans. Faraday Soc.* (1963) **59**, 2820.
9. Rochester, C. H., *Trans. Faraday Soc.* (1963) **59**, 2829.
10. Gold, V., Rochester, C. H., *J. Chem. Soc.* (1964) 1727.
11. Gold, V., Haves, B. W. V., *J. Chem. Soc.* (1951) 2102.
12. Gold, V., Rochester, C. H., *J. Chem. Soc.* (1964) 1722.
13. Schaal, R., *C.R. Acad. Sci.* (1954) **239** 1036.
14. Gold, V., Rochester, C. H., *J. Chem. Soc.* (1964) 1710, 1717.
15. Rochester, C. H., *Trans. Faraday Soc.* (1963) **59**, 2826.
16. Zucker, L., Hammett, L. P., *J. Am. Chem. Soc.* (1939) **61**, 2791.
17. Rochester, C. H., *J. Chem. Soc., B* (1967) 1076.
18. Kroeger, D. J., Stewart, R., *Can. J. Chem.* (1967) **45**, 2163.
19. Bover, W. J., Zuman, P., *J. Am. Chem. Soc.* (1973) **95**, 2531.
20. Hammett, L. P., Deyrup, A. J., *J. Am. Chem. Soc.* (1973) **54**, 2721.
21. Terrier, F., *Ann. Chim. (France)* (1969) **4**, 153.
22. Terrier, F., *Bull. Soc. Chim. France* (1969) 1894.
23. Terrier, F., Millot, F., Schaal, R., *Bull. Soc. Chim. France* (1969) 3002.
24. Crampton, M. R., *J. Chem. Soc., Perkins II* (1975) 185.
25. Bowden, K., *Can. J. Chem.* (1965) **43**, 2624.
26. Dolman, D., Stewart, R., *Can. J. Chem.* (1967) **45**, 911.
27. Buncel, E., Symons, E. A., Dolman, D., Stewart, R., *Can. J. Chem.* (1970) **48**, 3354.
28. Grunwald, E., Berkowitz, B. J., *J. Am. Chem. Soc.* (1951), **73**, 4939.
29. Gutbezahl, B., Grunwald, E., *J. Am. Chem. Soc.* (1953) **75**, 559.
30. Caldin, E. F., Long, G., *J. Chem. Soc.*, (1954) 3737.
31. Yagil, G., Anbar, M., *J. Am. Chem. Soc.* (1963) **85**, 2376.
32. Con, B. C., McTique, P. T., *Aust. J. Chem.* (1967) **20**, 1815.

RECEIVED July 21, 1975.

Activity Coefficients of Electrolytes in Water–Polyethylene Glycol Mixed Solvent by Isopiestic Method

Z. ADAMCOVÁ

Department of Physical Chemistry, Institute of Chemical Technology, Prague, Czechoslovakia

With the use of thermodynamic relations and numerical procedure, the activity coefficients of the solutes in a ternary system are expressed as a function of binary data and the water activity of the ternary system. The isopiestic method was used to obtain water activity data. The systems KCl–H₂O–PEG-200 and KBr–H₂O–PEG-200 were measured. The activity coefficient of potassium chloride is higher in the mixed solvent than in pure water. The activity coefficient of potassium bromide is smaller and changes very little with the increasing nonelectrolyte concentration. PEG-200 is salted out from the system with KCl, but it is salted in in the system with KBr within a certain concentration range.

The determination of the thermodynamic functions of individual components in a multicomponent system is based either on the total molar thermodynamic quantity (1) as in Equation 1

$$G = \sum_i N_i \bar{G}_i \qquad (1)$$

or on the use of the Gibbs–Duhem equation (Equation 2).

$$\sum_i n_i d\mu_i = 0. \qquad (2)$$

McKay (2) showed the possibility of using the Euler cross differentiation equation for the thermodynamic treatment of a three-component system.

$$\left(\frac{\partial \ln a_1}{\partial n_2}\right)_{n_1,n_3} = \left(\frac{\partial \ln a_2}{\partial n_1}\right)_{n_2,n_3} \qquad (3)$$

His procedure was used for the calculation of the activity coefficients in the aqueous solution of two electrolytes with a common ion from isopiestic data (3). Kelly, Robinson, and Stokes (4) proposed a treatment of isopiestic data of ternary systems with two electrolytes by a procedure based on the assumption that at all values of molal concentrations, m_1, m_2, the partial derivatives may be expressed by a sum of two functions in their differential form as follows:

$$\left(\frac{\partial \ln \gamma_1}{\partial m_2}\right)_{m_1} + \left(\frac{\partial \ln \gamma_2}{\partial m_1}\right)_{m_2} = f'(m_1) + F'(m_2) \tag{4}$$

In the system with one electrolyte and one nonelectrolyte, where one-parameter terms are insufficient to describe the dependence mentioned above, the mixed terms with both variables were added (5).

We used an analogous procedure (4) to derive the relations for activity coefficients assuming that the partial derivatives in Equation 4 can be expressed by means of a function $f(m_1, m_2)$. Then for the activity coefficients of the first and second component in a three-component system, the following integrated relations are valid:

$$\ln \gamma_1 = \ln \gamma_1^0 + \left[\int_{m_2=0}^{m_2} f(m_1, m_2)dm_2 \right]_{m_1} \tag{5}$$

$$\ln \gamma_2 = \ln \gamma_2^0 + \left[\int_{m_1=0}^{m_1} f(m_1, m_2)dm_1 \right]_{m_2} \tag{6}$$

If the function $f(m_1, m_2)$ is a polynomial of the general form

$$f(m_1, m_2) = \sum_{p=0,1,2,\ldots,q=0,1,2,\ldots,} B_{p,q}(m_1)^p(m_2)^q \tag{7}$$

then

$$\ln \gamma_1 = \ln \gamma_1^0 + \Sigma B_{p,q}(m_1)^p(m_2)^{q+1}(1/q + 1) \tag{8}$$

$$\ln \gamma_2 = \ln \gamma_2^0 + \Sigma B_{p,q}(m_1)^{p+1}(m_2)^q(1/p + 1) \tag{9}$$

The differentials of Equations 5 and 6 are introduced in the Gibbs–Duhem equation, the terms $m_i d \ln (m_i \gamma_i^0)$ are replaced by Bjerrum's terms $d(m_i \phi_i^0)$, and the integration is carried out. The resulting equation is

$$-55,51 \ln a_w = m_1 \gamma_1^0 + m_2 \gamma_2^0 + \left[m_1 \int_{m_2=0}^{m_2} f(m_1, m_2)dm_2 \right]_{m_1} + \left[m_2 \int_{m_1=0}^{m_1} f(m_1, m_2)dm_1 \right]_{m_2} \tag{10}$$

The sum of the two last terms is replaced by the quantity Δ (4). After the integration of $f(m_1, m_2)$ under the conditions indicated in Equation 10, the quantity Δ is equal to

$$\Delta = m_1 \Sigma B_{p,q}(m_1)^p(m_2)^{q+1}(1/q+1) + m_2 \Sigma B_{p,q}(m_1)^{p+1}(m_2)^q(1/p+1) \qquad (11)$$

which can be rearranged into the form

$$\Delta/m_1 \cdot m_2 = \Sigma B_{p,q}(m_1)^p(m_2)^q \frac{p+q+2}{(q+1)(p+1)} = \Sigma A_{p,q}(m_1)^p(m_2)^q \qquad (12)$$

The quantities Δ or $\Delta/m_1 m_2$ are available from isopiestic and binary data. Then the constants $A_{p,q}$ of Equation 12 are calculated by means of the nonlinear correlation and if transferred into $B_{p,q}$

$$B_{p,q} = A_{p,q} \frac{(q+1)(p+1)}{p+q+2} \qquad (13)$$

the final forms of Equations 8 and 9 are obtained

$$\ln \gamma_1 = \ln \gamma_1^0 + \sum_{p=0,1,2,\ldots,q=0,1,2,\ldots} A_{p,q}(m_1)^p(m_2)^{q+1} \left(\frac{p+1}{p+q+2} \right)$$

$$\ln \gamma_2 = \ln \gamma_2^0 + 1/\nu \sum_{p=0,1,2,\ldots,q=0,1,2,\ldots} A_{p,q}(m_1)^{p+1}(m_2)^q \left(\frac{p+1}{p+q+2} \right) \qquad (15)$$

where ν is the number of ions in one molecule of the present electrolyte.

Equations 14 and 15 differ from the work by Robinson, Stokes, and Marsh (6) in the coefficients $(q+1)/(p+q+2)$ and $(p+1)/(p+q+2)$. Nevertheless, as no physical meaning can be ascribed to the individual terms in the function $f(m_1, m_2)$, their choice and their analytical form are somewhat arbitrary. Thus the most convenient set (the smallest number of empirical constants) of terms can be found to reproduce the experimental quantity with an accuracy equal to that of the experimental measurement.

In this work the ternary systems water–KCl–polyethylene glycol-200 were studied. Polyethylene glycols (PEG) are used widely as water soluble polymers (7). Their behavior in the aqueous solution with an electrolyte is a subject of interest from the practical as well as theoretical viewpoint.

Experimental

The isopiestic method described elsewhere (8, 9, 10) was used to obtain the experimental data of water activity of the aqueous solutions. In our arrangement, eight gold-plated silver crucibles were embedded into hollows of a copper block and placed in the glass vacuum desiccator. The crucibles were provided with covers which can be handled from inside. Three desiccators were placed in a water bath (maintained at $25 \pm 0.01\,^\circ C$) on a brass construction which moved from one side to the other. This movement, aided by a moving glass bead in each crucible, mixed the solutions. The desiccators with the samples were evacuated to the pressure of about 20 torr and the evacuation after a time interval was repeated 2–3 times. The time needed for the establishment of equilibrium was

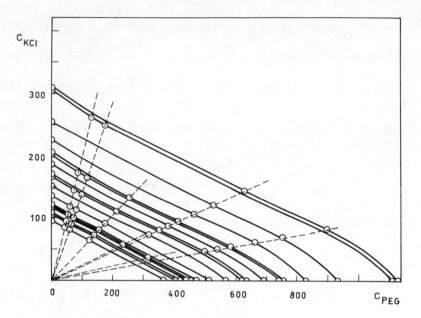

Figure 1.   *The curves of the same water activity in the system KCl–*
H_2O–PEG-200
c_{KCl}, c_{PEG} ... the concentration of the solute [g/1000 g H_2O]

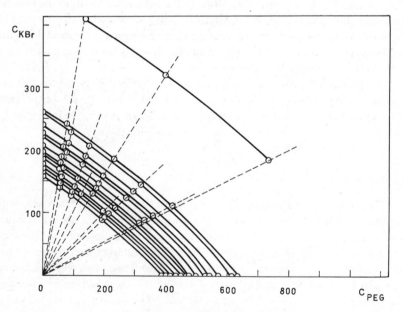

Figure 2.   *The curves of the same water activity in the system KBr–*
H_2O–PEG-200
c_{KBr}, c_{PEG} ... the concentration of the solute [g/1000 g H_2O]

72–120 hr. The initial weight of the solutions was 1–2 g. The weighing before and after reaching the equilibrium was carried out with covers on the crucibles.

Because of solvent loss, which occurred even with covered crucibles and which was of the same order as the weighing errors ($\pm 5 \times 10^{-4}$ g), the correction of weighing for the vacuum was not considered. Then the mean error in the molality determination was 0.3%. In each desiccator, 2–3 standard solutions (KCl or $CaCl_2$) were present. Their osmotic coefficient was taken from the literature (*11*). In the system considered equilibrated, the molal concentrations of the standard solutions differed by no more than 0.5%.

With the same samples, several determinations were made with either increasing or decreasing concentration so that the results could be plotted on a graph where each line represents the points of one equilibrated system with the same water activity and different ratio of the solutes. The points on the coordinates are the molalities of isotonic binary solutions. The experimental points must lie simultaneously on the straight line of the constant ratio of the concentrations of the solutes (Figures 1 and 2).

The homogeneous region of the ternary system is limited by the solubility curve (*12*). The area with isopiestic measured points is dashed in Figures 3 and 4.

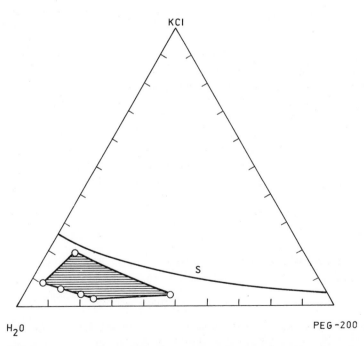

Figure 3.   The phase diagram of the system KCl–H$_2$O–PEG-200

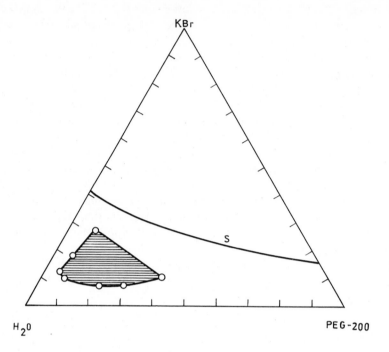

Figure 4. Phase diagram of the system KBr–H₂O–PEG-200

KCl and KBr of analytical reagent grade were recrystalized from distilled water and dried to constant weight at 150°C.

PEG-200 was a commercial product (Chemische Werke Hüls, A. G.) with mean molecular weight 182 as determined osmometrically in the vapor phase (with an error of ±5%) and with water content of 4.42% as determined by K. Fischer's method. (All weights were corrected with respect to this moisture.) $CaCl_2$ was of analytical reagent grade. For the standard solutions, the stock saturated solution was prepared and maintained at the constant temperature of 25°C. The concentration of $CaCl_2$ was tested by the chloride content argento-metrically.

Results

The Calculation of Activity Coefficients in the Binary System. The experimental quantity Δ was calculated from the relation:

$$\Delta = \nu_{ref}m_{ref}\phi_{ref} - m_1\phi_1^0 - \nu_2 m_2\phi_2^0 \tag{16}$$

The osmotic coefficients of the binary electrolyte solutions, ϕ_2^0 and ϕ_{ref} were calculated by means of the Lietzke–Stoughton relation (*11*) enlarging the Debye–Hückel equation with empirical terms:

$$\phi_2^0 = 1 - \frac{S}{A^3 I}\left[(1 + A\sqrt{I}) - 2\ln(1 + A\sqrt{I}) - \frac{1}{1 + A\sqrt{I}}\right] +$$
$$BI + CI^2 + DI^3 \quad (17)$$

The corresponding equation for the activity coefficient required in Equation 15 is then

$$\ln\gamma_2^0 = \frac{S\sqrt{I}}{1 + A\sqrt{I}} + 2BI + \frac{3}{2}CI^2 + \frac{4}{3}DI^3 \quad (18)$$

The constant S is the Debye–Hückel constant, the value of which is for 25°C and 1:1 electrolyte 1.17202 and for 2:1 electrolyte 2.34404. The other constants are listed in Table I.

The osmotic coefficient of the binary nonelectrolyte solution was obtained from isopiestic measurements

$$\phi_1^0 = \frac{\nu_{ref} m_{ref} \phi_{ref}}{m_1} \quad (19)$$

The three-parameter equation was chosen for the analytical description of the dependence $\phi_1^0 = f(m_1)$:

$$\phi_1^0 = 1 + A_1 m_1 + A_2 m_1^2 + A_3 m_1^3 \quad (20)$$

Table I.   The Values of Constants from Equation 17

	A	$B \times 10^2$	$C \times 10^3$	$D \times 10^4$	Limits of validity
KCl	1.30752	−0.359188	7.17091	−5.6750	0.106–4.81
CaCl$_2$	1.61291	4.56577	8.57310	−2.7380	0.002–6.00
KBr	1.29231	0.994831	4.34095	−3.50742	0.100–5.00

Table II.   Constants of Equation 20 and Osmotic and Activity Coefficients in the Aqueous Solution of PEG-200

A_1	A_2	A_3	Dev. (%)
0.0796	0.0333	−0.0089	0.854
m_1	ϕ_1^0	γ_1^0	
0.2	1.017	1.034	
0.4	1.036	1.071	
0.6	1.057	1.117	
0.8	1.081	1.166	
1.0	1.104	1.218	
1.2	1.128	1.274	
1.4	1.152	1.334	
1.6	1.176	1.397	
1.8	1.199	1.461	
2.0	1.221	1.527	

Table III. The Experimental and Calculated Data of the System KCl–H$_2$O–PEG–200[a]

m_{ref}	m_{PEG}	m_{KCl}	Δ	$\dfrac{\Delta}{m_{PEG}m_{KCl}}$	Dev. m_{ref} (%)
1.2810	0.2134	1.1268	0.0632	0.263	−0.22
	0.2882	1.0775	0.0745	0.240	−0.14
	0.6330	0.8664	0.0784	0.143	0.79
1.4620	0.2434	1.2851	0.0760	0.243	−0.01
	0.3282	1.2270	0.0930	0.231	−0.05
	0.7181	0.9836	0.1018	0.144	0.55
1.4139	0.2359	1.2457	0.0673	0.229	0.15
	0.3184	1.1904	0.0815	0.215	0.19
	0.6980	0.9553	0.0880	0.132	0.92
1.5550	1.1823	0.7632	0.1083	0.120	−0.67
	1.5964	0.4879	0.0537	0.069	−0.70
1.5909	1.2080	0.7798	0.1121	0.119	−0.78
	1.6345	0.4995	0.4735	0.058	−0.56
1.6008	0.2650	1.3991	0.1016	0.274	−0.39
	0.3564	1.3228	0.1447	0.307	−1.61
	0.7834	1.0721	0.1243	0.148	0.27
1.6321	0.2720	1.4359	0.0848	0.217	0.37
	0.3662	1.3700	0.1059	0.211	0.27
	0.8041	1.1005	0.1062	0.120	1.02
1.8221	0.3010	1.5890	0.1267	0.265	−0.27
	0.4060	1.5181	0.1461	0.237	−0.17
	0.8946	1.2243	0.1303	0.119	0.88
2.0258	1.6302	0.9802	0.0112	0.007	1.70
2.2361	0.3679	1.9427	0.1815	0.254	0.09
	0.4955	1.8525	0.2139	0.233	−0.20
	1.0892	1.4907	0.1932	0.119	0.39
2.2811	1.8084	1.0873	0.0610	0.031	0.02
	2.5564	0.6248	−0.1917	−0.120	2.12
2.5034	1.9514	1.1733	0.1374	0.060	−1.85
	2.7646	0.6757	−0.1513	−0.081	0.20
2.6916	0.4406	2.3260	0.2603	0.254	0.28
	0.5931	2.2177	0.3012	0.229	0.14
	1.3048	1.7857	0.2633	0.113	0.09
2.7427	2.1162	1.2724	0.1939	0.072	−3.01
	2.9939	0.7311	−0.1051	−0.048	−1.55
3.0150	2.3836	1.4332	0.0581	0.017	−0.40
	3.3910	0.8288	−0.2726	−0.097	0.75
3.4050	3.8620	0.9439	−0.2369	−0.065	0.81
4.0709	0.6657	3.5149	0.4773	0.204	0.28
	0.8946	3.3456	0.5627	0.188	−0.57
	1.9932	2.7278	0.3643	0.067	−0.20
4.1412	3.2182	1.9351	0.2927	0.047	1.16
	4.5989	1.1240	0.4549	0.088	−0.84

[a] $A_{00} = 0.3224$, $A_{10} = -0.1454$, $A_{01} = -0.1397$, $A_{20} = -0.01884$, $A_{02} = 0.1048$, $A_{11} = 0.01378$, $A_{30} = 0.00459$, $A_{03} = -0.0165$, $A_{21} = 0.01934$, $A_{12} = -0.01872$.

Table IV. The Experimental and Calculated Data of the System KBr–H₂O–PEG-200

m_{ref}	m_{PEG}	m_{KBr}	Δ	$\dfrac{\Delta}{m_{PEG}m_{KBr}}$	Dev. m_{ref} (%)
0.8223	0.2934	1.1778	0.0142	0.041	−0.18
	0.4825	1.0645	0.0205	0.040	−0.09
	1.0093	0.7438	−0.0030	−0.004	0.09
0.8548	0.2920	1.2344	0.0299	0.083	−0.73
	0.5059	1.1161	0.0190	0.034	−0.18
	1.0572	0.7791	−0.0066	−0.008	−0.14
0.8777	0.2874	1.3094	−0.0196	−0.052	1.05
	0.6485	1.0931	−0.0121	−0.017	0.80
	1.0251	0.8607	−0.0300	−0.034	0.68
0.8999	0.3270	1.3129	0.0159	0.037	−0.29
	0.5381	1.1872	0.0211	0.033	−0.47
	1.1252	0.8292	−0.0131	−0.014	−0.41
0.9394	0.2920	1.4234	−0.0029	−0.007	0.24
	0.8502	1.0989	−0.0206	−0.022	0.50
0.9676	0.3573	1.4346	0.0169	0.033	−0.46
	0.5879	1.2970	0.0229	0.030	−0.70
	1.2311	0.9073	−0.0279	−0.025	−0.59
0.9908	0.3111	1.5162	0.0038	0.008	−0.07
	0.9037	1.1681	−0.0055	−0.010	−0.10
	1.6283	0.7024	−0.0904	−0.079	−0.62
1.0203	0.3271	1.5947	−0.0438	−0.084	1.36
	0.9515	1.2299	−0.0655	−0.056	1.41
	1.7286	0.7457	−0.1882	−0.146	1.66
1.0821	0.4112	1.6512	0.0122	0.018	−0.39
	0.6766	1.4927	0.0192	0.019	−0.68
	1.4159	1.0434	−0.0547	−0.037	−0.69
1.0992	0.3543	1.7270	0.0000	0.000	−0.03
	1.0314	1.3332	−0.0303	−0.022	0.15
	1.8616	0.8031	−0.1540	−0.103	−0.40
1.1509	0.4420	1.7746	0.0345	0.044	−0.92
	0.7303	1.6110	0.0247	0.021	−0.75
	1.5260	1.1246	−0.0601	−0.035	−0.84
1.2201	0.4784	1.9209	0.0156	0.017	−0.27
	0.7888	1.7702	−0.0447	0.032	1.25
	1.6523	1.2176	−0.1006	−0.050	0.06
1.2491	0.4162	2.0288	0.0034	0.004	0.08
	1.2139	1.5691	−0.0495	−0.026	0.89
	2.1886	0.9442	−0.2252	−0.109	−0.18

[a] $A_{00} = 0.7531$, $A_{10} = -0.6519$, $A_{01} = -0.974$, $A_{20} = -0.09711$, $A_{02} = 0.3964$, $A_{11} = 0.5854$, $A_{30} = -0.0041$, $A_{03} = -0.05008$, $A_{21} = -0.05018$, $A_{12} = -0.1193$

The activity coefficient is as follows:

$$\ln \gamma_1^0 = 2A_1 m_1 + 3/2 A_2 m_1^2 + 4/3 A_3 m_1^3 \tag{21}$$

The calculated constants A_1, A_2, A_3, the relative deviation of the measured osmotic coefficients from calculated ones, and a set of calculated values of osmotic and activity coefficients for chosen concentrations of PEG-200 are given in Table II.

The Calculation of Activity Coefficients in the Ternary System. The polynomial with 10 constants was used to describe the experimental quantity $\Delta(m_1, m_2)$ in the whole concentration range

$$\Delta(m_1,m_2) = A_{00} + A_{10}m_1 + A_{01}m_2 + A_{20}m_1^2 + A_{02}m_2^2 + A_{11}m_1m_2 +$$
$$A_{30}m_1^3 + A_{03}m_2^3 + A_{21}m_1^2m_2 + A_{12}m_1m_2^2 \tag{22}$$

The ability of Equation 22 to reproduce the experimental data was checked by means of the calculated molal concentration of reference solution from the relation

$$m_{\text{ref,calc.}} = \frac{\phi_1^0 m_1 + \nu_2\phi_2^0 m_2 + \Delta_{\text{calc.}}}{\nu_{\text{ref}}\phi_{\text{ref}}^0} \tag{23}$$

for each experimental point.

The mean relative deviation was 0.64% in the system with KCl and 0.49% in the system with KBr. If we consider that the experimental error is of the same value, then the result can be taken as a satisfactory one.

We tried to reduce the number of the polynomial terms, but the reproducibility always became worse. For example, the six-parameter polynomial gave the mean relative deviation 1.30% in the system with KCl, which is twice as high as that from Equation 22.

The constants of Equation 22 were calculated by means of nonlinear correlation. The results are summarized in Tables III and IV.

Table V. Activity Coefficient of KCl in the Ternary System

m_{KCl}	m_{PEG} 0.5	1.0	1.5	2.0	2.5	3.0	3.5
0	0.6495	0.6034	0.5822	0.5717	0.5674	0.5672	0.5700
0.2	0.657	0.611	0.590	0.581	0.578	0.579	0.582
0.4	0.664	0.617	0.597	0.590	0.588	0.589	0.590
0.6	0.670	0.622	0.603	0.596	0.595	0.596	0.595
0.8	0.674	0.626	0.608	0.602	0.601	0.600	0.597
1.0	0.678	0.630	0.611	0.606	0.605	0.603	0.597
1.2	0.680	0.632	0.614	0.609	0.608	0.605	
1.4	0.680	0.633	0.616	0.611	0.610	0.604	
1.6	0.680	0.633	0.618	0.613	0.611		
1.8	0.679	0.633	0.619	0.615	0.612		

Table VI. Activity Coefficient of PEG-200 in the Ternary System

m_{PEG}

m_{KCl}	0	0.5	1.0	1.5	2.0	2.5	3.0	3.5
0.2	1.0344	1.101	1.164	1.236	1.326	1.440	1.575	1.723
0.4	1.071	1.131	1.184	1.244	1.319	1.409	1.512	1.612
0.6	1.117	1.166	1.209	1.258	1.318	1.389	1.464	1.527
0.8	1.166	1.204	1.237	1.276	1.324	1.379	1.431	1.462
1.0	1.218	1.245	1.269	1.299	1.336	1.377	1.410	1.417
1.2	1.274	1.290	1.304	1.325	1.354	1.383	1.402	
1.4	1.334	1.337	1.342	1.356	1.377	1.399	1.406	
1.6	1.397	1.386	1.383	1.390	1.407	1.423		
1.8	1.461	1.437	1.425	1.429	1.443	1.457		

Table VII. Activity Coefficient of KBr in the Ternary System

m_{PEG}

m_{KBr}	0.5	1.0	1.5	2.0	2.5	3.0	3.5
0	0.6663	0.6453	0.6522	0.6733	0.7048	0.7455	0.7953
0.2	0.663	0.615	0.596	0.590	0.591	0.595	0.600
0.4	0.668	0.614	0.594	0.589	0.592	0.597	0.598
0.6	0.671	0.613	0.594	0.590	0.594	0.597	
0.8	0.672	0.613	0.596	0.592	0.596	0.597	
1.0	0.671	0.613	0.599	0.596	0.599		
1.2	0.669	0.613	0.602	0.599	0.601		
1.4	0.666	0.614	0.606	0.604			
1.6	0.662	0.614	0.611	0.608			
1.8	0.657	0.615	0.615	0.612			

Table VIII. Activity Coefficient of PEG-200 in the Ternary System

m_{PEG}

m_{KBr}	0	0.5	1.0	1.5	2.0	2.5	3.0
0.2	1.034	1.132	1.149	1.130	1.112	1.108	1.119
0.4	1.071	1.144	1.153	1.143	1.138	1.148	1.163
0.6	1.117	1.162	1.165	1.162	1.171	1.191	1.206
0.8	1.166	1.185	1.183	1.188	1.208	1.237	1.247
1.0	1.218	1.213	1.207	1.218	1.249	1.284	
1.2	1.274	1.246	1.236	1.254	1.294	1.330	
1.4	1.334	1.283	1.269	1.293	1.341		
1.6	1.397	1.323	1.306	1.336	1.390		
1.8	1.461	1.365	1.346	1.382	1.438		

The activity coefficients of the nonelectrolyte and electrolyte were calculated by means of Equations 14 and 15 for chosen concentrations, and are listed in Tables V–VIII.

Discussion

The binary aqueous solution of PEG-200 has the osmotic coefficient greater than one, and its value rises with increasing concentration. From Bjerrum's definition of osmotic coefficient,

$$-55.51 \ln a_w = \phi_1^0 m_1^0 \tag{24}$$

it is evident that the water activity falls with increasing molal concentration of the solute. The more the osmotic coefficient increases over one, the greater is the drop. Considerable water activity lowering is caused by the interaction of solvent molecules with ethylene oxide segments and PEG end groups, as was also assumed in the discussion about the structure of polyethylene oxide chain and in the interpretation of viscosity data of PEG aqueous solution (13). The activity coefficient of PEG-200 (greater than one) indicates the association of the non-electrolyte molecules in aqueous solution.

In the ternary solution KCl–H₂O–PEG-200, the quantity Δ, which can be considered a type of excess function, is positive except in the range of higher nonelectrolyte concentration, where it becomes negative. Its contribution to the binary data is 5–10%.

In the system KBr–H₂O–PEG-200, the quantity is negative and, on the average, its value is smaller than that in the system with KCl.

The trend of activity coefficients of potassium chloride and potassium bromide is different in measured mixed solvent. The activity coefficient of potassium chloride is higher in the mixed solvent than in the pure water and rises smoothly with the nonelectrolyte content. The minimum value, about 2.0–3.0m in pure water, can be observed in the mixed solvent also. Because of the activity coefficient of the nonelectrolyte in the ternary system (also higher than that in pure water), both components are mutually salted out.

The activity coefficient of potassium bromide is smaller in the mixed solvent than in pure water and changes very little with the increasing nonelectrolyte concentration.

The activity coefficient of the PEG-200 with the concentration up to 0.8m is higher than in pure water. Thus, in this range, PEG is salted out. But above 1m, its value is lower than in pure water with a gap at the KBr concentration about 1m; both solutes are salted in.

Nomenclature

N_i	mole fraction of a component
n_i	number of moles of a component

μ_i chemical potential of a component

a_i activity of a component

m_i molal concentration of a component in ternary system

γ_i molal activity coefficient of a component in ternary system

ν_i number of ions from one mol of electrolyte

m_1^0 molal concentration of nonelectrolyte in pure water

m_2^0 molal concentration of electrolyte in pure water

γ_1^0 molal activity coefficient of nonelectrolyte in pure water

γ_2^0 molal activity coefficient of electrolyte in pure water

m_{ref} molal concentration of reference electrolyte

γ_{ref} molal mean activity coefficient of reference electrolyte

ν_{ref} number of ions from one mol of reference electrolyte

Φ_i^0 osmotic coefficient in binary solution

I ionic strength

Literature Cited

1. Darhen L. S., *J. Am. Chem. Soc.* (1950) **72**, 2909.
2. McKay, H. A. C., *Nature* (1952) **169**, 464.
3. McKay, H. A. C., Perring J. K., *Trans. Faraday Soc.* (1953) **49**, 163
4. Kelly, F. J., Robinson, R. A., Stokes, R. H., *J. Phys. Chem.* (1961) **65**, 1958.
5. Bower, V. E., Robinson, R. A., *J. Phys. Chem.* (1963) **67**, 1524.
6. Robinson, R. A., Stokes, R. H., Marsh, K. N., *J. Chem. Thermodyn.* (1970) **2**, 745.
7. Gaylord, N. G., "Polyethers," New York, Interscience, 1963.
8. Sinclair, D. A., *J. Phys. Chem.* (1933) **37**, 495.
9. Rush, R. M., Johnson, J. S., *J. Chem. Eng. Data* (1966) **11**, 590.
10. Adamcová, Z., *Chem.* (1972) **66**, 1233.
11. Lietzke, M. H., Stoughton, R. W., *J. Phys. Chem.* (1962) **66**, 508.
12. Adamcová, Z., Dang, Duc Tao, *Sci. Pap. Inst., Chem. Technol., Prague* (1973) **B17**, 217.
13. Doolittle, A. K., *J. Paint Technol.* (1969) **41**, 483.

RECEIVED July 1, 1975.

24

Salt Effect on Carbon Dioxide Solubility in Mixture of Methanol and Water

EIZO SADA, SHIGEHARU KITO, and YOSHITAKA ITO

Department of Chemical Engineering, Nagoya University, Furo-cho, Chikusa-ku, Nagoya, 464 Japan

Salt effects of lithium chloride and calcium chloride on the solubility of carbon dioxide in a mixture of methanol and water were observed at 25°C and 1 atm. Experimental results can be correlated by the Setschenow equation for a fixed solvent composition of salt-free basis. The salting-out parameter is not linear with solvent composition, which is opposite to the results obtained when a mixed salt is used.

The salt effect of single or mixed electrolytes on the solubility of a gas in water is of considerable industrial and theoretical interest. Methods to predict or correlate these effects have been presented by various workers and have been reviewed briefly (1). With the exception of a study by Clever and Reddy (2), previous investigations found no salt effect data on gas solubility in non-aqueous or mixed solvents. Clever and Reddy (2) observed the solubilities of helium and argon in methanol solutions of sodium iodide at 30°C and showed that the following Setschenow equation is not always applicable to such a system.

$$\log \frac{\alpha_0}{\alpha} = k \cdot I \tag{1}$$

In the present investigation, the salt effects of lithium chloride and calcium chloride on the solubility of carbon dioxide in the mixed solvents of methanol and water were experimentally studied at 25°C and 1 atm.

Experimental Apparatus and Procedure

Solubilities of carbon dioxide were determined as in the previous investigations (3, 4, 5, 7). Full descriptions of the apparatus and the mode of operation

374

are given elsewhere (3). The principle of this method is that a known volume of gas in liquid-saturated vapor state is brought into contact with a measured amount of gas-free liquid. Equilibrium is established by agitation, and the volume of remaining gas is measured. After correction for vapor pressure, the change in gas volume gives the solubility. Here, the vapor pressures of salt solutions at 25°C were calculated by use of Raoult's law.

A gas-free solution was prepared by mixing methanol, water, and salt, each of which was separately degassed or desorbed. The methanol and water were each boiled in a flask with a reflux condenser under a reduced pressure to remove dissolved gases in the solvent. Then they were introduced into constant-temperature measuring devices out of contact with air. The amounts of both solvents were adjusted to accommodate a desired composition of the mixture.

Salt was degassed (desorbed) as follows. A measured amount of sample salt was put into a Pyrex degassing vessel and heated to about 200°C for about one hour, with strong shaking in vacuo. It was then allowed to cool to room temperature. Thus, water and adsorbed gas in salt were removed. The gas-free water and methanol were introduced into the degassing vessel without exposure to air.

The composition of mixed solvent (free of salt) was calculated from the amounts of methanol and water before mixing. The reproducibility of the preparation of the mixed solvent was within 0.001 mole fraction over the whole composition range.

The concentration of salt in a solution was determined by evaporation to dryness and measurement of the density of the solvent remaining in the degassing vessel.

As a check on the degassing procedure, solubilities of carbon dioxide in aqueous calcium chloride solutions prepared in the same manner were determined at 25°C and 1 atm. The results agreed with the previous results (4), in which the aqueous solutions were prepared by another method. Thus the degassing procedure adopted in this work can be considered satisfactory.

Reagent-grade methanol and salts manufactured by Nakarai Chemicals, Ltd. (Tokyo) were used without further purifications; distilled water was used. The physical properties of methanol and water were compared with the published data (8). Carbon dioxide was supplied from a commercial cylinder and the purity was confirmed by gas chromatogram to be greater than 99.8%.

A series of gas solubility measurements was carried out by varying salt concentration at a fixed solvent composition. Such a measurement was repeated at various solvent compositions.

Experimental Results and Discussion

In this study, the solubilities of carbon dioxide were represented by the Bunsen absorption coefficients and their values given in Table I for lithium chloride and in Table II for calcium chloride. The densities of salt solutions are

Table I. Experimental Results of Carbon Dioxide Solubility in
Methanol–Water–Lithium Chloride at 25°C and 1 atm

x_{MeOH}	$I[mole/l.]$	$\alpha[-]$	$\rho^{25}[g/cc]$
0.0	0	0.7597	—
	0.7707	0.6565	1.0157
	1.0798	0.6207	1.0229
	1.1883	0.6050	1.0254
0.280	0	0.9695	—
	0.7433	0.8078	0.9493
	0.9897	0.7714	0.9553
	1.3854	0.7066	0.9661
0.389	0	1.1775	—
	0.7288	0.9875	0.9271
	1.0750	0.9010	0.9366
	1.1291	0.8955	0.9386
0.587	0	1.8161	—
	0.8652	1.4639	0.8899
	1.1876	1.3525	0.8998
	2.2293	1.0496	0.9253
0.800	0	2.7813	—
	0.4316	2.4566	0.8367
	0.5525	2.4073	0.8478
	0.8636	2.2003	0.8518
	1.2905	1.9650	0.8661
1.0	0	3.9003	—
	0.5930	3.2643	0.8104
	0.8000	3.0662	0.8182
	2.2550	2.0391	0.8707

Table II. Experimental Results of Carbon Dioxide Solubility in
Methanol–Water–Calcium Chloride at 25°C and 1 atm

x_{MeOH}	$I[mole/l.]$	$\alpha[-]$	$\rho^{25}[g/cc]$
0.280	0	0.9695	—
	0.6072	0.8885	0.9472
	0.9154	0.8486	0.9565
	1.4510	0.7897	0.9721
	2.0969	0.7228	0.9912
0.389	0	1.1775	—
	0.6389	1.0707	0.9260
	0.7978	1.0546	0.9308
	0.9624	1.0249	0.9358
	1.7948	0.9180	0.9610
	2.2506	0.8694	0.9743
0.587	0	1.8161	—
	0.7021	1.6361	0.8857
	0.9454	1.5963	0.8931
	2.1877	1.3187	0.9330

Table II (Continued)

$x\,MeOH$	$I[mole/l.]$	$\alpha[-]$	$\rho^{25}[g/cc]$
0.800	0	2.7813	—
	0.9771	2.3869	0.8562
	1.7551	2.1193	0.8827
	2.6001	1.8634	0.9112
1.0	0	3.9003	—
	0.8805	3.3315	—
	1.9928	2.7794	0.8637
	2.8868	2.4118	0.8952

also shown. Here the concentration of salt is represented by an ionic strength calculated from the following equation to diminish the effect caused by the difference of ionic charges.

$$I = \frac{1}{2}(C_c z_c^2 + C_a z_a^2) \tag{2}$$

The variations of carbon dioxide solubilities with salt concentrations are shown in Figures 1 and 2 as forms of the modified Setschenow plot for each solvent

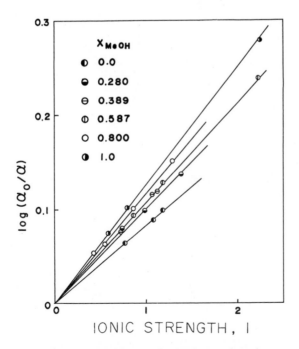

Figure 1. Solubilities of carbon dioxide in methanol–water–lithium chloride at 25°C

composition.   As can be seen in these figures, the effects of these salts on the solubility of carbon dioxide in the mixed solvents was found to correlate relatively well with the Setschenow equation for every salt species at each solvent composition.   These features are the same as in aqueous salt solutions.

Salting-out parameters for each salt and each solvent composition were calculated by a linear regression analysis.   In order to examine the dependence of the salting-out parameter on solvent composition, the salting-out parameters, k, are shown as a function of solvent composition in Figure 3.   In this figure, the salting-out parameter corresponding to aqueous calcium chloride solution was calculated from the data reported in the previous paper (3).   From this figure it can be seen that when the solvent is a mixture, the salting-out parameter does not always vary linearly with solvent composition.   This trend contradicts the results seen when a mixed salt is used.

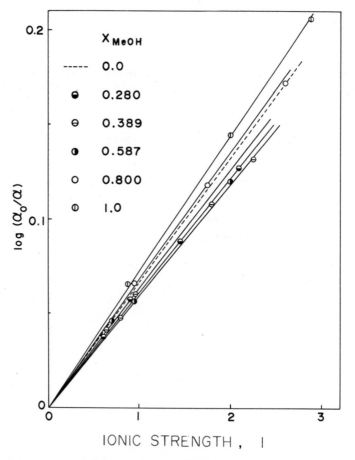

Figure 2.   Solubilities of carbon dioxide in methanol–water–calcium chloride at 25°C

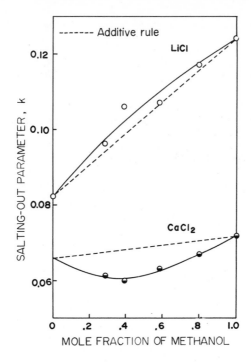

Figure 3. Variations of salting out parameters with solvent composition

Bocklis et al. (*1*) and Sergeeva et al. (*9*) have observed the salt effects on solubilities of benzoic acid in a few mixed solvents. However, upon review of their results and ours, no unified features of the dependence of the salting-out parameter on solvent composition were found.

Nomenclature

C_i = molarity of ionic species i, mole/l.
I = ionic strength defined in Equation 2, mole/l.
k = salting-out parameter in Equation 1, l./mole
x = mole fraction (salt-free basis)
z_i = valency of ionic species i

Greek letters
α = the Bunsen absorption coefficient
ρ^{25} = density at 25°C, g/cc

Subscripts
0   = free of salt
a   = anion
c   = cation
MeOH = methanol

Literature Cited

1. Bockris, J. O'M., Bowler-Reed, J., Kitchener, J. A., *Trans. Faraday Soc.* (1951) **47**, 184.
2. Clever, H. L., Reddy, G. S., *J. Chem. Eng. Data* (1963) **8**, 191.
3. Onda, K., Sada, E., Kobayashi, T., Kito, S., Ito, K., *J. Chem. Eng. Japan* (1970) **3**, 18.
4. *Ibid.*, 137.
5. Sada, E., Kito, S., *Kagaku Kogaku* (1972) **36**, 218.
6. Sada, E., Kito, S., Morisue, T., *Kagaku Kogaku* (1973) **37**, 983.
7. Sada, E., Kito, S., Ito, Y., *J. Chem. Eng. Japan* (1974) **7**, 57.
8. Sada, E., Kito, S., Oda, T., Ito, Y., *Chem. Eng. J.* (1975) **10**, 155.
9. Sergeeva, V. F., *Russ. Chem. Rev.* (1965) **34**, 309.

RECEIVED August 1, 1975.

The Removal of Organic Lead from Aqueous Effluents by a Combined Chemical Complexing–Solvent Extraction Technique

A. J. BARKER and S. R. M. ELLIS

Department of Chemical Engineering, The University of Birmingham, P.O. Box 363, Birmingham B15 2TT England

A. B. CLARKE

Colworth Welwyn Laboratory, Unilever Ltd., Colworth House, Sharnbrook, Bedford MK44 1LQ England

The removal of organic lead from aqueous solution can be achieved by a method of chemical complexing followed by solvent extraction. The mechanism of removal depends first on the formation of a coordinate complex; this is determined by the distribution of organic lead species in the aqueous phase in the presence of NaCl. The second step in the process is the transfer of the coordinate or neutral species from the aqueous to the solvent phase. The extent of this transfer is determined by the distribution of these species between the two phases. The continued formation of the complex depends on the law of mass action, and the rate of extraction is determined by factors controlling the physical distribution in the process. The major operating parameters determining the level of organic lead removal are the organic-to-aqueous phase ratio and the complexing reagent-to-organic lead ratio.

Waste waters containing low concentrations of soluble organic lead in the presence of high concentrations of other diverse ions such as Cl⁻ pose a particularly difficult treatment problem. Generally, organic lead exists in solution as the tri- or dialkyl lead chloride species. These salts are not amenable to the conventional methods used to remove inorganic lead, viz., those of pH adjustment followed by settling. The technique of chemical conversion of the organic lead salts to the inorganic state, while possible, is precluded on economic grounds because of the very high capital and energy costs of effective processes. An alternative process which did not possess these major disadvantages was therefore sought.

The method of chemical complexing followed by solvent extraction has been used for some time as an analytical technique for metal determination by chelate extraction (1). The chelating agent diethyl dithiocarbamic acid sodium salt is an established reagent for the determination of inorganic lead. The chelating power of the dithiocarbamates and other potentially interesting reagents such as the thiazoles and xanthates is based upon the affinity of the sulfur-bearing groups of the chelating agent for heavy metals.

Initially the application of this analytical technique to an effluent treatment problem would not appear to be very favorable. In the analytical technique high reagent concentrations relative to the metal salt are employed, and the extraction is dependent upon pH level of the aqueous phase (1, 2). Furthermore, organic lead salts in the presence of high concentrations of ions such as Cl^- will form a series of complexes. As a consequence, any method of extraction must take into account the equilibrium of these species. However, these apparent objections to the use of the analytical technique can be resolved and so the process of combined chemical complexing–solvent extraction does have the potential of successful deployment in the field of the large scale treatment of organic-lead-contaminated waste waters.

Experimental

For the investigation of the potential of a chemical complexing–solvent extraction technique for the treatment of waste waters containing organic lead, a sequence of experiments was performed using synthetic effluents. As the major organic lead contaminant in these waste waters is generally trialkyl lead chloride in the presence of a high concentration of chloride ions, the synthetic effluents made usually contained up to 100 ppm of trialkyl lead chloride and 0.83 m sodium chloride.

The chemical complexing–solvent extraction technique employed in this work involved the formation of a neutral complex in the aqueous phase between trialkyl lead chloride and a dithiocarbamate reagent such as sodium diethyl dithiocarbamate. The complex was subsequently removed either as a precipitate or by extraction into an organic solvent. The extent of lead removal was traced by analysis of the aqueous phase for residual trialkyl lead using a Pye-Unicam 8000 spectrophotometer.

Any dialkyl lead present was determined spectrophotometrically as the dialkyllead 4(2-pyridylazo)resorcinol (P.A.R.) complex at pH 9. Trialkyl lead does not form a complex with P.A.R.; therefore its concentration was obtained by conversion to the dialkyl lead form in iodine monochloride solution followed by determination as the P.A.R. complex. Any inorganic lead in solution was masked by the addition of excess ethylenediaminetetraacetic acid, disodium salt.

Characterization of the Complex

Of the complexing reagents investigated, sodium diethyl dithiocarbamate proved on a number of counts to be the most effective. The bulk of experiments therefore used this compound. The nature of this complex between the reagent and trialkyl lead chloride was first characterized in the absence of sodium chloride in the aqueous phase.

A sample of complex sufficient for elemental analysis was obtained by adding 0.5 g $(C_2H_5)_2NCSSNa\cdot3\ H_2O$ in 50 ml distilled water to 0.5 g of solid $(C_2H_5)_3PbCl$ dissolved in 50 ml distilled water. A substantial precipitate was formed which flocculated sufficiently for a large percentage of it to be separated in a laboratory filter. The precipitate was washed with distilled water and dried in a vacuum desiccator for 30 hr.

Examination of the precipitate showed that it consisted of flocs of irregularly shaped particles 1–5 μ in size. Accordingly, at low organic lead concentrations of, e.g., 10 ppm, the use of Centriflo membrane filters was required for efficient separation of the precipitate from solution. The filters, holding 7-ml aliquots of the aqueous phase were centrifuged for 3 min at 1500 rpm.

Elemental analysis of the product formed between $(C_2H_5)_3PbCl$ and $(C_2H_5)_2NCSSNa\cdot3H_2O$ showed the product to contain stoichiometric quantities of reagent and organic lead corresponding to complexing on a 1:1 molar basis; that is the formation of $(C_2H_5)_3PbSCSN(C_2H_5)_2\cdot3H_2O$. The results of the analysis were compared with those calculated for a 1:1 and 2:1 complex as shown in Table I.

The formation of a 1:1 complex was confirmed by examination of the removal of organic lead from the aqueous phase as a function of the reagent-to-organic lead ratio C_R/C_L at a temperature of 30°C. Figure 1 shows that removal of organic lead corresponds to what is calculated. For a molar reagent-to-lead ratio $(C_R/C_L = 1)$, complete removal of organic lead is achieved. This demonstrates that, in the absence of NaCl, the complex is essentially insoluble in the aqueous phase, at least to within the accuracy of analysis, ±0.1 ppm.

From the results of a factorially designed experiment, variance analysis of the complexing reaction showed that over the range 15°–60°C, temperature had

Table I. Analysis of Complex of Triethyl Lead Chloride and Sodium Diethyl Dithiocarbamate

Element	Calculated (2:1)	Calculated (1:1)	Found
C	27.5	26.6	26.9
H	6.7	6.3	5.8
S	18.6	12.7	12.9

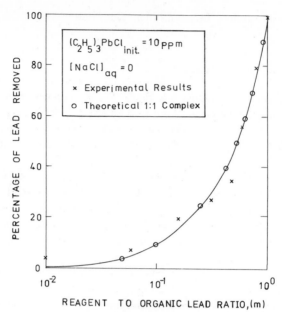

Figure 1.   *Complex of* $(C_2H_5)_3PbCl$ *with*
$(C_2H_5)_2NCSSNa$

no significant effect on either the level or rate of organic lead removal, reaction being virtually instantaneous in all cases.

Characterization of the complexing reaction was then conducted in the presence of 0.83 m sodium chloride in the aqueous phase over a temperature range 15°–60°C. In addition, removal of $(C_2H_5)_3PbCl$ as a function of the reagent-to-organic lead ratio C_R/C_L in the presence of 5 wt % NaCl was examined at 30°C using in turn three other dithiocarbamate reagents besides that of sodium diethyl dithiocarbamate.

In all cases, the results demonstrate that in the presence of sodium chloride in the aqueous phase, the level of organic lead removed from solution is significantly reduced. Figure 2 shows that for complete precipitation of organic lead a ratio, C_R/C_L, of at least 350 is necessary. This is over two orders of magnitude greater than in the absence of sodium chloride. Clearly such an excess of reagent is undesirable in both economic and environmental terms.

The high reagent-to-organic lead ratio required is attributed to the formation of the series of chloro complexes, $(C_2H_5)_3PbCl_n{}^{1-n}$, $n = 0, 1, 2, 3$, in the presence of the chloride ion. A high excess of reagent is required to cause a swing in the equilibrium, viz.

$$
\begin{matrix}
(C_2H_5)_3Pb^+ \\
\text{complex with} \\
(C_2H_5)_3 \; NCSSNa \cdot 3H_2O
\end{matrix}
\Bigg\downarrow
\quad + \; n \; Cl^- \; \rightleftharpoons \; (C_2H_5)_3PbCl_n{}^{1-n}
$$

The existence of the anionic species $(C_2H_5)_3PbCl_2^-$ and $(C_2H_5)_3PbCl_3^{2-}$ has previously been established by paper chromatography (3, 4), ion exchange (5, 6, 7), and amine extraction techniques (8).

The work of Barker and Clarke (9) has demonstrated the influence of the chloride ion on the relative concentration of individual chloro species. Figure 3 shows the concentration of the species $(C_2H_5)_3Pb^+$ and $(C_2H_5)_3PbCl_2^-$ remaining in the aqueous phase after amine extraction, assuming $(C_2H_5)_3PbCl_2^-$ to be the dominant species (as suggested by values of the stability constants) and assuming the neutral species $(C_2H_5)_3PbCl^0$ to be absent from the aqueous phase or in a concentration too low to measure. The latter assumption is consistent with a high distribution of the neutral species between organic and aqueous phases.

Owing to the inadequacy of the mathematical model available for analysis of the amine extraction system (7), accurate values of the stability constants could not be evaluated for the $(C_2H_5)_3Pb_n^{1-n}$ system in the presence of NaCl. However, using the values of stability constants obtained by Bertazzi for the system $(C_2H_5)_3PbCl_n^{1-n}$ in LiCl at 8.0 m (10), viz. $\beta_1 = 3.5$, $\beta_2 = 1.0$, $\beta_3 = 0.1$, the neutral species $(C_2H_5)_3PbCl^0$ ($n = 1$) is seen to be dominant. Therefore a simple solvent extraction would be expected to remove a certain amount of triethyl lead from solution. As shown in Table II, this is seen to be so. However,

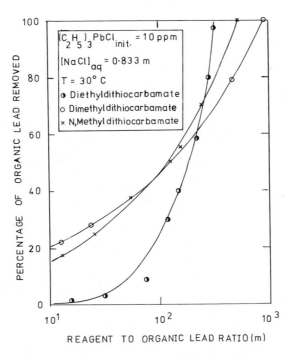

Figure 2.   Complex of $(C_2H_5)_3PbCl$ with various dithiocarbamate reagents

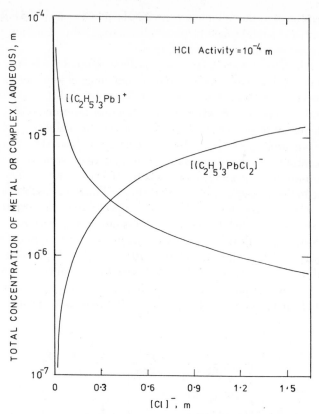

Figure 3.   Concentration of $(C_2H_5)_3PbCl$ vs. aqueous chloride concentration

Table II.   Solvent Extraction of $(C_2H_5)_3Cl$ from Synthetic Effluent

Initial Concentration of $(C_2H_5)_3PbCl$ = 10 ppm
Concentration of NaCl = 0.833m
V_{aq}/V_{org} = 5
No reagent

Solvent	Final Concentration $(C_2H_5)_3PbCl$ (ppm)	Solvent Solubility (per 100 parts water)
Benzene	6.5	0.07
Chloroform	2.5	0.82
CCl_4	6.6	0.08
Xylene	7.5	0.05
Trichloroethylene		0.10
Diethylether	6.5	7.50
n-Octane	10	0.002
Isooctane	10	insignificant
$C_{14}H_{30}$	10	insignificant
$C_{15}H_{32}$	10	insignificant

this is true for only certain solvents which in general are those having a high solubility in the aqueous phase (relative to desirable limits for toxic solvents such as benzene, chloroform, carbon tetrachloride).

Formation and subsequent extraction of a neutral species such as $(C_2H_5)_3PbCl^0$ suggest that a combined chemical complexing–solvent extraction technique might be more effective in terms of a lower C_R/C_L ratio than direct precipitation. This is confirmed by the results of the chemical complexing solvent–extraction studies.

Chemical Complexing–Solvent Extraction Studies

Chemical complexing–solvent extraction studies were conducted using aliquots of synthetic effluents which contained varying quantities of $(C_2H_5)_3PbCl$ and $(C_2H_5)_3NCSSNa \cdot 3H_2O$ with the sodium chloride concentration maintained at 0.83 m.

The ratio of reagent-to-organic lead in the effluent was varied between 0.1 and 10.0. For an initial $[(C_2H_5)_3PbCl] = 10$ ppm, 50-ml aliquots of effluent were shaken for 15 sec with 10 ml of an organic solvent in a 250-ml separating funnel. The choice of the phase ratio, V_{aq}/V_{org}, was an arbitrary one. After phase separation, the aqueous phase was run off and analyzed for organic lead. This

Table III. Chemical Complexing–Solvent Extraction of $(C_2H_5)_3PbCl$ from Synthetic Effluent

Initial Concentrations of $(C_2H_5)_3PbCl$ = 67 ppm
Concentration of NaCl = 0.833 m
V_{aq}/V_{org} = 1.0
Reagent $(C_2H_5)_2NCSSNa \cdot 3H_2O$

| | C_R/C_L = 0.75 | | C_R/C_L = 1.00 | |
| | Final concentration $(C_2H_5)_3PbCl$ | | Final concentration $(C_2H_5)_3PbCl$ | |
Solvent	(ppm)	% Removal	(ppm)	% Removal
Kerosene	21.0	70.0	9.6	81.0
Petroleum Ether	17.2	74.6	10.0	80.0
Diethyl Ether	4.8	93.0	—	—
Xylene	6.4	90.3	—	—
Benzene	4.6	93.2	0.0	100.0
n-Pentane	27.0	59.7	7.0	86.0
Isooctane	20.0	69.0	12.0	76.0
CCl_4	6.6	90.0	—	—
Chloroform	0.0	100.0	—	—
n-Nonane, 99%	12.0	82.0	10.2	79.0
$C_{14}H_{30}$	20.3	70.0	—	—
$C_{15}H_{32}$	18.1	72.0	12.0	76.0
Toluene	9.6	86.0	—	—
Cyclohexane	12.6	82.3	8.4	83.0

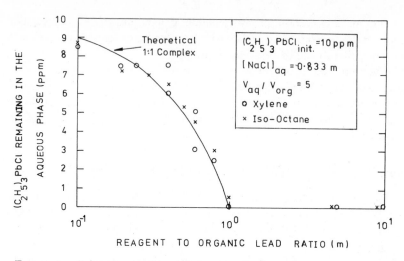

Figure 4.   Solvent extraction of the complex by xylene and isooctane

procedure was repeated for the solvents diethyl ether, isooctane, xylene, kerosene, $C_{14}H_{30}$ and $C_{15}H_{32}$, all at ambient temperature $16°C \pm 2°C$.

Further experiments were conducted with a wide range of organic solvent for the initial conditions: $[(C_2H_5)_3PbCl] = 67$ ppm, $V_{aq}/V_{org} = 1.0$. $C_R/C_L = 0.75$ m, $[NaCl] = 0.83$ m.

Reference to Table III shows that in the presence of 0.83 m sodium chloride and for an aqueous-to-organic phase ratio $V_{aq}/V_{org} = 1.0$, a ratio C_R/C_L of 1.0 is sufficient to remove at least 75% organic lead.   For solvents such as benzene and chloroform this ratio is sufficient to achieve complete organic lead extraction.

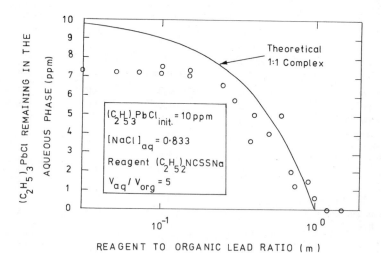

Figure 5.   Solvent extraction of the complex by diethyl ether

Table IV.   Comparison of Solvent Strength

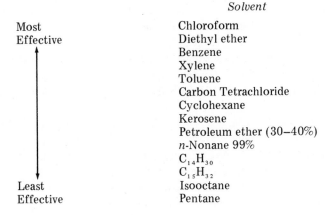

	Solvent
Most	Chloroform
Effective	Diethyl ether
↑	Benzene
	Xylene
	Toluene
	Carbon Tetrachloride
	Cyclohexane
	Kerosene
	Petroleum ether (30–40%)
	n-Nonane 99%
	$C_{14}H_{30}$
	$C_{15}H_{32}$
Least	Isooctane
Effective	Pentane

Table V.   Removal of $(C_2H_5)_3PbCl$ as a Function of the Ratios V_{aq}/V_{org}, C_R/C_L

Initial concentration of $(C_2H_5)_3PbCl$ = 27 ppm
Concentration of NaCl = 0.833m
Reagent $(C_2H_5)_2NCSSNa \cdot 3H_2O$
Solvent $C_{15}H_{32}$
pH 7.2
Aqueous phase shaken with two aliquots of organic phase to give residual lead concentrations of Pb_{f_1} and Pb_{f_2}

C_R/C_L (m)	V_{aq}/V_{org}	Pb_{f_1} (ppm)	Pb_{f_2} (ppm)
0.75	25	9.0	9.0
0.75	50	11.5	11.4
0.75	75	13.5	11.2
0.75	100	19.0	11.2
1.5	25	3.0	1.7
1.5	50	2.7	2.7
1.5	75	4.4	2.8
1.5	100	3.5	3.3
2.2	25	0.1	0.0
2.2	50	1.3	2.7
2.2	75	1.3	1.3
2.2	100	2.7	1.3
4.0	25	0.3	0.0
4.0	50	0.3	0.0
4.0	75	1.6	0.5
4.0	100	2.4	0.5
4.0	150	4.5	2.7
4.0	200	7.4	5.3
4.0	300	9.1	6.1

Figures 4 and 5 for $V_{aq}/V_{org} = 5.0$ show that removal of organic lead corresponds closely to that which would be obtained for a theoretical 1:1 complex in the absence of sodium chloride. Variations in extraction efficiency are observed for different solvents, but for all of the solvents employed a ratio $C_R/C_L = 1.0$ is sufficient to reduce an initial triethyl lead chloride level of 10 ppm to <1 ppm. A comparison of solvents used suggests an approximate order of effectiveness (corresponding to solubility of the organo-lead complex $(C_2H_5)_3PbSCSN(C_2H_5)_2$ and also the neutral species $(C_2H_5)_3PbCl^0$ in the solvent) of the form shown in Table IV.

While organo-lead removal is a function of the solvent employed, the level to which the concentration can be reduced depends upon the interaction of the variables V_{aq}/V_{org} and C_R/C_L. This effect is clearly illustrated in Table V for the solvent $C_{15}H_{32}$ at an initial triethyl lead chloride level of 27 ppm.

From the results of the more extensive experiments using a range of organic solvents, typical values of operating ratios required to reduce an organic lead level of 10 ppm to less than 1 ppm are $10 < V_{aq}/V_{org} < 20, 2 < C_R/C_L < 3$.

Stirred Cell Studies for the Investigation of Extraction Mechanism

For ascertaining whether removal of organic lead by chemical complexing–solvent extraction was chemical reaction or physical diffusion controlled, stirred cell studies were conducted.

The stirred cell of Davies (11) was used to investigate the possible interfacial mechanism of extraction. Transfer of the complex $(C_2H_5)_3PbSCSN(C_2H_5)_2$ from the aqueous to the organic phase was studied as a function of the stirring rate in the aqueous phase.

The initial aqueous phase concentration of triethyl lead chloride was approximately 20 ppm, determined accurately in each case. The sodium chloride concentration was maintained at 0.83 m and the reagent-to-lead ratio, C_R/C_L at 3.0, thus ensuring complexing at the phase ratio employed, $V_{aq}/V_{org} = 2.0$. The volume of the aqueous phase was 400 ml, and that of the organic phase, xylene, 200 ml. The extraction, conducted at ambient conditions of 16–18°C, was followed by a determination of the concentration of triethyl lead in the aqueous phase at time intervals up to 150 min.

The results are shown in Figure 6, which gives a plot of mass transfer coefficient, k_t, at time t, for the transfer of the complex $(C_2H_5)_3PbSCSN(C_2H_5)_2$ from the aqueous phase as a function of the stirring speed in the aqueous phase, N. The mass transfer coefficient, k_t, was calculated from a relationship of the type

$$Q/t = k_t \cdot A_c \cdot (\Delta C)_i$$

The plot of k_t vs. N has a functional form which closely resembles that found by Austin and Sawistowski (12) for the interdiffusion of pairs of liquids by physical absorption (for which values of k ranged from 2×10^{-6} to 50×10^{-6} m/s). Austin

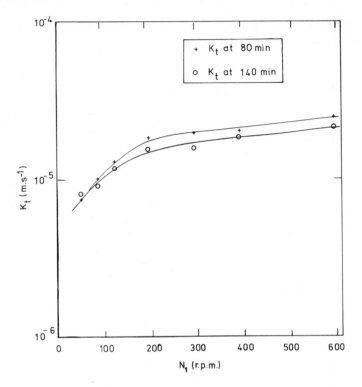

Figure 6.   Mass transfer coefficient K_t vs. stirring speed in aqueous phase N_1

and Sawistowski observed a change of slope for k vs. N between 150 and 200 rpm signifying the onset of interfacial renewal. At lower stirring speeds, the density gradients associated with diffusion opposed interfacial renewal. Such a change of slope is observed in the present work and would appear to be consistent with mass transfer by physical absorption. Had the rate controlling step been a chemical reaction, the mass transfer coefficient would have been independent of stirring speed (*13*).

Distribution Studies

In order to examine the influence of sodium chloride concentration on the distribution of triethyl lead chloride between an organic and aqueous phase, distribution studies were initiated. The solvent chosen for the initial studies was benzene because it had been shown that up to 30% of triethyl lead chloride is transferred to the organic phase as the neutral species $(C_2H_5)_3PbCl^0$ when using the chemical complexing–solvent extraction technique.

Aliquots of 60 ml of an aqueous phase containing varying concentrations of triethyl lead chloride and sodium chloride were shaken with 10 ml of benzene in a 250-ml separatory funnel. After phase separation, the aqueous phase was analyzed for residual triethyl lead chloride. The results of these studies are shown in Figure 7.

It is assumed that transfer of triethyl lead chloride between an organic phase of low dielectric constant and water will reduce to the distribution of the neutral species, an equilibrium being established for this species between the aqueous and organic phases. In the absence of sodium chloride in the aqueous phase, $(C_2H_5)_3PbCl^0$ will ionize to form the positive species $(C_2H_5)_3Pb^+$, thus affecting the equilibrium between the two phases and lowering the distribution between them. Although a simple distribution law (based on saturation solubilities of triethyl lead chloride in each phase) would suggest a value of 40:1 for distribution between organic and aqueous phases in the benzene solvent chosen, such a value

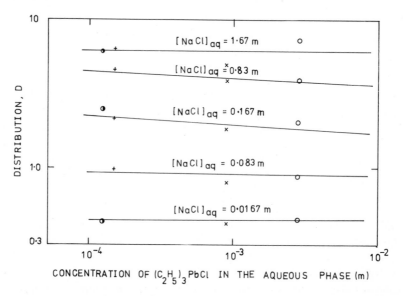

Figure 7.   Distribution of $(C_2H_5)_3PbCl$ between water and benzene

is not attained. In fact, as shown by Figure 7, the distribution tends toward zero as the sodium chloride concentration decreases. In the presence of sodium chloride the series of chloro species $(C_2H_5)_3PbCl_n{}^{1-n}$ is established, $(C_2H_5)_3PbCl^0$ being the dominant species. An equilibrium is considered to exist between the different species in the aqueous phase. The overall distribution is therefore a function of the equilibrium between both the species within the aqueous phase, and the aqueous and organic phases.

Mechanism of Extraction

From the evidence of the foregoing studies, the following mechanism of extraction of triethyl lead chloride from aqueous solution using an organic solvent may be propounded. In chloride solutions, a horizontal equilibrium concentration of the neutral species is formed. This, for the majority of solvents, immediately transfers to the organic phase by virtue of its greater solubility in that phase. At the same time there exists in the aqueous phase an equilibrium concentration of the positive species $(C_2H_5)_3Pb^+$. This complexes with the negative dithiocarbamate reagent to form the complex $(C_2H_5)_3PbSCSN(C_2H_5)_2$ which is insoluble in aqueous solution but is readily soluble in the majority of common organic solvents, and, therefore, is extracted from the aqueous phase. Accordingly, triethyl lead is removed from the aqueous phase by simultaneous extraction of the neutral species, $(C_2H_5)_3PbCl^0$ and the complex $(C_2H_5)_3PbSCSN(C_2H_5)_2$.

Because there exists an equilibrium distribution of species in the aqueous phase, removal of certain species will lead, by the law of mass action, to a readjustment of the equilibrium. The concentration of negative species will decrease to compensate for continued loss of the neutral and positive species to the organic phase (loss of the positive species is, of course, caused by complexing). This continued adjustment of equilibrium will be maintained until all of the organic lead has been removed from the aqueous phase if the organic phase is not saturated.

Experimental results indicate that the loadings of the organo lead complex which may be obtained for the solvent kerosene are approximately 300 g/l. If such an order of magnitude can be taken as typical of what might be obtained for common organic solvents, considerable recycle potential exists for the solvent, particularly for effluents containing only trace quantities of organic lead.

Discussion

From initial studies on the removal of triethyl lead chloride from aqueous chloride solution it was possible to establish the mechanism of the chemical complexing–solvent extraction technique. Furthermore, working values of the operating parameters V_{aq}/V_{org} and C_R/C_L suggested that the technique might provide the basis of an economic industrial process. However, initial investigations were conducted largely with synthetic effluent and were limited to laboratory-scale batch operations. Therefore, in order to assess the validity of the results in terms of design data for an industrial operation, a continuous pilot scale process was developed. The chemical complexing–solvent extraction technique was applied to a continuous mixer–centrifugal separator unit using a synthetic effluent at flow rates of 50 l./hr. The solvent was kerosene. This was continually recycled within the unit in order to obtain the maximum loading of organic lead.

The pilot scale work confirmed that levels of the operating parameters, V_{aq}/V_{org} and C_R/C_L which might typically be used to achieve a reduction of 90% or greater in organic lead levels (for initial lead levels up to 50 ppm) were identical to those observed for the laboratory-scale batch experiments, viz. $2 < C_R/C_L < 3$ for $10 < V_{aq}/V_{org} < 20$, and in all cases a C_R/C_L ratio of five reduced the organic lead level to <1 ppm. Since the performance of laboratory centrifugal separators is representative of what would be obtained for industrial models (14), the levels of operating parameters suggests a viable industrial process.

Clearly the running costs of extraction based on a chemical complexing–solvent extraction technique would be dictated largely by the quantities of solvent and reagent employed. Therefore, the use of a cheaper solvent such as kerosene would be economically attractive. However, it was observed from the laboratory experiments that kerosene was one of the least effective solvents. This was illustrated particularly in the extraction of trimethyl lead chloride; for solvents such as benzene and chloroform, using the reagent sodium diethyl dithiocarbamate extraction of trimethyl lead chloride was completely analogous to using triethyl lead chloride on both laboratory and pilot scale. Using kerosene, the loadings of the trimethyl lead complex in the organic phase were approximately one fifth of those obtained for the triethyl lead complex, implying that up to five times the quantity of solvent would be required to achieve the same levels of removal (excluding the possibility of regenerating the solvent).

Extraction efficiency is not the only factor to be examined in the choice of solvent or reagent for a particular application. Environmental, as well as economic considerations must be taken into account. Solvents such as benzene and chloroform (which have solubilities of 0.07 and 0.82 parts per 100 parts of water) might be preferred for extractive efficiency, but their use would result in large losses to the aqueous phase. Not only would this be expensive, but it would be undesirable for reasons of health; the toxic organo lead salts would be removed, but an equally toxic organic solvent would be added to the effluent. Addition of a solvent recovery unit subsequent to the extraction step might render the technique uneconomic (relative to alternative effluent treatment techniques).

Similarly, consideration must be given to the reagent employed. Sodium diethyl dithiocarbamate is a relatively cheap and readily available reagent. Unfortunately, it is toxic and the level which may be added to an effluent stream is limited. However, it should be stressed that the extraction technique is not restricted to the use of this one reagent. Screening tests conducted with a number of other reagents (8, 15) have shown the extraction of trialkyl lead to be analogous to that obtained for $(C_2H_5)_3NCSSNa \cdot 3H_2O$.

The established use of $(C_2H_5)_3NCSSNa \cdot 3H_2O$ as an analytical reagent for the determination of metals in solution suggests that the chemical complexing–solvent extraction technique could be applied to the removal of a wide range of metals from solution. Whereas analytical determination (by solvent extraction of metal chelates) are generally pH dependent and require high loadings of reagent in the organic phase, the chemical complexing–solvent extraction technique offers flexibility in its application for any chloride (anion) concentration, from zero to saturation levels, and at reagent levels directly related to concentrations of solute. The only restriction is that the reagent stability may be pH dependent, as is the case with $(C_2H_5)_3NCSSNa \cdot 3H_2O$ which is subject to decomposition in strongly acidic media.

In conclusion, the chemical complexing–solvent extraction technique is particularly suited to the polishing of effluents containing trace quantities of metals or organo metallic compounds, since this would minimize the amount of reagent added to the effluent (as excess) and the ratio of organic phase to aqueous phase. Choice of solvent and reagent for a particular application will be dictated by both economic and environmental considerations.

Acknowledgment

The authors would like to thank the Science Research Council for a research grant and the Associated Octel Co., Ltd., for the provision of analytical equipment and methods of analysis. In particular we are indebted to J. R. Grove of the Associated Octel Co. Ltd., for the interest he has taken in the work.

Nomenclature

C_R/C_L	= Molar ratio of complexing reagent to organic lead
V_{aq}/V_{org}	= Volume ratio of the aqueous phase to the organic phase
β	= Stability constant for a chlorocomplex
Q	= Quantity of complex transferred to the organic phase in time t
t	= Time
K_t	= Mass transfer coefficient at time t
A_c	= Interfacial area of the stirred cell
(ΔC)	= Interfacial concentration driving force

Literature Cited

1. Stary, J., "Solvent Extraction of Metal Chelates," Pergamon, New York, 1964.
2. Freiser, H., *Anal. Chem.* (1966) **38**, 131R.
3. Barbieri, R., Belluro, U., Tagliavini, G., *Ann. Chim.* (*Rome*) (1958) **48**, 940
4. Giustiniani, M., Faraglia, G., Barbieri, R., *J. Chromatog.* (1964) **15**, 207.
5. Barbieri, R., *Ric. Sci.* (1963) **33**, 635.
6. Barbieri, R., Giustiniani, M., Faraglia, G., Tagliavini, G., *Ric. Sci.* (1963) **33**, 975.
7. Barbieri, R., Faraglia, G., Giustiniani, M., *Ric. Sci.* (1964) **34**, 109.
8. Clarke, A. B., "The Treatment of Organo Lead Effluents," Ph.D. Thesis, University of Birmingham, 1974.
9. Barker, A. J., Clarke, A. B., *J. Inorg. Nucl. Chem.* (1974) **36**, 921.
10. Bertazzi, N., Alonzo, G., Silvestri, A., *J. Inorg. Nucl. Chem.* (1972) **34**, 1943.
11. Davies, J. T., "Turbulence Phenomena," Academic, New York, 1972. Cell used for Surface Renewal Studies, Chemical Engineering Department, University of Birmingham.
12. Austin, L. J., Sawistowski, H., *Inst. Chem. Eng. Symp. Ser.* (*London*) (1967) **26**, 3.
13. Sankholkar, D. S., Sharma, M. M., *Chem. Eng. Sci.* (1973) **28**, 2091.
14. "Centrifugals for the Chemical and Process Industries," Alpha Laval Co., Publication No. 1B, 40200E.
15. Edmondson, T., The Associated Octel Co. Ltd., private communication.

RECEIVED June 13, 1975.

INDEX

INDEX

The text of this book is set in 10 point Highland with two points of leading. The chapter numerals are set in 26 point Times Roman; the chapter titles are set in 18 point Baskerville Bold.

The book is printed offset on Text White Opaque, 50-pound. The cover is Joanna Book Binding blue linen.

Jacket design by Linda McKnight.
Editing and production by Joan Comstock.

The book was composed by the Mack Printing Co., Easton, Pa., printed and bound by The Maple Press Co., York, Pa.